Hilbert's Fifth Problem and Related Topics

Hilbert's Fifth Problem and Related Topics

Terence Tao

Graduate Studies
in Mathematics

Volume 153

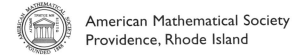

American Mathematical Society
Providence, Rhode Island

2010 *Mathematics Subject Classification.* Primary 22D05, 22E05, 22E15, 11B30, 20F65.

For additional information and updates on this book, visit
www.ams.org/bookpages/gsm-153

Library of Congress Cataloging-in-Publication Data

Tao, Terence, 1975–
    Hilbert's fifth problem and related topics / Terence Tao.
      pages cm. – (Graduate studies in mathematics ; volume 153)
    Includes bibliographical references and index.
    ISBN 978-1-4704-1564-8 (alk. paper)
    1. Hilbert, David, 1862–1943. 2. Lie groups. 3. Lie algebras. Characteristic functions.
I. Title.

QA387.T36   2014
512′.482–dc23   2014009022

To Garth Gaudry, who set me on the road;
To my family, for their constant support;
And to the readers of my blog, for their feedback and contributions.

Contents

Preface

Hilbert's fifth problem, from his famous list of twenty-three problems in mathematics from 1900, asks for a topological description of Lie groups, without any direct reference to smooth structure. As with many of Hilbert's problems, this question can be formalised in a number of ways, but one commonly accepted formulation asks whether any locally Euclidean topological group is necessarily a Lie group. This question was answered affirmatively by Montgomery and Zippin [**MoZi1952**] and Gleason [**Gl1952**]; see Theorem 1.1.9. As a byproduct of the machinery developed to solve this problem, the structure of locally compact groups was greatly clarified, leading in particular to the very useful *Gleason-Yamabe theorem* (Theorem 1.1.13) describing such groups. This theorem (and related results) have since had a number of applications, most strikingly in Gromov's celebrated theorem [**Gr1981**] on groups of polynomial growth (Theorem 1.3.1), and in the classification of finite approximate groups (Theorem 1.2.12). These results in turn have applications to the geometry of manifolds, and on related topics in geometric group theory.

In the fall of 2011, I taught a graduate topics course covering these topics, developed the machinery needed to solve Hilbert's fifth problem, and then used it to classify approximate groups and then finally to develop applications such as Gromov's theorem. Along the way, one needs to develop a number of standard mathematical tools, such as the Baker-Campbell-Hausdorff formula relating the group law of a Lie group to the associated Lie algebra, the Peter-Weyl theorem concerning the representation-theoretic structure of a compact group, or the basic facts about ultrafilters and ultraproducts that underlie nonstandard analysis.

This text is based on the lecture notes from that course, as well as from some additional posts on my blog at `terrytao.wordpress.com` on further topics related to Hilbert's fifth problem. Part 1 of this text can thus serve as the basis for a one-quarter or one-semester advanced graduate course, depending on how much of the optional material one wishes to cover. The material here assumes familiarity with basic graduate real analysis (such as measure theory and point set topology), as covered for instance in my texts [**Ta2011**], [**Ta2010**], and including topics such as the Riesz representation theorem, the Arzelá-Ascoli theorem, Tychonoff's theorem, and Urysohn's lemma. A basic understanding of linear algebra (including, for instance, the spectral theorem for unitary matrices) is also assumed.

The core of the text is Part 1. The first part of this section of the book is devoted to the theory surrounding Hilbert's fifth problem, and in particular in fleshing out the long road from locally compact groups to Lie groups. First, the theory of Lie groups and Lie algebras is reviewed, and it is shown that a Lie group structure can be built from a special type of metric known as a *Gleason metric*, thanks to tools such as the Baker-Campbell-Hausdorff formula. Some representation theory (and in particular, the Peter-Weyl theorem) is introduced next, in order to classify compact groups. The two tools are then combined to prove the fundamental Gleason-Yamabe theorem, which among other things leads to a positive solution to Hilbert's fifth problem.

After this, the focus turns from the "soft analysis" of locally compact groups to the "hard analysis" of approximate groups, with the useful tool of *ultraproducts* serving as the key bridge between the two topics. By using this bridge, one can start imposing approximate Lie structure on approximate groups, which ultimately leads to a satisfactory classification of approximate groups as well. Finally, Part 1 ends with applications of this classification to geometric group theory and the geometry of manifolds, and in particular in reproving Gromov's theorem on groups of polynomial growth.

Part 2 contains a variety of additional material that is related to one or more of the topics covered in Part 1, but which can be omitted for the purposes of teaching a graduate course on the subject.

Notation

For reasons of space, we will not be able to define every single mathematical term that we use in this book. If a term is italicised for reasons other than emphasis or for definition, then it denotes a standard mathematical object, result, or concept, which can be easily looked up in any number of references. (In the blog version of the book, many of these terms were linked to their Wikipedia pages, or other on-line reference pages.)

Given a subset E of a space X, the *indicator function* $1_E : X \to \mathbf{R}$ is defined by setting $1_E(x)$ equal to 1 for $x \in E$ and equal to 0 for $x \notin E$.

The cardinality of a finite set E will be denoted $|E|$. We will use the asymptotic notation $X = O(Y)$, $X \ll Y$, or $Y \gg X$ to denote the estimate $|X| \le CY$ for some absolute constant $C > 0$. In some cases we will need this constant C to depend on a parameter (e.g., d), in which case we shall indicate this dependence by subscripts, e.g., $X = O_d(Y)$ or $X \ll_d Y$. We also sometimes use $X \sim Y$ as a synonym for $X \ll Y \ll X$. (Note though that once we deploy the machinery of nonstandard analysis in Chapter 7, we will use a closely related, but slightly different, asymptotic notation.)

Acknowledgments

I am greatly indebted to my students of the course on which this text was based, as well as many further commenters on my blog, including Marius Buliga, Tony Carbery, Nick Cook, Alin Galatan, Pierre de la Harpe, Ben Hayes, Richard Hevener, Vitali Kapovitch, E. Mehmet Kiral, Allen Knutson, Mateusz Kwasnicki, Fred Lunnon, Peter McNamara, William Meyerson, Joel Moreira, John Pardon, Ravi Raghunathan, David Roberts, David Ross, Olof Sisask, David Speyer, Benjamin Steinberg, Neil Strickland, Lou van den Dries, Joshua Zelinsky, Pavel Zorin, and several anonymous commenters. These comments can be viewed online at:

`terrytao.wordpress.com/category/teaching/254a-hilberts-fifth-problem/`

The author was supported by a grant from the MacArthur Foundation, by NSF grant DMS-0649473, and by the NSF Waterman award. Last, but not least, I thank Emmanuel Breuillard and Ben Green for introducing me to the beautiful interplay between geometric group theory, additive combinatorics, and topological group theory that arises in this text.

Part 1

Hilbert's Fifth Problem

Chapter 1

Introduction

This text focuses on three related topics:

- *Hilbert's fifth problem* on the topological description of Lie groups, as well as the closely related (local) classification of *locally compact groups* (the *Gleason-Yamabe theorem*, see Theorem 1.1.13);

- Approximate groups in nonabelian groups, and their classification [**Hr2012**], [**BrGrTa2011**] via the Gleason-Yamabe theorem; and

- Gromov's theorem [**Gr1981**] on groups of polynomial growth, as proven via the classification of approximate groups (as well as some consequences to fundamental groups of Riemannian manifolds).

These three families of results exemplify two broad principles (part of what I like to call the *the dichotomy between structure and randomness* [**Ta2008**]):

- (Rigidity) If a group-like object exhibits a weak amount of regularity, then it (or a large portion thereof) often automatically exhibits a strong amount of regularity as well.

- (Structure) Furthermore, this strong regularity manifests itself either as Lie type structure (in continuous settings) or *nilpotent* type structure (in discrete settings). (In some cases, "nilpotent" should be replaced by sister properties such as *"abelian"*, *"solvable"*, or *"polycyclic"*.)

Let us illustrate these two principles with two simple examples, one in the continuous setting and one in the discrete setting. We begin with a continuous example. Given an $n \times n$ complex matrix $A \in M_n(\mathbf{C})$, define the *matrix exponential* $\exp(A)$ of A by the formula

$$\exp(A) := \sum_{k=0}^{\infty} \frac{A^k}{k!} = 1 + A + \frac{1}{2!}A^2 + \frac{1}{3!}A^3 + \dots$$

which can easily be verified to be an absolutely convergent series.

Exercise 1.0.1. Show that the map $A \mapsto \exp(A)$ is a real analytic (and even complex analytic) map from $M_n(\mathbf{C})$ to $M_n(\mathbf{C})$, and obeys the restricted homomorphism property

$$(1.1) \qquad\qquad \exp(sA)\exp(tA) = \exp((s+t)A)$$

for all $A \in M_n(\mathbf{C})$ and $s, t \in \mathbf{C}$.

Proposition 1.0.1 (Rigidity and structure of matrix homomorphisms). *Let n be a natural number. Let $\mathrm{GL}_n(\mathbf{C})$ be the group of invertible $n \times n$ complex matrices. Let $\Phi : \mathbf{R} \to \mathrm{GL}_n(\mathbf{C})$ be a map obeying two properties:*

   (1) *(Group-like object) Φ is a homomorphism, thus $\Phi(s)\Phi(t) = \Phi(s+t)$ for all $s, t \in \mathbf{R}$.*

   (2) *(Weak regularity) The map $t \mapsto \Phi(t)$ is continuous.*

Then:

   (i) *(Strong regularity) The map $t \mapsto \Phi(t)$ is smooth (i.e., infinitely differentiable). In fact it is even real analytic.*

   (ii) *(Lie-type structure) There exists a (unique) complex $n \times n$ matrix A such that $\Phi(t) = \exp(tA)$ for all $t \in \mathbf{R}$.*

Proof. Let Φ be as above. Let $\varepsilon > 0$ be a small number (depending only on n). By the homomorphism property, $\Phi(0) = 1$ (where we use 1 here to denote the identity element of $\mathrm{GL}_n(\mathbf{C})$), and so by continuity we may find a small $t_0 > 0$ such that $\Phi(t) = 1 + O(\varepsilon)$ for all $t \in [-t_0, t_0]$ (we use some arbitrary norm here on the space of $n \times n$ matrices, and allow implied constants in the $O()$ notation to depend on n).

The map $A \mapsto \exp(A)$ is real analytic and (by the *inverse function theorem*) is a diffeomorphism near 0. Thus, by the inverse function theorem, we can (if ε is small enough) find a matrix B of size $B = O(\varepsilon)$ such that $\Phi(t_0) = \exp(B)$. By the homomorphism property and (1.1), we thus have

$$\Phi(t_0/2)^2 = \Phi(t_0) = \exp(B) = \exp(B/2)^2.$$

On the other hand, by another application of the inverse function theorem we see that the squaring map $A \mapsto A^2$ is a diffeomorphism near 1 in $\mathrm{GL}_n(\mathbf{C})$, and thus (if ε is small enough)

$$\Phi(t_0/2) = \exp(B/2).$$

We may iterate this argument (for a fixed, but small, value of ε) and conclude that

$$\Phi(t_0/2^k) = \exp(B/2^k)$$

for all $k = 0, 1, 2, \ldots$. By the homomorphism property and (1.1) we thus have

$$\Phi(qt_0) = \exp(qB)$$

whenever q is a dyadic rational, i.e., a rational of the form $a/2^k$ for some integer a and natural number k. By continuity we thus have

$$\Phi(st_0) = \exp(sB)$$

for all real s. Setting $A := B/t_0$ we conclude that

$$\Phi(t) = \exp(tA)$$

for all real t, which gives existence of the representation and also real analyticity and smoothness. Finally, uniqueness of the representation $\Phi(t) = \exp(tA)$ follows from the identity

$$A = \frac{d}{dt} \exp(tA)|_{t=0}. \qquad \square$$

Exercise 1.0.2. Generalise Proposition 1.0.1 by replacing the hypothesis that Φ is continuous with the hypothesis that Φ is Lebesgue measurable. (*Hint:* Use the *Steinhaus theorem*, see e.g. [**Ta2011**, Exercise 1.6.8].) Show that the proposition fails (assuming the axiom of choice) if this hypothesis is omitted entirely.

Note how one needs both the group-like structure and the weak regularity in combination in order to ensure the strong regularity; neither is sufficient on its own. We will see variants of the above basic argument throughout the course. Here, the task of obtaining smooth (or real analytic structure) was relatively easy, because we could borrow the smooth (or real analytic) structure of the domain $\mathbf{R}$ and range $M_n(\mathbf{C})$; but, somewhat remarkably, we shall see that one can still build such smooth or analytic structures even when none of the original objects have any such structure to begin with.

Now we turn to a second illustration of the above principles, namely *Jordan's theorem* [**Jo1878**], which uses a discreteness hypothesis to upgrade Lie type structure to nilpotent (and in this case, abelian) structure. We shall formulate Jordan's theorem in a slightly stilted fashion in order to emphasise the adherence to the above-mentioned principles.

Theorem 1.0.2 (Jordan's theorem). *Let G be an object with the following properties:*

(1) *(Group-like object) G is a group.*

(2) *(Discreteness) G is finite.*

(3) *(Lie-type structure) G is a subgroup of* $U_n(\mathbf{C})$ *(the group of unitary* $n \times n$ *matrices) for some* n.

Then there is a subgroup G' *of* G *such that*

(i) *(G' is close to G) The index* $|G/G'|$ *of* G' *in* G *is* $O_n(1)$ *(i.e., bounded by* C_n *for some quantity* C_n *depending only on* n).

(ii) *(Nilpotent-type structure) G' is abelian.*

A key observation in the proof of Jordan's theorem is that if two unitary elements $g, h \in U_n(\mathbf{C})$ are close to the identity, then their *commutator* $[g, h] = g^{-1}h^{-1}gh$ is even closer to the identity (in, say, the operator norm $\|\|_{\mathrm{op}}$). Indeed, since multiplication on the left or right by unitary elements does not affect the operator norm, we have

$$\|[g,h] - 1\|_{\mathrm{op}} = \|gh - hg\|_{\mathrm{op}}$$
$$= \|(g-1)(h-1) - (h-1)(g-1)\|_{\mathrm{op}}$$

and so by the triangle inequality

(1.2) $$\|[g,h] - 1\|_{\mathrm{op}} \le 2\|g - 1\|_{\mathrm{op}}\|h - 1\|_{\mathrm{op}}.$$

Now we can prove Jordan's theorem.

Proof. We induct on n, the case $n = 1$ being trivial. Suppose first that G contains a *central element* g (i.e., an element that commutes with every element in G) which is not a multiple of the identity. Then, by definition, G is contained in the *centraliser* $Z(g) := \{h \in U_n(\mathbf{C}) : gh = hg\}$ of g, which by the spectral theorem is isomorphic to a product $U_{n_1}(\mathbf{C}) \times \cdots \times U_{n_k}(\mathbf{C})$ of smaller unitary groups. Projecting G to each of these factor groups and applying the induction hypothesis, we obtain the claim.

Thus we may assume that G contains no central elements other than multiples of the identity. Now pick a small $\varepsilon > 0$ (one could take $\varepsilon = \frac{1}{10n}$ in fact) and consider the subgroup G' of G generated by those elements of G that are within ε of the identity (in the operator norm). By considering a maximal ε-net of G we see that G' has index at most $O_{n,\varepsilon}(1)$ in G. By arguing as before, we may assume that G' has no central elements other than multiples of the identity.

If G' consists only of multiples of the identity, then we are done. If not, take an element g of G' that is not a multiple of the identity, and which is as close as possible to the identity (here is where we crucially use that G is finite). Note that g is within ε of the identity. By (1.2), we see that if ε is sufficiently small depending on n, and if h is one of the generators of G', then $[g, h]$ lies in G' and is closer to the identity than g, and is thus a multiple of the identity. On the other hand, $[g, h]$ has determinant 1. Given that it is so close to the identity, it must therefore be the identity (if ε is small enough). In other words, g is central in G', and is thus a multiple of the identity. But this contradicts the hypothesis that there are no central elements other than multiples of the identity, and we are done. $\square$

Commutator estimates such as (1.2) will play a fundamental role in many of the arguments we will see in this text; as we saw above, such estimates combine very well with a discreteness hypothesis, but will also be very useful in the continuous setting.

Exercise 1.0.3. Generalise Jordan's theorem to the case when G is a finite subgroup of $\mathrm{GL}_n(\mathbf{C})$ rather than of $\mathrm{U}_n(\mathbf{C})$. (*Hint:* The elements of G are not necessarily unitary, and thus do not necessarily preserve the standard Hilbert inner product of $\mathbf{C}^n$. However, if one averages that inner product by the finite group G, one obtains a new inner product on $\mathbf{C}^n$ that is preserved by G, which allows one to conjugate G to a subgroup of $\mathrm{U}_n(\mathbf{C})$. This averaging trick is (a small) part of *Weyl's unitary trick* in representation theory.)

Remark 1.0.3. We remark that one can strengthen Jordan's theorem further by relaxing the finiteness assumption on G to a periodicity assumption; see Chapter 11.

Exercise 1.0.4 (Inability to discretise nonabelian Lie groups)**.** Show that if $n \geq 3$, then the orthogonal group $\mathrm{O}_n(\mathbf{R})$ cannot contain arbitrarily dense finite subgroups, in the sense that there exists an $\varepsilon = \varepsilon_n > 0$ depending only on n such that for every finite subgroup G of $\mathrm{O}_n(\mathbf{R})$, there exists a ball of radius ε in $\mathrm{O}_n(\mathbf{R})$ (with, say, the operator norm metric) that is disjoint from G. What happens in the $n = 2$ case?

Remark 1.0.4. More precise classifications of the finite subgroups of $\mathrm{U}_n(\mathbf{C})$ are known, particularly in low dimensions. For instance, it is a classical result that the only finite subgroups of $\mathrm{SO}_3(\mathbf{R})$ (which $\mathrm{SU}_2(\mathbf{C})$ is a double cover of) are isomorphic to either a cyclic group, a *dihedral group*, or the symmetry group of one of the *Platonic solids*.

1.1. Hilbert's fifth problem

One of the fundamental categories of objects in modern mathematics is the category of *Lie groups*, which are rich in both algebraic and analytic structure. Let us now briefly recall the precise definition of what a Lie group is.

Definition 1.1.1 (Smooth manifold)**.** Let $d \geq 0$ be a natural number. A d-dimensional *topological manifold* is a *Hausdorff* topological space M which is *locally Euclidean*, thus every point in M has a neighbourhood which is homeomorphic to an open subset of $\mathbf{R}^d$.

A *smooth atlas* on a d-dimensional topological manifold M is a family $(\phi_\alpha)_{\alpha \in A}$ of homeomorphisms $\phi_\alpha : U_\alpha \to V_\alpha$ from open subsets U_α of M to open subsets V_α of $\mathbf{R}^d$, such that the U_α form an open cover of M, and for any $\alpha, \beta \in A$, the map $\phi_\beta \circ \phi_\alpha^{-1}$ is smooth (i.e., infinitely differentiable) on the domain of definition $\phi_\alpha(U_\alpha \cap U_\beta)$. Two smooth atlases are *equivalent* if their union is also a smooth atlas; this is easily seen to be an equivalence relation. An equivalence class of smooth atlases is a *smooth structure*. A *smooth manifold* is a topological manifold equipped with a smooth structure.

A map $\psi : M \to M'$ from one smooth manifold to another is said to be smooth if $\phi'_\alpha \circ \psi \circ \phi_\beta^{-1}$ is a smooth function on the domain of definition $V_\beta \cap \phi_\beta^{-1}(U_\beta \cap \psi^{-1}(U_\alpha))$ for any smooth charts ϕ_β, ϕ'_α in any the smooth atlases of M, M' respectively (one easily verifies that this definition is independent of the choice of smooth atlas in the smooth structure).

Note that we do not require manifolds to be connected, nor do we require them to be embeddable inside an ambient Euclidean space such as $\mathbf{R}^n$, although certainly many key examples of manifolds are of this form. The requirement that the manifold be Hausdorff is a technical one, in order to exclude pathological examples such as the line with a doubled point (formally, consider the double line $\mathbf{R} \times \{0, 1\}$ after identifying $(x, 0)$ with $(x, 1)$ for all $x \in \mathbf{R} \backslash \{0\}$), which is locally Euclidean but not Hausdorff[1].

Remark 1.1.2. It is a plausible, but nontrivial, fact that a (nonempty) topological manifold can have at most one dimension d associated to it; thus a manifold M cannot both be locally homeomorphic to $\mathbf{R}^d$ and locally homeomorphic to $\mathbf{R}^{d'}$ unless $d = d'$. This fact is a consequence of Brouwer's *invariance of domain theorem*; see Exercise 6.0.4. On the other hand, it is an easy consequence of the *rank-nullity theorem* that a *smooth* manifold can have at most one dimension, without the need to invoke invariance of domain; we leave this as an exercise.

Definition 1.1.3 (Lie group). A *Lie group* is a group $G = (G, \cdot)$ which is also a smooth manifold, such that the group operations $\cdot : G \times G \to G$ and $()^{-1} : G \to G$ are smooth maps. (Note that the Cartesian product of two smooth manifolds can be given the structure of a smooth manifold in the obvious manner.) We will also use additive notation $G = (G, +)$ to describe some Lie groups, but only in the case when the Lie group is abelian.

---

[1] In some literature, additional technical assumptions such as *paracompactness*, *second countability*, or *metrisability* are imposed to remove pathological examples of topological manifolds such as the *long line*, but it will not be necessary to do so in this text, because (as we shall see later) we can essentially get such properties "for free" for locally Euclidean groups.

Remark 1.1.4. In some literature, Lie groups are required to be connected (and occasionally, are even required to be simply connected), but we will not adopt this convention here. One can also define infinite-dimensional Lie groups, but in this text all Lie groups are understood to be finite dimensional.

Example 1.1.5. Every group can be viewed as a Lie group if given the discrete topology (and the discrete smooth structure). (Note that we are not requiring Lie groups to be connected.)

Example 1.1.6. Finite-dimensional vector spaces such as $\mathbf{R}^d$ are (additive) Lie groups, as are sublattices such as $\mathbf{Z}^d$ or quotients such as $\mathbf{R}^d/\mathbf{Z}^d$. However, nonclosed subgroups such as $\mathbf{Q}^d$ are not manifolds (at least with the topology induced from $\mathbf{R}^d$) and are thus not Lie groups; similarly, quotients such as $\mathbf{R}^d/\mathbf{Q}^d$ are not Lie groups either (they are not even Hausdorff). Also, infinite-dimensional topological vector spaces (such as $\mathbf{R}^{\mathbf{N}}$ with the product topology) will not be Lie groups.

Example 1.1.7. The *general linear group* $\mathrm{GL}_n(\mathbf{C})$ of invertible $n \times n$ complex matrices is a Lie group. A theorem of Cartan (Theorem 3.0.14) asserts that any closed subgroup of a Lie group is a smooth submanifold of that Lie group and is in particular also a Lie group. In particular, closed linear groups (i.e., closed subgroups of a general linear group) are Lie groups; examples include the real general linear group $\mathrm{GL}_n(\mathbf{R})$, the *unitary group* $\mathrm{U}_n(\mathbf{C})$, the *special unitary group* $\mathrm{SU}_n(\mathbf{C})$, the *orthogonal group* $\mathrm{O}_n(\mathbf{R})$, the *special orthogonal group* $\mathrm{SO}_n(\mathbf{R})$, and the *Heisenberg group*

$$\begin{pmatrix} 1 & \mathbf{R} & \mathbf{R} \\ 0 & 1 & \mathbf{R} \\ 0 & 0 & 1 \end{pmatrix}$$

of unipotent upper triangular 3×3 real matrices. Many Lie groups are isomorphic to closed linear groups; for instance, the additive group $\mathbf{R}$ can be identified with the closed linear group

$$\begin{pmatrix} 1 & \mathbf{R} \\ 0 & 1 \end{pmatrix},$$

the circle $\mathbf{R}/\mathbf{Z}$ can be identified with $\mathrm{SO}_2(\mathbf{R})$ (or $\mathrm{U}_1(\mathbf{C})$), and so forth. However, not all Lie groups are isomorphic to closed linear groups. A somewhat trivial example is that of a discrete group with cardinality larger than the continuum, which is simply too large to fit inside any linear group. A less pathological example is provided by the Weil-Heisenberg group

$$(1.3) \qquad G := \begin{pmatrix} 1 & \mathbf{R} & \mathbf{R}/\mathbf{Z} \\ 0 & 1 & \mathbf{R} \\ 0 & 0 & 1 \end{pmatrix} := \begin{pmatrix} 1 & \mathbf{R} & \mathbf{R} \\ 0 & 1 & \mathbf{R} \\ 0 & 0 & 1 \end{pmatrix} / \begin{pmatrix} 1 & 0 & \mathbf{Z} \\ 0 & 1 & 0 \\ 0 & 0 & 1 \end{pmatrix}$$

which is isomorphic to the image of the Heisenberg group under the *Weil representation*, or equivalently the group of isometries of $L^2(\mathbf{R})$ generated by translations and modulations. Despite this, though, it is helpful to think of closed linear groups and Lie groups as being almost the same concept as a first approximation. For instance, one can show using *Ado's theorem* (Theorem 13.0.4) that every Lie group is *locally* isomorphic to a linear *local group* (a concept we will discuss in Section 2.1).

An important subclass of the closed linear groups are the *linear algebraic groups*, in which the group is also a real or complex *algebraic variety* (or at least an algebraically constructible set). All of the examples of closed linear groups given above are linear algebraic groups, although there exist closed linear groups that are not isomorphic to any algebraic group; see Proposition 21.0.4.

Exercise 1.1.1 (Weil-Heisenberg group is not linear)**.** Show that there is no injective homomorphism $\rho : G \to \mathrm{GL}_n(\mathbf{C})$ from the Weil-Heisenberg group (1.3) to a general linear group $\mathrm{GL}_n(\mathbf{C})$ for any finite n. (*Hint:* The centre $[G, G]$ maps via ρ to a circle subgroup of $\mathrm{GL}_n(\mathbf{C})$; diagonalise this subgroup and reduce to the case when the image of the centre consists of multiples of the identity. Now, use the fact that commutators in $\mathrm{GL}_n(\mathbf{C})$ have determinant one.) This fact was first observed by Birkhoff.

Hilbert's fifth problem, like many of Hilbert's problems, does not have a unique interpretation, but one of the most commonly accepted interpretations of the question posed by Hilbert is to determine if the requirement of smoothness in the definition of a Lie group is redundant. (There is also an analogue of Hilbert's fifth problem for group *actions*, known as the *Hilbert-Smith conjecture*; see Chapter 17.) To answer this question, we need to relax the notion of a Lie group to that of a *topological group*.

Definition 1.1.8 (Topological group)**.** A *topological group* is a group $G = (G, \cdot)$ that is also a topological space, in such a way that the group operations $\cdot : G \times G \to G$ and $()^{-1} : G \to G$ are continuous. (As before, we also consider additive topological groups $G = (G, +)$ provided that they are abelian.)

Clearly, every Lie group is a topological group if one simply forgets the smooth structure, retaining only the topological and group structures. Furthermore, such topological groups remain locally Euclidean. It was established by Montgomery-Zippin [**MoZi1952**] and Gleason [**Gl1952**] that the converse statement holds, thus solving at least one formulation of Hilbert's fifth problem:

Theorem 1.1.9 (Hilbert's fifth problem)**.** *Let G be an object with the following properties:*

(1) *(Group-like object) G is a topological group.*

(2) *(Weak regularity) G is locally Euclidean.*

Then

(i) *(Lie-type structure) G is isomorphic to a Lie group.*

Exercise 1.1.2. Show that a locally Euclidean topological group is necessarily Hausdorff (without invoking Theorem 1.1.9).

We will prove this theorem in Section 6. As it turns out, Theorem 1.1.9 is not directly useful for many applications, because it is often difficult to verify that a given topological group is locally Euclidean. On the other hand, the weaker property of *local compactness*, which is clearly implied by the locally Euclidean property, is much easier to verify in practice. One can then ask the more general question of whether every locally compact group is isomorphic to a Lie group. Unfortunately, the answer to this question is easily seen to be no, as the following examples show:

Example 1.1.10 (Trivial topology). A group equipped with the trivial topology is a compact (hence locally compact) group, but will not be Hausdorff (and thus not Lie) unless the group is also trivial. Of course, this is a rather degenerate counterexample and can be easily eliminated in practice. For instance, we will see later that any topological group can be made Hausdorff by quotienting out the closure of the identity.

Example 1.1.11 (Infinite-dimensional torus). The infinite-dimensional torus $(\mathbf{R}/\mathbf{Z})^{\mathbf{N}}$ (with the product topology) is an (additive) topological group, which is compact (and thus locally compact) by *Tychonoff's theorem*. However, it is not a Lie group.

Example 1.1.12 (p-adics). Let p be a prime. We define the *p-adic norm* $\|\|_p$ on the integers $\mathbf{Z}$ by defining $\|n\|_p := p^{-j}$, where p^j is the largest power of p that divides n (with the convention $\|0\|_p := 0$). This is easily verified to generate a metric (and even an *ultrametric*) on $\mathbf{Z}$; the p-adic integers $\mathbf{Z}_p$ are then defined as the *metric completion* of $\mathbf{Z}$ under this metric. This is easily seen to be a compact (hence locally compact) additive group (topologically, it is homeomorphic to a *Cantor set*). However, it is not locally Euclidean (or even *locally connected*), and so is not isomorphic to a Lie group.

One can also extend the p-adic norm to the ring $\mathbf{Z}[\frac{1}{p}]$ of rationals of the form a/p^j for some integers a, j in the obvious manner; the metric completion of this space is then the p-adic rationals $\mathbf{Q}_p$. This is now a locally compact additive group rather than a compact one ($\mathbf{Z}_p$ is a compact open neighbourhood of the identity); it is still not locally connected, so it is still not a Lie group.

One can also define algebraic groups such as GL_n over the p-adic rationals $\mathbf{Q}_p$; thus for instance $GL_n(\mathbf{Q}_p)$ is the group of invertible $n \times n$ matrices with entries in the p-adics. This is still a locally compact group, and is certainly not Lie.

Exercise 1.1.3 (Solenoid). Let p be a prime. Let G be the *solenoid group* $G := (\mathbf{Z}_p \times \mathbf{R})/\mathbf{Z}^\Delta$, where $\mathbf{Z}^\Delta := \{(n,n) : n \in \mathbf{Z}\}$ is the diagonally embedded copy of the integers in $\mathbf{Z}_p \times \mathbf{R}$. (Topologically, G can be viewed as the set $\mathbf{Z}_p \times [0,1]$ after identifying $(x+1,1)$ with $(x,0)$ for all $x \in \mathbf{Z}_p$.) Show that G is a compact additive group that is connected but not locally connected (and thus not a Lie group). Thus one cannot eliminate p-adic type behaviour from locally compact groups simply by restricting attention to the connected case (although we will see later that one can do so by restricting to the *locally connected* case).

We have now seen several examples of locally compact groups that are not Lie groups. However, all of these examples are "almost" Lie groups in that they can be turned into Lie groups by quotienting out a small compact normal subgroup. (It is easy to see that the quotient of a locally compact group by a compact normal subgroup is again a locally compact group.) For instance, a group with the trivial topology becomes Lie after quotienting out the entire group (which is "small" in the sense that it is contained in every open neighbourhood of the origin). The infinite-dimensional torus $(\mathbf{R}/\mathbf{Z})^{\mathbf{N}}$ can be quotiented into a finite-dimensional torus $(\mathbf{R}/\mathbf{Z})^d$ (which is of course a Lie group) by quotienting out the compact subgroup $\{0\}^d \times (\mathbf{R}/\mathbf{Z})^{\mathbf{N}}$; note from the definition of the product topology that these compact subgroups shrink to zero in the sense that every neighbourhood of the group identity contains at least one (and in fact all but finitely many) of these subgroups. Similarly, with the p-adic group $\mathbf{Z}_p$, one can quotient out by the compact (and open) subgroups $p^j\mathbf{Z}_p$ (which also shrink to zero, as discussed above) to obtain the cyclic groups $\mathbf{Z}/p^j\mathbf{Z}$, which are discrete and thus Lie. Quotienting out $\mathbf{Q}_p$ by the same compact open subgroups $p^j\mathbf{Z}_p$ also leads to discrete (hence Lie) quotients; similarly for algebraic groups defined over $\mathbf{Q}_p$, such as $GL_n(\mathbf{Q}_p)$. Finally, with the solenoid group $G := (\mathbf{Z}_p \times \mathbf{R})/\mathbf{Z}^\Delta$, one can quotient out the copy of $p^j\mathbf{Z}_p \times \{0\}$ in G for $j = 0, 1, 2, \ldots$ (which are another sequence of compact subgroups shrinking to zero) to obtain the quotient group $(\mathbf{Z}/p^j\mathbf{Z} \times \mathbf{R})/\mathbf{Z}^\Delta$, which is isomorphic to a (highly twisted) circle $\mathbf{R}/\mathbf{Z}$ and is thus Lie.

Inspired by these examples, we might be led to the following conjecture: if G is a locally compact group, and U is a neighbourhood of the identity, then there exists a compact normal subgroup K of G contained in U such that G/K is a Lie group. In the event that G is Hausdorff, this is equivalent to asserting that G is the *projective limit* (or *inverse limit*) of Lie groups.

This conjecture is true in several cases; for instance, one can show using the *Peter-Weyl theorem* (which we will discuss in Chapter 4) that it is true for compact groups, and we will later see that it is also true for connected locally compact groups (see Theorem 6.0.11). However, it is not quite true in general, as the following example shows.

Exercise 1.1.4. Let p be a prime, and let $T : \mathbf{Q}_p \to \mathbf{Q}_p$ be the automorphism $Tx := px$. Let $G := \mathbf{Q}_p \rtimes_T \mathbf{Z}$ be the semidirect product of $\mathbf{Q}_p$ and $\mathbf{Z}$ twisted by T; more precisely, G is the Cartesian product $\mathbf{Q}_p \times \mathbf{Z}$ with the product topology and the group law

$$(x, n)(y, m) := (x + T^n y, n + m).$$

Show that G is a locally compact group which is not isomorphic to a Lie group, and that $\mathbf{Z}_p \times \{0\}$ is an open neighbourhood of the identity that contains no nontrivial normal subgroups of G. Conclude that the conjecture stated above is false.

The difficulty in the above example was that it was not easy to keep a subgroup normal with respect to the entire group $\mathbf{Q}_p \rtimes_T \mathbf{Z}$. Note however that G contains a "large" (and more precisely, *open*) subgroup $\mathbf{Q}_p \times \{0\}$ which *is* the projective limit of Lie groups. So the above examples do not rule out that the conjecture can still be salvaged if one passes from a group G to an open subgroup G'. This is indeed the case:

Theorem 1.1.13 (Gleason-Yamabe theorem [**Gl1951, Ya1953b**]). *Let G obey the following hypotheses:*

(1) *(Group-like object) G is a topological group.*

(2) *(Weak regularity) G is locally compact.*

Then for every open neighbourhood U of the identity, there exists a subgroup G' of G and a compact normal subgroup K of G' with the following properties:

(i) *(G'/K is close to G) G' is an open subgroup of G, and K is contained in U.*

(ii) *(Lie-type structure) G'/K is isomorphic to a Lie group.*

The proof of this theorem will occupy the next few sections of this text, being finally proven in Chapter 5. As stated, G' may depend on U, but one can in fact take the open subgroup G' to be uniform in the choice of U; we will show this in later sections. Theorem 1.1.9 can in fact be deduced from Theorem 1.1.13 and some topological arguments involving the invariance of domain theorem; this will be shown in Chapter 6.

The Gleason-Yamabe theorem asserts that locally compact groups are "essentially" Lie groups, after ignoring the very large scales (by restricting

to an open subgroup) and also ignoring the very small scales (by allowing
one to quotient out by a small group). In special cases, the conclusion of the
theorem can be simplified. For instance, it is easy to see that an open subgroup G' of a topological group G is also closed (since the complement $G\backslash G'$
is a union of cosets of G'), and so if G is connected, there are no open subgroups other than G itself. Thus, in the connected case of Theorem 1.1.13,
one can take $G = G'$. In a similar spirit, if G has the *no small subgroups*
(NSS) property, that is to say that there exists an open neighbourhood of
the identity that contains no nontrivial subgroups of G, then we can take
K to be trivial. Thus, as a special case of the Gleason-Yamabe theorem, we
see that all connected NSS locally compact groups are Lie; in fact it is not
difficult to then conclude that any locally compact NSS group (regardless of
connectedness) is Lie. Conversely, this claim (which we isolate as Corollary
5.3.3) turns out to be a key step in the *proof* of Theorem 1.1.13, as we shall
see later. (It is also not difficult to show that all Lie groups are NSS; see
Exercise 5.3.1.)

The proof of the Gleason-Yamabe theorem proceeds in a somewhat
lengthy series of steps in which the initial regularity (local compactness)
on the group G is gradually upgraded to increasingly stronger regularity
(e.g. metrisability, the NSS property, or the locally Euclidean property) until one eventually obtains Lie structure; see Figure 1. A key turning point
in the argument will be the construction of a metric (which we call a *Gleason metric*) on (a large portion of) G which obeys a commutator estimate
similar to (1.2).

While the Gleason-Yamabe theorem does not completely classify all locally compact groups (as mentioned earlier, it primarily controls the medium-
scale behaviour, and not the very fine-scale or very coarse-scale behaviour,
of such groups), it is still powerful enough for a number of applications, to
which we now turn.

1.2. Approximate groups

We now discuss what appears at first glance to be an unrelated topic, namely
that of *additive combinatorics* (and its noncommutative counterpart, multiplicative combinatorics). One of the main objects of study in either additive
or multiplicative combinatorics are *approximate groups* — sets A (typically
finite) contained in an additive or multiplicative ambient group G that are
"almost groups" in the sense that they are "almost" closed under either
addition or multiplication. (One can also consider abstract approximate
groups that are not contained in an ambient genuine group, but we will not
do so here.)

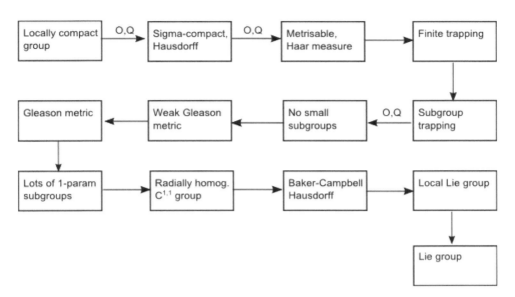

Figure 1. A schematic description of the steps needed to establish the Gleason-Yamabe theorem. The annotation O, Q on an arrow indicates that one has to pass to an open subgroup, and then quotient out a compact normal subgroup, in order to obtain the additional structure at the end of the arrow.

There are several ways to quantify what it means for a set A to be "almost" closed under addition or multiplication. Here are some common formulations of this idea (phrased in multiplicative notation, for the sake of concreteness):

(1) (Statistical multiplicative structure) For a "large" proportion of pairs $(a, b) \in A \times A$, the product ab also lies in A.

(2) (Small product set) The "size" of the product set $A \cdot A := \{ab : a, b \in A\}$ is "comparable" to the "size" of the original set A. (For technical reasons, one sometimes uses the triple product $A \cdot A \cdot A := \{abc : a, b, c \in A\}$ instead of the double product.)

(3) (Covering property) The product set $A \cdot A$ can be covered by a "bounded" number of (left or right) translates of the original set A.

Of course, to make these notions precise one would have to precisely quantify the various terms in quotes. Fortunately, the basic theory of additive combinatorics (and multiplicative combinatorics) can be used to show that all these different notions of additive or multiplicative structure are "essentially" equivalent; see [**TaVu2006**, Chapter 2] or [**Ta2008b**] for more discussion.

For the purposes of this text, it will be convenient to focus on the use of covering to describe approximate multiplicative structure. More precisely:

Definition 1.2.1 (Approximate groups). Let G be a multiplicative group, and let $K \geq 1$ be a real number. A *K-approximate subgroup* of G, or *K-approximate group* for short, is a subset A of G which contains the identity, is symmetric (thus $A^{-1} := \{a^{-1} : a \in A\}$ is equal to A) and is such that $A \cdot A$ can be covered by at most K left-translates (or equivalently by symmetry, right translates) of A, thus there exists a subset X of G of cardinality at most K such that $A \cdot A \subset X \cdot A$.

In most combinatorial applications, one only considers approximate groups that are finite sets, but one could certainly also consider countably or uncountably infinite approximate groups. We remark that this definition is essentially from [**Ta2008b**] (although the definition in [**Ta2008b**] places some additional minor constraints on the set X which have turned out not to be terribly important in practice).

Example 1.2.2. A 1-approximate subgroup of G is the same thing as a genuine subgroup of G.

Example 1.2.3. In the additive group of the integers $\mathbf{Z}$, the symmetric arithmetic progression $\{-N, \ldots, N\}$ is a 2-approximate group for any $N \geq 1$. More generally, in any additive group G, the *symmetric generalised arithmetic progression*

$$\{a_1 v_1 + \cdots + a_r v_r : a_1, \ldots, a_r \in \mathbf{Z}, |a_i| \leq N_i \forall i = 1, \ldots, r\}$$

with $v_1, \ldots, v_r \in G$ and $N_1, \ldots, N_r > 0$, is a 2^r-approximate group.

Exercise 1.2.1. Let A be a convex symmetric subset of $\mathbf{R}^d$. Show that A is a 5^d-approximate group. (*Hint:* Greedily pack $2A$ with disjoint translates of $\frac{1}{2}A$.)

Example 1.2.4. If A is an open *precompact*[2] symmetric neighbourhood of the identity in a locally compact group G, then A is a K-approximate group for some finite K. Thus we see some connection between locally compact groups and approximate groups; we will see a deeper connection involving *ultraproducts* in Chapter 7.

Example 1.2.5. Let G be a d-dimensional Lie group. Then G is a smooth manifold, and can thus be (nonuniquely) given the structure of a *Riemannian manifold*. If one does so, then for sufficiently small radii r, the ball $B(1, r)$ around the identity 1 will be a $O_d(1)$-approximate group.

---

[2]A subset of a topological space is said to be precompact if its closure is compact.

Example 1.2.6 (Extensions). Let $\phi : G \to H$ be a surjective group homomorphism (thus G is a *group extension* of H by the kernel $\ker(\phi)$ of ϕ). If A is a K-approximate subgroup of H, then $\phi^{-1}(A)$ is a K-approximate subgroup of G. One can think of $\phi^{-1}(A)$ as an extension of the approximate group A by $\ker(\phi)$.

The classification of approximate groups is of importance in additive combinatorics, and has connections with number theory, geometric group theory, and the theory of expander graphs. One can ask for a quantitative classification, in which one has explicit dependence of constants on the approximate group parameter K, or one can settle for a qualitative classification in which one does not attempt to control this dependence of constants. In this text we will focus on the latter question, as this allows us to bring in qualitative tools such as the Gleason-Yamabe theorem to bear on the problem.

In the abelian case when the ambient group G is additive, approximate groups are classified by *Freiman's theorem for abelian groups*[3] [**GrRu2007**]. As before, we phrase this theorem in a slightly stilted fashion (and in a qualitative, rather than quantitative, manner) in order to demonstrate its alignment with the general principles stated in the introduction.

Theorem 1.2.7 (Freiman's theorem in an abelian group). *Let A be an object with the following properties:*

(1) *(Group-like object) A is a subset of an additive group G.*

(2) *(Weak regularity) A is a K-approximate group.*

(3) *(Discreteness) A is finite.*

Then there exists a finite subgroup H of G, and a subset P of G/H, with the following properties:

(i) *(P is close to A/H) $\pi^{-1}(P)$ is contained in $4A := A + A + A + A$, where $\pi : G \to G/H$ is the quotient map, and $|P| \gg_K |A|/|H|$.*

(ii) *(Nilpotent type structure) P is a symmetric generalised arithmetic progression of rank $O_K(1)$ (see Example 1.2.3).*

Informally, this theorem asserts that in the abelian setting, discrete approximate groups are essentially bounded rank symmetric generalised arithmetic progressions, extended by finite groups (such extensions are also known as *coset progressions*). The theorem has a simpler conclusion (and is simpler to prove) in the case when G is a torsion-free abelian group (such as $\mathbf{Z}$), since in this case H is trivial.

---

[3]The original theorem of Freiman [**Fr1973**] obtained an analogous classification in the case when G was a torsion-free abelian group, such as the integers $\mathbf{Z}$.

We will not discuss the proof of Theorem 1.2.7 from [**GrRu2007**] here, save to say that it relies heavily on Fourier-analytic methods, and as such, does not seem to easily extend to a general nonabelian setting. To state the nonabelian analogue of Theorem 1.2.7, one needs multiplicative analogues of the concept of a generalised arithmetic progression. An ordinary (symmetric) arithmetic progression $\{-Nv, \ldots, Nv\}$ has an obvious multiplicative analogue, namely a (symmetric) geometric progression $\{a^{-N}, \ldots, a^N\}$ for some generator $a \in G$. In a similar vein, if one has r *commuting* generators $a_1, \ldots, a_r$ and some dimensions $N_1, \ldots, N_r > 0$, one can form a symmetric generalised geometric progression

$$(1.4) \qquad\qquad P := \{a_1^{n_1} \ldots a_r^{n_r} : |n_i| \leq N_i \forall 1 \leq i \leq r\},$$

which will still be a 3^r-approximate group. However, if the $a_1, \ldots, a_r$ do not commute, then the set P defined in (1.4) is not quite the right concept to use here; for instance, there it is no reason for P to be symmetric. However, it can be modified as follows:

Definition 1.2.8 (noncommutative progression). Let $a_1, \ldots, a_r$ be elements of a (not necessarily abelian) group $G = (G, \cdot)$, and let $N_1, \ldots, N_r > 0$. We define the *noncommutative progression* $P = P(a_1, \ldots, a_r; N_1, \ldots, N_r)$ of rank r with generators $a_1, \ldots, a_r$ and dimensions $N_1, \ldots, N_r$ to be the collection of all words w composed using the alphabet $a_1, a_1^{-1}, \ldots, a_r, a_r^{-1}$, such that for each $1 \leq i \leq r$, the total number of occurrences of a_i and a_i^{-1} combined in w is at most N_i.

Example 1.2.9. $P(a, b; 1, 2)$ consists of the elements

$$1, a^{\pm}, b^{\pm}, b^{\pm}a^{\pm}, a^{\pm}b^{\pm}, b^{\pm2}, b^{\pm2}a^{\pm}, b^{\pm}a^{\pm}b^{\pm}, a^{\pm}b^{\pm2},$$

where each occurrence of $\pm$ can independently be set to $+$ or $-$; thus $P(a, b; 1, 2)$ can have as many as 31 elements.

Example 1.2.10. If the $a_1, \ldots, a_r$ commute, then the noncommutative progression $P(a_1, \ldots, a_r; N_1, \ldots, N_r)$ simplifies to (1.4).

Exercise 1.2.2. Let G be the discrete Heisenberg group

$$(1.5) \qquad\qquad G = \begin{pmatrix} 1 & \mathbf{Z} & \mathbf{Z} \\ 0 & 1 & \mathbf{Z} \\ 0 & 0 & 1 \end{pmatrix}$$

and let

$$e_1 := \begin{pmatrix} 1 & 1 & 0 \\ 0 & 1 & 0 \\ 0 & 0 & 1 \end{pmatrix}, e_2 := \begin{pmatrix} 1 & 0 & 0 \\ 0 & 1 & 1 \\ 0 & 0 & 1 \end{pmatrix}$$

be the two generators of G. Let $N \geq 1$ be a sufficiently large natural number. Show that the noncommutative progression $P(e_1, e_2; N, N)$ contains all the group elements $\begin{pmatrix} 1 & a & c \\ 0 & 1 & b \\ 0 & 0 & 1 \end{pmatrix}$ of G with $|a|, |b| \leq \delta N$ and $|c| \leq \delta N^2$ for a sufficiently small absolute constant $\delta > 0$; conversely, show that all elements of $P(e_1, e_2; N, N)$ are of the form $\begin{pmatrix} 1 & a & c \\ 0 & 1 & b \\ 0 & 0 & 1 \end{pmatrix}$ with $|a|, |b| \leq CN$ and $|c| \leq CN^2$ for some sufficiently large absolute constant $C > 0$. Thus, informally, we have

$$P(e_1, e_2; N, N) = \begin{pmatrix} 1 & O(N) & O(N^2) \\ 0 & 1 & O(N) \\ 0 & 0 & 1 \end{pmatrix}.$$

It is clear that noncommutative progressions $P(a_1, \ldots, a_r; N_1, \ldots, N_r)$ are symmetric and contain the identity. However, if the $a_1, \ldots, a_r$ do not have any commutative properties, then the size of these progressions can grow exponentially in $N_1, \ldots, N_r$ and will not be approximate groups with any reasonable parameter K. However, the situation changes when the $a_1, \ldots, a_r$ generate a *nilpotent group*[4]:

Proposition 1.2.11. *Suppose that $a_1, \ldots, a_r \in G$ generate a nilpotent group of step s, and suppose that $N_1, \ldots, N_r$ are all sufficiently large depending on r, s. Then $P(a_1, \ldots, a_r; N_1, \ldots, N_r)$ is an $O_{r,s}(1)$-approximate group.*

We will prove this proposition in Chapter 12.

We can now state the noncommutative analogue of Theorem 1.2.7, proven in [**BrGrTa2011**]:

Theorem 1.2.12 (Freiman's theorem in an arbitrary group). *Let A be an object with the following properties:*

(1) *(Group-like object) A is a subset of a multiplicative group G.*

(2) *(Weak regularity) A is a K-approximate group.*

(3) *(Discreteness) A is finite.*

Then there exists a finite subgroup H of G, and a subset P of $N(H)/H$ (where $N(H) := \{g \in G : gH = Hg\}$ is the normaliser of H), with the following properties:

(i) *(P is close to A/H) $\pi^{-1}(P)$ is contained in $A^4 := A \cdot A \cdot A \cdot A$, where $\pi : N(H) \to N(H)/H$ is the quotient map, and $|P| \gg_K |A|/|H|$.*

(ii) *(Nilpotent type structure) P is a noncommutative progression of rank $O_K(1)$, whose generators generate a nilpotent group of step $O_K(1)$.*

---

[4] A group G is nilpotent if the lower central series $G_1 := G, G_2 := [G, G_1], G_3 := [G, G_2], \ldots,$ etc. eventually becomes trivial.

The proof of this theorem relies on the Gleason-Yamabe theorem (Theorem 1.1.13), and will be discussed in Chapter 8. The key connection will take some time to explain properly, but roughly speaking, it comes from the fact that the *ultraproduct* of a sequence of K-approximate groups can be used to generate a locally compact group, to which the Gleason-Yamabe theorem can be applied. This in turn can be used to place a metric on approximate groups that obeys a commutator estimate similar to (1.2), which allows one to run an argument similar to that used to prove Theorem 1.0.2.

1.3. Gromov's theorem

The final topic of this chapter will be *Gromov's theorem on groups of polynomial growth* [**Gr1981**]. This theorem is analogous to Theorem 1.1.13 or Theorem 1.2.12, but in the category of *finitely generated groups* rather than locally compact groups or approximate groups.

Let G be a group that is generated by a finite set S of generators; for notational simplicity we will assume that S is symmetric and contains the origin. Then S defines a (right-invariant) *word metric* on G, defined by setting $d(x, y)$ for $x, y \in G$ to be the least natural number n such that $x \in S^n y$. One easily verifies that this is indeed a metric that is right-invariant (thus $d(xg, yg) = d(x, y)$ for all $x, y, g \in G$). Geometrically, this metric describes the geometry of the *Cayley graph* on G formed by connecting x to sx for each $x \in G$ and $s \in S$. (See [**Ta2011c**, §2.3] for more discussion of using Cayley graphs to study groups geometrically.)

Let us now consider the growth of the balls $B(1, R) = S^{\lfloor R \rfloor}$ as $R \to \infty$, where $\lfloor R \rfloor$ is the integer part of R. On the one hand, we have the trivial upper bound

$$|B(1, R)| \leq |S|^R$$

that shows that such balls can grow at most exponentially. And for "typical" nonabelian groups, this exponential growth actually occurs; consider the case for instance when S consists of the generators of a free group (together with their inverses, and the group identity). However, there are some groups for which the balls grow at a much slower rate. A somewhat trivial example is that of a finite group G, since clearly $|B(1, R)|$ will top out at $|G|$ (when R reaches the *diameter* of the Cayley graph) and stop growing after that point. Another key example is the abelian case:

Exercise 1.3.1. If G is an abelian group generated by a finite symmetric set S containing the identity, show that

$$|B(1, R)| \leq (1 + R)^{|S|}.$$

In particular, $B(1, R)$ grows at a polynomial rate in R.

Let us say that a finite group G is a *group of polynomial growth* if one has $|B(1, R)| \leq CR^d$ for all $R \geq 1$ and some constants $C, d > 0$.

Exercise 1.3.2. Show that the notion of a group of polynomial growth (as well as the rate d of growth) does not depend on the choice of generators S; thus if S' is another set of generators for G, show that G has polynomial growth with respect to S with rate d if and only if it has polynomial growth with respect to S' with rate d.

Exercise 1.3.3. Let G be a finitely generated group, and let G' be a finite index subgroup of G.

  (i) Show that G' is also finitely generated. (*Hint:* Let S be a symmetric set of generators for G containing the identity, and locate a finite integer n such that $S^{n+1}G' = S^n G'$. Then show that the set $S' := G' \cap S^{2n+1}$ is such that $S^{n+1} \subset S^n S'$. Conclude that S^n meets every coset of $\langle S' \rangle$ (or equivalently that $G = S^n \langle S' \rangle$), and use this to show that S' generates G'.)

  (ii) Show that G has polynomial growth if and only if G' has polynomial growth.

  (iii) More generally, show that any finitely generated subgroup of a group of polynomial growth also has polynomial growth. Conclude in particular that a group of polynomial growth cannot contain the free group on two generators.

From Exercise 1.2.2 we see that the discrete Heisenberg group (1.5) is of polynomial growth. It is in fact not difficult to show that, more generally, any nilpotent finitely generated group is of polynomial growth. By Exercise 1.3.3, this implies that any *virtually nilpotent* finitely generated group is of polynomial growth.

Gromov's theorem asserts the converse statement:

Theorem 1.3.1 (Gromov's theorem [**Gr1981**]). *Let G be an object with the following properties:*

  (1) *(Group-like object) G is a finitely generated group.*

  (2) *(Weak regularity) G is of polynomial growth.*

Then there exists a subgroup G' of G such that

  (i) *(G' is close to G) The index $|G/G'|$ is finite.*

  (ii) *(Nilpotent type structure) G' is nilpotent.*

More succinctly: A finitely generated group is of polynomial growth if and only if it is virtually nilpotent.

Groups of polynomial growth are related to approximate groups by the following observation.

Exercise 1.3.4 (Pigeonhole principle). Let G be a finitely generated group of polynomial growth, and let S be a symmetric set of generators for G containing the identity.

(i) Show that there exists a $C > 1$ such that $|B(1, 5R/2)| \leq C|B(1, R/2)|$ for a sequence $R = R_n$ of radii going to infinity.

(ii) Show that there exists a $K > 1$ such that $B(1, R)$ is a K-approximate group for a sequence $R = R_n$ of radii going to infinity. (*Hint:* Argue as in Exercise 1.2.1.)

In Chapter 9 we will use this connection to deduce Theorem 1.3.1 from Theorem 1.2.12. From a historical perspective, this was not the first proof of Gromov's theorem; Gromov's original proof in [**Gr1981**] relied instead on a variant of Theorem 1.1.9 (as did some subsequent variants of Gromov's argument, such as the nonstandard analysis variant in [**vdDrWi1984**]), and a subsequent proof of Kleiner [**Kl2010**] went by a rather different route, based on earlier work of Colding and Minicozzi [**CoMi1997**] on harmonic functions of polynomial growth. (This latter proof is discussed in [**Ta2009**, §1.2] and [**Ta2011c**, §2.5].) The proof we will give in this text is more recent, based on an argument of Hrushovski [**Hr2012**]. We remark that the strategy used to prove Theorem 1.2.12 — namely taking an ultralimit of a sequence of approximate groups — also appears in Gromov's original argument[5]. We will discuss these sorts of limits more carefully in Chapter 7, but an informal example to keep in mind for now is the following: If one takes a discrete group (such as $\mathbf{Z}^d$) and rescales it (say to $\frac{1}{N}\mathbf{Z}^d$ for a large parameter N), then intuitively this rescaled group "converges" to a continuous group (in this case $\mathbf{R}^d$). More generally, one can generate locally compact groups (or at least locally compact spaces) out of the limits of (suitably normalised) groups of polynomial growth or approximate groups, which is one of the basic observations that tie the three different topics discussed above together.

As we shall see in Chapter 10, finitely generated groups arise naturally as the *fundamental groups* of compact manifolds. Using the tools of Riemannian geometry (such as the *Bishop-Gromov inequality*), one can relate the growth of such groups to the curvature of a metric on such a manifold. As a consequence, Gromov's theorem and its variants can lead to some nontrivial conclusions about the relationship between the topology of a manifold and its geometry. The following simple consequence is typical:

---

[5]Strictly speaking, he uses *Gromov-Hausdorff limits* instead of ultralimits, but the two types of limits are closely related, as we shall see in Chapter 7.

Proposition 1.3.2. *Let M be a compact Riemannian manifold of nonnegative Ricci curvature. Then the fundamental group $\pi_1(M)$ of M is virtually nilpotent.*

We will discuss this result and some related results (such as a relaxation of the nonnegative curvature hypothesis to an almost nonnegative curvature hypothesis) in Section 10. We also remark that the above proposition can also be proven (with stronger conclusions) by more geometric means, but there are some results of the above type which currently have no known proof that does not employ some version of Gromov's theorem at some point.

Chapter 2

Lie groups, Lie algebras, and the Baker-Campbell-Hausdorff formula

In this chapter, we describe the basic analytic structure theory of Lie groups, by relating them to the simpler concept of a *Lie algebra*. Roughly speaking, the Lie algebra encodes the "infinitesimal" structure of a Lie group, but is a simpler object, being a vector space rather than a nonlinear manifold. Nevertheless, thanks to the fundamental theorems of Lie, the Lie algebra can be used to reconstruct the Lie group (at a local level, at least), by means of the *exponential map* and the *Baker-Campbell-Hausdorff formula*. As such, the local theory of Lie groups is completely described (in principle, at least) by the theory of Lie algebras, which leads to a number of useful consequences, such as the following:

(1) (Local Lie implies Lie) A topological group G is Lie (i.e., it is isomorphic to a Lie group) if and only if it is locally Lie (i.e., the group operations are smooth near the origin).

(2) (Uniqueness of Lie structure) A topological group has at most one smooth structure on it that makes it Lie.

(3) (Weak regularity implies strong regularity, I) Lie groups are automatically real analytic. (In fact one only needs a "local $C^{1,1}$" regularity on the group structure to obtain real analyticity.)

(4) (Weak regularity implies strong regularity, II) A continuous homomorphism from one Lie group to another is automatically smooth (and real analytic).

The connection between Lie groups and Lie algebras also highlights the role of *one-parameter subgroups* of a topological group, which will play a central role in the solution of Hilbert's fifth problem (cf. Figure 1).

Remark 2.0.3. There is also a very important *algebraic* structure theory of Lie groups and Lie algebras, in which the Lie algebra is split into *solvable* and *semisimple* components, with the latter being decomposed further into *simple components*, which can then be completely classified using *Dynkin diagrams*. This classification is of fundamental importance in many areas of mathematics (e.g., representation theory, arithmetic geometry, and group theory), and many of the deeper facts about Lie groups and Lie algebras are proven via this classification (although in such cases it can be of interest to also find alternate proofs that avoid the classification). However, it turns out that we will not need this theory here, and so we will not discuss it further (though it can of course be found in any graduate text on Lie groups and Lie algebras, e.g., [**Bo1968**]).

2.1. Local groups

The connection between Lie groups and Lie algebras will be *local* in nature — the only portion of the Lie group that will be of importance will be the portion that is close to the group identity 1. To formalise this locality, it is convenient to introduce the notion of a *local group* and a *local Lie group*, which are local versions of the concept of a topological group and a Lie group respectively. We will only set up the barest bones of the theory of local groups here; a more detailed discussion is given in Chapter 15.

Definition 2.1.1 (Local group). A *local topological group*

$$G = (G, \Omega, \Lambda, 1, \cdot, ()^{-1}),$$

or *local group* for short, is a topological space G equipped with an identity element $1 \in G$, a partially defined but continuous multiplication operation $\cdot : \Omega \to G$ for some domain $\Omega \subset G \times G$, and a partially defined but continuous inversion operation $()^{-1} : \Lambda \to G$, where $\Lambda \subset G$, obeying the following axioms:

(1) (Local closure) Ω is an open neighbourhood of $G \times \{1\} \cup \{1\} \times G$, and Λ is an open neighbourhood of 1.

(2) (Local associativity) If $g, h, k \in G$ are such that $(g \cdot h) \cdot k$ and $g \cdot (h \cdot k)$ are both well-defined in G, then they are equal. (Note however that

it may be possible for one of these products to be defined but not the other.)

(3) (Identity) For all $g \in G$, $g \cdot 1 = 1 \cdot g = g$.

(4) (Local inverse) If $g \in G$ and g^{-1} are well-defined in G, then[1] $g \cdot g^{-1} = g^{-1} \cdot g = 1$. (In particular this, together with the other axioms, forces $1^{-1} = 1$.)

We will sometimes use additive notation for local groups if the groups are abelian (by which we mean the statement that if $g + h$ is defined, then $h + g$ is also defined and equal to $g + h$.)

A local group is said to be *symmetric* if $\Lambda = G$, i.e., if every element g in G has an inverse g^{-1} that is also in G.

A *local Lie group* is a local group that is also a smooth manifold, in such a fashion that the partially defined group operations $\cdot, ()^{-1}$ are smooth on their domain of definition.

Clearly, every topological group is a local group, and every Lie group is a local Lie group. We will sometimes refer to the former concepts as *global* topological groups and *global* Lie groups in order to distinguish them from their local counterparts. One could also consider local discrete groups, in which the topological structure is just the discrete topology, but we will not need to study such objects in here.

A model class of examples of a local (Lie) group comes from *restricting* a global (Lie) group to an open neighbourhood of the identity. Let us formalise this concept:

Definition 2.1.2 (Restriction). If G is a local group, and U is an open neighbourhood of the identity in G, then we define the *restriction* $G \!\restriction_U$ of G to U to be the topological space U with domains $\Omega \!\restriction_U := \{(g, h) \in \Omega : g, h, g \cdot h \in U\}$ and $\Lambda \!\restriction_U := \{g \in \Lambda : g, g^{-1} \in U\}$, and with the group operations $\cdot, ()^{-1}$ being the restriction of the group operations of G to $\Omega \!\restriction_U$, $\Lambda \!\restriction_U$ respectively. If U is symmetric (in the sense that g^{-1} is well-defined and lies in U for all $g \in U$), then this restriction $G \!\restriction_U$ will also be symmetric. If G is a global or local Lie group, then $G \!\restriction_U$ will also be a local Lie group. We will sometimes abuse notation and refer to the local group $G \!\restriction_U$ simply as U.

Thus, for instance, one can take the Euclidean space $\mathbf{R}^d$, and restrict it to a ball B centred at the origin, to obtain an additive local group $\mathbf{R}^d \!\restriction_B$. In this group, two elements x, y in B have a well-defined sum $x + y$ only when

---

[1] Here we adopt the convention that any mathematical sentence involving an undefined operation is automatically false; thus, for instance, $g \cdot g^{-1} = 1$ is false unless $g \cdot g^{-1}$ is well-defined, so that $(g, g^{-1}) \in \Omega$.

their sum in $\mathbf{R}^d$ stays inside B. Intuitively, this local group behaves like the global group $\mathbf{R}^d$ as long as one is close enough to the identity element 0, but as one gets closer to the boundary of B, the group structure begins to break down.

It is natural to ask the question as to whether *every* local group arises as the restriction of a global group. The answer to this question is somewhat complicated, and can be summarised as "essentially yes in certain circumstances, but not in general"; see Chapter 15.

A key example of a local Lie group arises from pushing forward a Lie group via a coordinate chart near the origin:

Example 2.1.3. Let G be a global or local Lie group of some dimension d, and let $\phi : U \to V$ be a smooth coordinate chart from a neighbourhood U of the identity 1 in G to a neighbourhood V of the origin 0 in $\mathbf{R}^d$, such that ϕ maps 1 to 0. Then we can define a local group $\phi_* G \mid_U$ which is the set V (viewed as a smooth submanifold of $\mathbf{R}^d$) with the local group identity 0, the local group multiplication law $*$ defined by the formula

$$x * y := \phi(\phi^{-1}(x) \cdot \phi^{-1}(y))$$

defined whenever $\phi^{-1}(x), \phi^{-1}(y), \phi^{-1}(x) \cdot \phi^{-1}(y)$ are well-defined and lie in U, and the local group inversion law $()^{*-1}$ defined by the formula

$$x^{*-1} := \phi(\phi^{-1}(x)^{-1})$$

defined whenever $\phi^{-1}(x), \phi^{-1}(x)^{-1}$ are well-defined and lie in U. One easily verifies that $\phi_* G \mid_U$ is a local Lie group. We will sometimes denote this local Lie group as $(V, *)$, to distinguish it from the additive local Lie group $(V, +)$ arising by restriction of $(\mathbf{R}^d, +)$ to V. The precise distinction between the two local Lie groups will in fact be a major focus of this section.

Example 2.1.4. Let G be the Lie group $\mathrm{GL}_n(\mathbf{R})$, and let U be the ball $U := \{g \in \mathrm{GL}_n(\mathbf{R}) : \|g - 1\|_{\mathrm{op}} < 1\}$. If we then let $V \subset M_n(\mathbf{R})$ be the ball $V := \{x \in M_n(\mathbf{R}) : \|x\|_{\mathrm{op}} < 1\}$ and ϕ be the map $\phi(g) := g - 1$, then ϕ is a smooth coordinate chart (after identifying $M_n(\mathbf{R})$ with $\mathbf{R}^{n \times n}$), and by the construction in the preceding exercise, $V = \phi_* G \mid_U$ becomes a local Lie group with the operations

$$x * y := x + y + xy$$

(defined whenever $x, y, x + y + xy$ all lie in V) and

$$x^{*-1} := (1 + x)^{-1} - 1 = x - x^2 + x^3 - \dots$$

(defined whenever x and $(1+x)^{-1} - 1$ both lie in V). Note that this Lie group structure is not equal to the additive structure $(V, +)$ on V, nor is it equal to the multiplicative structure $(V, \cdot)$ on V given by matrix multiplication, which is one of the reasons why we use the symbol $*$ instead of $+$ or $\cdot$ for such structures.

Many (though not all) of the familiar constructions in group theory can be generalised to the local setting, though often with some slight additional subtleties. We will not systematically do so here, but we give a single such generalisation for now:

Definition 2.1.5 (Homomorphism). A *continuous homomorphism* $\phi : G \to H$ between two local groups G, H is a continuous map from G to H with the following properties:

(i) ϕ maps the identity 1_G of G to the identity 1_H of H: $\phi(1_G) = 1_H$.

(ii) If $g \in G$ is such that g^{-1} is well-defined in G, then $\phi(g)^{-1}$ is well-defined in H and is equal to $\phi(g^{-1})$.

(iii) If $g, h \in G$ are such that $g \cdot h$ is well-defined in G, then $\phi(g) \cdot \phi(h)$ is well-defined and equal to $\phi(g \cdot h)$.

A *smooth homomorphism* $\phi : G \to H$ between two local Lie groups G, H is a continuous homomorphism that is also smooth.

A *(continuous) local homomorphism* $\phi : U \to H$ between two local groups G, H is a continuous homomorphism from an open neighbourhood U of the identity in G to H. Two local homomorphisms are said to be *equivalent* if they agree on a (possibly smaller) open neighbourhood of the identity. One can of course define the notion of a smooth local homomorphism similarly.

It is easy to see that the composition of two continuous homomorphisms is again a continuous homomorphism, and that the identity map on a local group is automatically a continuous homomorphism; this gives the class of local groups the structure of a *category*. Similarly, the class of local Lie groups with their smooth homomorphisms is also a category.

Example 2.1.6. With the notation of Example 2.1.3, $\phi : U \to V$ is a smooth homomorphism from the local Lie group $G \downharpoonright_U$ to the local Lie group $\phi_* G \downharpoonright_U$. In fact, it is a smooth isomorphism, since $\phi^{-1} : V \to U$ provides the inverse homomorphism.

Let us say that a word $g_1 \ldots g_n$ in a local group G is *well-defined in G* (or *well-defined*, for short) if every possible way of associating this word using parentheses is well-defined from applying the product operation. For instance, in order for $abcd$ to be well-defined, $((ab)c)d$, $(a(bc))d$, $(ab)(cd)$, $a(b(cd))$, and $a((bc)d)$ must all be well-defined. For instance, in the additive local group $\{-9, \ldots, 9\}$ (with the group structure restricted from that of the integers $\mathbf{Z}$), $-2 + 6 + 5$ is not well-defined because one of the ways of associating this sum, namely $-2 + (6 + 5)$, is not well-defined (even though $(-2 + 6) + 5$ is well-defined).

Exercise 2.1.1 (Iterating the associative law)**.**

(i) Show that if a word $g_1 \ldots g_n$ in a local group G is well-defined, then all ways of associating this word give the same answer, and so we can uniquely evaluate $g_1 \ldots g_n$ as an element in G.

(ii) Give an example of a word $g_1 \ldots g_n$ in a local group G which has two ways of being associated that are both well-defined, but give *different* answers. (*Hint:* The local associativity axiom prevents this from happening for $n \leq 3$, so try $n = 4$. A small discrete local group will already suffice to give a counterexample; verifying the local group axioms are easier if one makes the domain of definition of the group operations as small as one can get away with while still having the counterexample.)

Exercise 2.1.2. Show that the number of ways to associate a word $g_1 \ldots g_n$ is given by the *Catalan number* $C_{n-1} := \frac{1}{n}\binom{2n-2}{n-1}$.

Exercise 2.1.3. Let G be a local group, and let $m \geq 1$ be an integer. Show that there exists a symmetric open neighbourhood U_m of the identity such that every word of length m in U_m is well-defined in G (or more succinctly, U_m^m is well-defined). (Note, though, that these words will usually only take values in G, rather than in U_m, and also the sets U_m tend to become smaller as m increases.)

2.2. Some differential geometry

To define the Lie algebra of a Lie group, we must first quickly recall some basic notions from differential geometry associated to smooth manifolds (which are not necessarily embedded in some larger Euclidean space, but instead exist intrinsically as abstract geometric structures). This requires a certain amount of abstract formalism in order to define things rigorously, though for the purposes of visualisation, it is more intuitive to view these concepts from a more informal geometric perspective.

We begin with the concept of the tangent space and related structures.

Definition 2.2.1 (Tangent space)**.** Let M be a smooth d-dimensional manifold. At every point x of this manifold, we can define the *tangent space* $T_x M$ of M at x. Formally, this tangent space can be defined as the space of all continuously differentiable curves $\gamma : I \to G$ defined on an open interval I containing 0 with $\gamma(0) = x$, modulo the relation that two curves γ_1, γ_2 are considered equivalent if they have the same derivative at 0, in the sense that

$$\frac{d}{dt}\phi(\gamma_1(t))|_{t=0} = \frac{d}{dt}\phi(\gamma_2(t))|_{t=0}$$

where $\phi : U \to V$ is a coordinate chart of G defined in a neighbourhood of x; it is easy to see from the chain rule that this equivalence is independent of the actual choice of ϕ. Using such a coordinate chart, one can identify the tangent space $T_x M$ with the Euclidean space $\mathbf{R}^d$, by identifying γ with $\frac{d}{dt}\phi(\gamma(t))|_{t=0}$. One easily verifies that this gives $T_x M$ the structure of a d-dimensional vector space, in a manner which is independent of the choice of coordinate chart ϕ. Elements of $T_x M$ are called *tangent vectors* of M at x. If $\gamma : I \to G$ is a continuously differentiable curve with $\gamma(0) = x$, the equivalence class of γ in $T_x M$ will be denoted $\gamma'(0)$.

The space $TM := \bigcup_{x \in M} (\{x\} \times T_x M)$ of pairs (x, v), where x is a point in M and v is a tangent vector of M at x, is called the *tangent bundle*.

If $\Phi : M \to N$ is a smooth map between two manifolds, we define the *derivative map* $D\Phi : TM \to TN$ to be the map defined by setting

$$D\Phi((x, \gamma'(0))) := (\Phi(x), (\Phi \circ \gamma)'(0))$$

for all continuously differentiable curves $\gamma : I \to G$ with $\gamma(0) = x$ for some $x \in M$; one can check that this map is well-defined. We also write $(\Phi(x), D\Phi(x)(v))$ for $D\Phi(x, v)$, so that for each $x \in M$, $D\Phi(x)$ is a map from $T_x M$ to $T_{\Phi(x)} N$. One can easily verify that this latter map is linear. We observe the *chain rule*[2]

$$(2.1) \qquad\qquad D(\Psi \circ \Phi) = (D\Psi) \circ (D\Phi)$$

for any smooth maps $\Phi : M \to N$, $\Psi : N \to O$.

Observe that if V is an open subset of $\mathbf{R}^d$, then TV may be identified with $V \times \mathbf{R}^d$. In particular, every coordinate chart $\phi : U \to V$ of M gives rise to a coordinate chart $D\phi : TU \to V \times \mathbf{R}^d$ of TM, which gives TM the structure of a smooth $2d$-dimensional manifold.

Remark 2.2.2. Informally, one can think of a tangent vector (x, v) as an infinitesimal vector from the point x of M to a nearby point $x + \varepsilon v + O(\varepsilon^2)$ on M, where $\varepsilon > 0$ is infinitesimally small; a smooth map ϕ then sends $x + \varepsilon v + O(\varepsilon^2)$ to $\phi(x) + \varepsilon D\phi(x)(v) + O(\varepsilon^2)$. One can make this informal perspective rigorous by means of *nonstandard analysis*, but we will not do so here.

Once one has the notion of a tangent bundle, one can define the notion of a smooth vector field:

---

[2]Indeed, one can view the tangent operator T and the derivative operator D together as a single *covariant functor* from the category of smooth manifolds to itself, although we will not need to use this perspective here.

Definition 2.2.3 (Vector fields). A *smooth vector field* on M is a smooth map $X : M \to TM$ which is a right inverse for the projection map $\pi : TM \to M$, thus (by slight abuse of notation) X maps x to $(x, X(x))$ for some $X(x) \in T_x M$. The space of all smooth vector fields is denoted $\Gamma(TM)$. It is clearly a real vector space. In fact, it is a $C^\infty(M)$-module: given a smooth vector field $X \in \Gamma(TM)$ and a smooth function $f \in C^\infty(M)$ (i.e., a smooth map $f : M \to \mathbf{R}$), one can define the product fX in the obvious manner: $fX(x) := f(x)X(x)$, and one easily verifies the axioms for a module.

Given a smooth function $f \in C^\infty(M)$ and a smooth vector field $X \in \Gamma(TM)$, we define the *directional derivative* $\nabla_X f \in C^\infty(M)$ of f along X by the formula

$$\nabla_X f(x) := \frac{d}{dt} f(\gamma(t))|_{t=0}$$

whenever $\gamma : I \to M$ is a continuously differentiable function with $\gamma(0) = x$ and $\gamma'(0) = X(x)$; one easily verifies that $\nabla_X f$ is well-defined and is an element of $C^\infty(M)$.

Remark 2.2.4. One can define $\nabla_X f$ in a more "coordinate free" manner as

$$\nabla_X f = \eta \circ Df \circ X,$$

where $\eta : T\mathbf{R} \to \mathbf{R}$ is the projection map to the second coordinate of $T\mathbf{R} \equiv \mathbf{R} \times \mathbf{R}$; one can also view $\nabla_X f$ as the *Lie derivative* of f along X (although, in most texts, the latter definition would be circular, because the Lie derivative is usually defined using the directional derivative).

Remark 2.2.5. If V is an open subset of $\mathbf{R}^d$, a smooth vector field on V can be identified with a smooth map $X : V \to \mathbf{R}^d$ from V to $\mathbf{R}^d$. If $X : M \to TM$ is a smooth vector field on M and $\phi : U \to V$ is a coordinate chart of M, then the *pushforward* $\phi_* X := D\phi \circ X \circ \phi^{-1} : V \to TV$ of X by ϕ is a smooth vector field of V. Thus, in coordinates, one can view vector fields as maps from open subsets of $\mathbf{R}^d$ to $\mathbf{R}^d$. This perspective is convenient for quick and dirty calculations; for instance, in coordinates, the directional derivative $\nabla_X f$ is the same as the familiar directional derivative $X \cdot \nabla f$ from several variable calculus. If, however, one wishes to perform several changes of variable, then the more intrinsically geometric (and "coordinate-free") perspective outlined above can be more helpful.

There is a fundamental link between smooth vector fields and derivations of $C^\infty(M)$:

Exercise 2.2.1 (Correspondence between smooth vector fields and derivations). Let M be a smooth manifold.

(i) If $X \in \Gamma(TM)$ is a smooth vector field, show that $\nabla_X : C^\infty(M) \to C^\infty(M)$ is a *derivation* on the (real) algebra $C^\infty(M)$, i.e., a (real) linear map that obeys the Leibniz rule

(2.2) $$\nabla_X(fg) = f\nabla_X g + (\nabla_X f)g$$

for all $f, g \in C^\infty(M)$.

(ii) Conversely, if $d : C^\infty(M) \to C^\infty(M)$ is a derivation on $C^\infty(M)$, show that there exists a unique smooth vector field X such that $d = \nabla_X$.

We see from the above exercise that smooth vector fields can be interpreted as a purely algebraic construction associated to the real algebra $C^\infty(M)$, namely as the space of derivations on that vector space. This can be useful for analysing the algebraic structure of such vector fields. Indeed, we have the following basic algebraic observation:

Exercise 2.2.2 (Commutator of derivations is a derivation)**.** Let $d_1, d_2 : A \to A$ be two derivations on an algebra A. Show that the commutator $[d_1, d_2] := d_1 \circ d_2 - d_2 \circ d_1$ is also a derivation on A.

From the preceding two exercises, we can define the *Lie bracket* $[X, Y]$ of two vector fields $X, Y \in \Gamma(TM)$ by the formula

$$\nabla_{[X,Y]} := [\nabla_X, \nabla_Y].$$

This gives the space $\Gamma(TM)$ of smooth vector fields the structure of an (infinite-dimensional) *Lie algebra*:

Definition 2.2.6 (Lie algebra)**.** A (real) Lie algebra is a real vector space V (possibly infinite dimensional), together with a bilinear map $[,] : V \times V \to V$ which is anti-symmetric (thus $[X, Y] = -[Y, X]$ for all $X, Y \in V$, or equivalently $[X, X] = 0$ for all $X \in V$) and obeys the *Jacobi identity*

(2.3) $$[[X, Y], Z] + [[Y, Z], X] + [[Z, X], Y] = 0$$

for all $X, Y, Z \in V$.

Exercise 2.2.3. If M is a smooth manifold, show that $\Gamma(TM)$ (equipped with the Lie bracket) is a Lie algebra.

Remark 2.2.7. This is the abstract definition of a Lie algebra. A more concrete definition would be to let V be a subspace of an algebra of operators, and to define the Lie bracket as the commutator. The relation between the two notions of a Lie algebra is explored in Chapter 13.

2.3. The Lie algebra of a Lie group

Let G be a (global) Lie group. By definition, G is then a smooth manifold, so we can thus define the tangent bundle TG and smooth vector fields $X \in \Gamma(TG)$ as in the preceding section. In particular, we can define the tangent space T_1G of G at the identity element 1.

If $g \in G$, then the left multiplication operation $\rho_g^{\text{left}} : x \mapsto gx$ is, by definition of a Lie group, a smooth map from G to G. This creates a derivative map $D\rho_g^{\text{left}} : TG \to TG$ from the tangent bundle TG to itself. We say that a vector field $X \in \Gamma(TG)$ is *left-invariant* if one has $(\rho_g^{\text{left}})_* X = X$ for all $g \in G$, or equivalently if $(D\rho_g^{\text{left}}) \circ X = X \circ \rho_g^{\text{left}}$ for all $g \in G$.

Exercise 2.3.1. Let G be a (global) Lie group.

(i) Show that for every element x of T_1G there is a unique left-invariant vector field $X \in \Gamma(TG)$ such that $X(1) = x$.

(ii) Show that the commutator $[X, Y]$ of two left-invariant vector fields is again a left-invariant vector field.

From the above exercise, we can identify the tangent space T_1G with the left-invariant vector fields on TG, and the Lie bracket structure on the latter then induces a Lie bracket (which we also call $[,]$) on T_1G. The vector space T_1G together with this Lie bracket is then a (finite-dimensional) Lie algebra, which we call the *Lie algebra* of the Lie group G, and we write as $\mathfrak{g}$.

Remark 2.3.1. Informally, an element x of the Lie algebra $\mathfrak{g}$ is associated with an infinitesimal perturbation $1 + \varepsilon x + O(\varepsilon^2)$ of the identity in the Lie group G. This intuition can be formalised fairly easily in the case of matrix Lie groups such as $\text{GL}_n(\mathbf{C})$; for more abstract Lie groups, one can still formalise things using nonstandard analysis, but we will not do so here.

Exercise 2.3.2.

(i) Show that the Lie algebra $\mathfrak{gl}_n(\mathbf{C})$ of the general linear group $\text{GL}_n(\mathbf{C})$ can be identified with the space $M_n(\mathbf{C})$ of $n \times n$ complex matrices, with the Lie bracket $[A, B] := AB - BA$.

(ii) Describe the Lie algebra $\mathfrak{u}_n(\mathbf{C})$ of the unitary group $\text{U}_n(\mathbf{C})$.

(iii) Describe the Lie algebra $\mathfrak{su}_n(\mathbf{C})$ of the special unitary group $\text{SU}_n(\mathbf{C})$.

(iv) Describe the Lie algebra $\mathfrak{o}_n(\mathbf{R})$ of the orthogonal $\text{O}_n(\mathbf{R})$.

(v) Describe the Lie algebra $\mathfrak{so}_n(\mathbf{R})$ of the special orthogonal $\text{SO}_n(\mathbf{R})$.

(vi) Describe the Lie algebra of the Heisenberg group $\begin{pmatrix} 1 & \mathbf{R} & \mathbf{R} \\ 0 & 1 & \mathbf{R} \\ 0 & 0 & 1 \end{pmatrix}$.

Exercise 2.3.3. Let $\phi : G \to H$ be a smooth homomorphism between (global) Lie groups. Show that the derivative map $D\phi(1_G)$ at the identity element 1_G is then a Lie algebra homomorphism from the Lie algebra $\mathfrak{g}$ of G to the Lie algebra $\mathfrak{h}$ of H (thus this map is linear and preserves the Lie bracket). (From this and the chain rule (2.1), we see that the map $\phi \mapsto D\phi(1_G)$ creates a covariant functor from the category of Lie groups to the category of Lie algebras.)

We have seen that every global Lie group gives rise to a Lie algebra. One can also associate Lie algebras to *local* Lie groups as follows:

Exercise 2.3.4. Let G be a local Lie group. Let U be a symmetric neighbourhood of the identity in G. (It is not difficult to see that at least one such neighbourhood exists.) Call a vector field $X \in \Gamma(TU)$ *left-invariant* if, for every $g \in U$, one has $(\rho_g^{\text{left}})_* X(g) = X(g)$, where ρ_g^{left} is the left-multiplication map $x \mapsto gx$, defined on the open set $\{x \in U : gx \in U\}$ (where we adopt the convention that $gx \in U$ is shorthand for "$g \cdot x$ is well-defined and lies in U").

    (i) Establish the analogue of Exercise 2.3.1 in this setting. Conclude that one can give $T_1 G$ the structure of a Lie algebra, which is independent of the choice of U.

    (ii) Establish the analogue of Exercise 2.3.3 in this setting.

Remark 2.3.2. In the converse direction, it is also true that every finite-dimensional Lie algebra can be associated to either a local or a global Lie group; this is known as *Lie's third theorem*. However, this theorem is somewhat tricky to prove (particularly if one wants to associate the Lie algebra with a *global* Lie group), requiring the nontrivial algebraic tool of *Ado's theorem* (discussed in Section 13); see Exercise 2.5.6 below.

2.4. The exponential map

The *exponential map* $x \mapsto \exp(x)$ on the reals $\mathbf{R}$ (or its extension to the complex numbers $\mathbf{C}$) is of course fundamental to modern analysis. It can be defined in a variety of ways, such as the following:

    (i) $\exp : \mathbf{R} \to \mathbf{R}$ is the differentiable map obeying the ODE $\frac{d}{dx} \exp(x) = \exp(x)$ and the initial condition $\exp(0) = 1$.

    (ii) $\exp : \mathbf{R} \to \mathbf{R}$ is the differentiable map obeying the homomorphism property $\exp(x+y) = \exp(x)\exp(y)$ and the initial condition $\frac{d}{dx} \exp(x)|_{x=0} = 1$.

    (iii) $\exp : \mathbf{R} \to \mathbf{R}$ is the limit of the functions $x \mapsto (1 + \frac{x}{n})^n$ as $n \to \infty$.

    (iv) $\exp : \mathbf{R} \to \mathbf{R}$ is the limit of the infinite series $x \mapsto \sum_{n=0}^{\infty} \frac{x^n}{n!}$.

We will need to generalise this map to arbitrary Lie algebras and Lie groups. In the case of matrix Lie groups (and matrix Lie algebras), one can use the matrix exponential, which can be defined efficiently by modifying definition (iv) above, and which was already discussed in Section 1. It is however difficult to use this definition for abstract Lie algebras and Lie groups. The definition based on (ii) will ultimately be the best one to use for the purposes of this text, but for foundational purposes (i) or (iii) is initially easier to work with. In most of the foundational literature on Lie groups and Lie algebras, one uses (i), in which case the existence and basic properties of the exponential map can be provided by the *Picard existence theorem* from the theory of ordinary differential equations. However, we will use (iii), because it relies less heavily on the smooth structure of the Lie group, and will therefore be more aligned with the spirit of Hilbert's fifth problem (which seeks to minimise the reliance of smoothness hypotheses whenever possible). Actually, for minor technical reasons it is slightly more convenient to work with the limit of $(1 + \frac{x}{2^n})^{2^n}$ rather than $(1 + \frac{x}{n})^n$.

We turn to the details. It will be convenient to work in local coordinates, and for applications to Hilbert's fifth problem it will be useful to "forget" almost all of the smooth structures. We make the following definition:

Definition 2.4.1 ($C^{1,1}$ *local group*). A $C^{1,1}$ *local group* is a local group V that is an open neighbourhood of the origin 0 in a Euclidean space $\mathbf{R}^d$, with group identity 0, and whose group operation $*$ obeys the estimate

$$(2.4) \qquad\qquad x * y = x + y + O(|x||y|)$$

for all sufficiently small x, y, where the implied constant in the $O()$ notation can depend on V but is uniform in x, y.

Example 2.4.2. Let G be a local Lie group of some dimension d, and let $\phi : U \to V$ be a smooth coordinate chart that maps a neighbourhood U of the group identity 1 to a neighbourhood V of the origin 0 in $\mathbf{R}^d$, with $\phi(1) = 0$. Then, as explained in Example 2.1.3, $V = (V, *) = \phi_* G \restriction_U$ is a local Lie group with identity 0; in particular, one has

$$0 * x = x * 0 = x.$$

From Taylor expansion (using the smoothness of $*$) we thus have (2.4) for sufficiently small x, y. Thus we see that every local Lie group generates a $C^{1,1}$ local group when viewed in coordinates.

Remark 2.4.3. In real analysis, a (locally) $C^{1,1}$ function is a function $f : U \to \mathbf{R}^m$ on a domain $U \subset \mathbf{R}^n$ which is continuously differentiable (i.e., in the regularity class C^1), and whose first derivatives ∇f are (locally) Lipschitz (i.e., in the regularity class $C^{0,1}$) the $C^{1,1}$ regularity class is slightly weaker (i.e., larger) than the class C^2 of twice continuously differentiable functions, but much stronger than the class C^1 of singly continuously differentiable functions. See [**Ta2010**, §1.14] for more on these sorts of regularity classes. The reason for the terminology $C^{1,1}$ in the above definition is that $C^{1,1}$ regularity is essentially the minimal regularity for which one has the Taylor expansion

$$f(x) = f(x_0) + \nabla f(x_0) \cdot (x - x_0) + O(|x - x_0|^2)$$

for any x_0 in the domain of f, and any x sufficiently close to x_0; note that the asymptotic (2.4) is of this form.

We now estimate various expressions in a $C^{1,1}$ local group.

Exercise 2.4.1. Let V be a $C^{1,1}$ local group. Throughout this exercise, the implied constants in the $O()$ notation can depend on V, but not on parameters such as x, y, ε, k, n.

(i) Show that there exists an $\varepsilon > 0$ such that one has

$$(2.5) \qquad x_1 * \cdots * x_k = x_1 + \cdots + x_k + O\left(\sum_{1 \le i < j \le k} |x_i||x_j|\right)$$

whenever $k \ge 1$ and $x_1, \ldots, x_k \in V$ are such that $\sum_{i=1}^k |x_i| \le \varepsilon$, and the implied constant is uniform in k. Here and in the sequel we adopt the convention that a statement such as (2.5) is automatically false unless all expressions in that statement are well-defined. (*Hint:* Induct on k using (2.4). It is best to replace the asymptotic $O()$ notation by explicit constants C in order to ensure that such constants remain uniform in k.) In particular, one has the crude estimate

$$x_1 * \cdots * x_k = O(\sum_{i=1}^k |x_i|)$$

under the same hypotheses as above.

(ii) Show that one has

$$x^{*-1} = -x + O(|x|^2)$$

for x sufficiently close to the origin.

(iii) Show that

$$x * y * x^{*-1} * y^{*-1} = O(|x||y|)$$

for x, y sufficiently close to the origin. (*Hint:* First show that $x * y = y * x + O(|x||y|)$, then express $x * y$ as the product of $x * y * x^{*-1} * y^{*-1}$ and $y * x$.)

(iv) Show that
$$x * y * x^{*-1} = y + O(|x||y|)$$
whenever x, y are sufficiently close to the origin.

(v) Show that
$$y * x^{*-1}, x^{*-1} * y = O(|x - y|)$$
whenever x, y are sufficiently close to the origin.

(vi) Show that there exists an $\varepsilon > 0$ such that
$$x_1 * \cdots * x_k = y_1 * \cdots * y_k + O\left(\sum_{i=1}^{k} |x_i - y_i|\right)$$
whenever $k \geq 1$ and $x_1, \ldots, x_k, y_1, \ldots, y_k$ are such that
$$\sum_{i=1}^{k} |x_i|, \sum_{j=1}^{k} |y_i| \leq \varepsilon.$$

(vii) Show that there exists an $\varepsilon > 0$ such that
$$\frac{1}{2}|n||x - y| \leq |x^{*n} - y^{*n}| \leq 2|n||x - y|$$
for all $n \in \mathbf{Z}$ and $x, y \in \mathbf{R}^d$ such that $|nx|, |ny| \leq \varepsilon$, where $x^{*n} = x * \cdots * x$ is the product of n copies of x (assuming of course that this product is well-defined) for $n \geq 0$, and $x^{*-n} := (x^{*n})^{*-1}$.

(viii) Show that there exists an $\varepsilon > 0$ such that
$$(xy)^{*n} = x^{*n}y^{*n} + O\left(|n|^2|x||y|\right)$$
for all $n \in \mathbf{Z}$ and $x, y \in \mathbf{R}^d$ such that $|nx|, |ny| \leq \varepsilon$. (*Hint:* Do the case when n is positive first. In that case, express $x^{*-n} * (xy)^{*n}$ as the product of n conjugates of y by various powers of x.)

We can now define the *exponential map* $\exp : V' \to V$ on this $C^{1,1}$ local group by defining

(2.6)
$$\exp(x) := \lim_{n \to \infty} \left(\frac{1}{2^n}x\right)^{*2^n}$$

for any x in a sufficiently small neighbourhood V' of the origin in V.

Exercise 2.4.2. Let V be a local $C^{1,1}$ group.

(i) Show that if V' is a sufficiently small neighbourhood of the origin in V, then the limit in (2.6) exists for all $x \in V'$. (*Hint:* Use the previous exercise to estimate the distance between $(\frac{1}{2^n}x)^{*2^n}$ and $(\frac{1}{2^{n+1}}x)^{*2^{n+1}}$.) Establish the additional estimate

(2.7)
$$\exp(x) = x + O(|x|^2).$$

(ii) Show that if $\gamma : I \to G$ is a smooth curve with $\gamma(0) = 1$, and $\gamma'(0)$ is sufficiently small, then

$$\exp(\gamma'(0)) = \lim_{n \to \infty} \gamma(1/2^n)^{*2^n}.$$

(iii) Show that for all sufficiently small x, y, one has the bilipschitz property

$$|(\exp(x) - \exp(y)) - (x - y)| \leq \frac{1}{2}|x - y|.$$

Conclude, in particular, that for V' sufficiently small, exp is a homeomorphism between V' and an open neighbourhood $\exp(V')$ of the origin. (*Hint:* To show that $\exp(V')$ contains a neighbourhood of the origin, use (2.7) and the contraction mapping theorem.)

(iv) Show that

(2.8) $$\exp(sx) * \exp(tx) = \exp((s + t)x)$$

for $s, t \in \mathbf{R}$ and $x \in \mathbf{R}^d$ with sx, tx sufficiently small. (*Hint:* First handle the case when $s, t \in \mathbf{Z}[\frac{1}{2}]$ are dyadic numbers.)

(v) Show that for any sufficiently small $x, y \in \mathbf{R}^d$, one has

(2.9) $$\exp(x + y) = \lim_{n \to \infty} (\exp(x/2^n) * \exp(y/2^n))^{*2^n}.$$

Then conclude the stronger estimate

(2.10) $$\exp(x + y) = \lim_{n \to \infty} (\exp(x/n) * \exp(y/n))^{*n}.$$

(vi) Show that for any sufficiently small $x, y \in \mathbf{R}^d$, one has

$$\exp(x + y) = \exp(x) * \exp(y) + O(|x||y|).$$

(*Hint:* Use the previous part, as well as Exercise 2.4.1(viii).)

Let us say that a $C^{1,1}$ local group is *radially homogeneous* if one has

(2.11) $$sx * tx = (s + t)x$$

whenever $s, t \in \mathbf{R}$ and $x \in \mathbf{R}^d$ are such that sx, tx are sufficiently small. (In particular, this implies that $x^{*-1} = -x$ for sufficiently small x.) From the above exercise, we see that any $C^{1,1}$ local group V can be made into a radially homogeneous $C^{1,1}$ local group V' by first restricting to an open neighbourhood $\exp(V')$ of the identity, and then applying the logarithmic homeomorphism $\exp^{-1}$. Thus:

Corollary 2.4.4. *Every $C^{1,1}$ local group has a neighbourhood of the identity which is isomorphic (as a topological group) to a radially homogeneous $C^{1,1}$ local group.*

Now we study the exponential map on global Lie groups. If G is a global Lie group, and $\mathfrak{g}$ is its Lie algebra, we define the exponential map $\exp : \mathfrak{g} \to G$ on a global Lie group G by setting

$$\exp(\gamma'(0)) := \lim_{n \to \infty} \gamma(1/2^n)^{2^n}$$

whenever $\gamma : I \to G$ is a smooth curve with $\gamma(0) = 1$.

Exercise 2.4.3. Let G be a global Lie group.

(i) Show that the exponential map is well-defined. (*Hint:* First handle the case when $\gamma'(0)$ is small, using the previous exercise, then bootstrap to larger values of $\gamma'(0)$.)

(ii) Show that for all $x, y \in \mathfrak{g}$ and $s, t \in \mathbf{R}$, one has

$$\exp(sx)\exp(tx) = \exp((s+t)x) \tag{2.12}$$

and

$$\exp(x+y) = \lim_{n \to \infty} (\exp(x/n)\exp(y/n))^n. \tag{2.13}$$

(*Hint:* Again, begin with the case when x, y are small.)

(iii) Show that the exponential map is continuous.

(iv) Show that for each $x \in \mathfrak{g}$, the function $t \mapsto \exp(tx)$ is the unique homomorphism from $\mathbf{R}$ to G that is differentiable at $t = 0$ with derivative equal to x.

Proposition 2.4.5 (Lie's first theorem). *Let G be a Lie group. Then the exponential map is smooth. Furthermore, there is an open neighbourhood U of the origin in $\mathfrak{g}$ and an open neighbourhood V of the identity in G such that the exponential map $\exp$ is a diffeomorphism from U to V.*

Proof. We begin with the smoothness. From the homomorphism property we see that

$$\frac{d}{dt}\exp(tx) = \left(\rho_{\exp(tx)}^{\text{left}}\right)_* x$$

for all $x \in \mathfrak{g}$ and $t \in \mathbf{R}$. If x and t are sufficiently small, and one uses a coordinate chart ϕ near the origin, the function $f(t, x) := \phi(\exp(tx))$ then satisfies an ODE of the form

$$\frac{d}{dt}f(t, x) = F(f(t, x), x)$$

for some smooth function F, with initial condition $f(0, x) = 0$; thus by the fundamental theorem of calculus we have

$$f(t, x) = \int_0^t F(f(t', x), x) \, dt'. \tag{2.14}$$

Now let $k \geq 0$. An application of the contraction mapping theorem (in the function space $L_t^\infty C_x^k$ localised to small region of spacetime) then shows that f lies in $L_t^\infty C_x^k$ for small enough t, x, and by further iteration of the integral equation we then conclude that $f(t, x)$ is k times continuously differentiable for small enough t, x. By (2.8) we then conclude that exp is smooth everywhere.

Since
$$\frac{d}{dt} \exp(tx)|_{t=0} = x$$
we see that the derivative of the exponential map at the origin is the identity map on $\mathfrak{g}$. The second claim of the proposition thus follows from the inverse function theorem. $\square$

In view of this proposition, we see that given a vector space basis $X_1, \ldots, X_d$ for the Lie algebra $\mathfrak{g}$, we may obtain a smooth coordinate chart $\phi : U \to V$ for some neighbourhood U of the identity and neighbourhood V of the origin in $\mathbf{R}^d$ by defining
$$\tilde{\phi}(\exp(t_1 X_1 + \cdots + t_d X_d)) := (t_1, \ldots, t_d)$$
for sufficiently small $t_1, \ldots, t_d \in \mathbf{R}$. These are known as *exponential coordinates of the first kind*. Although we will not use them much here, we also note that there are *exponential coordinates of the second kind*, in which the expression $\exp(t_1 X_1 + \cdots + t_d X_d)$ is replaced by the slight variant $\exp(t_1 X_1) \ldots \exp(t_d X_d)$.

Using exponential coordinates of the first kind, we see that we may identify a local piece U of the Lie group G with the radially homogeneous $C^{1,1}$ local group V. In the next section, we will analyse such radially homogeneous $C^{1,1}$ groups further. For now, let us record some easy consequences of the existence of exponential coordinates. Define a *one-parameter subgroup* of a topological group G to be a continuous homomorphism $\phi : \mathbf{R} \to G$ from $\mathbf{R}$ to G.

Exercise 2.4.4 (Classification of one-parameter subgroups). Let G be a Lie group. For any $X \in \mathfrak{g}$, show that the map $t \mapsto \exp(tX)$ is a one-parameter subgroup. Conversely, if $\phi : \mathbf{R} \to G$ is a one-parameter subgroup, there exists a unique $X \in \mathfrak{g}$ such that $\phi(t) = \exp(tX)$ for all $t \in \mathbf{R}$. (*Hint:* Mimic the proof of Proposition 1.0.1.)

Proposition 2.4.6 (Weak regularity implies strong regularity). *Let G, H be global Lie groups, and let $\Phi : G \to H$ be a continuous homomorphism. Then Φ is smooth.*

Proof. Since Φ is a continuous homomorphism, it maps one-parameter subgroups of G to one-parameter subgroups of H. Thus, for every $X \in \mathfrak{g}$, there exists a unique element $L(X) \in \mathfrak{h}$ such that

$$\Phi(\exp(tX)) = \exp(tL(X))$$

for all $t \in \mathbf{R}$. In particular, we see that L is homogeneous: $L(sX) = sL(X)$ for all $X \in \mathfrak{g}$ and $s \in \mathbf{R}$. Next, we observe using (2.9) and the fact that Φ is a continuous homomorphism that for any $X, Y \in \mathfrak{g}$ and $t \in \mathbf{R}$, one has

$$
\begin{aligned}
\Phi(\exp(t(X+Y))) &= \Phi\left(\lim_{n\to\infty} (\exp(tX/2^n)\exp(tY/2^n))^{2^n}\right) \\
&= \lim_{n\to\infty} (\Phi(\exp(tX/2^n))\Phi(\exp(tY/2^n)))^{2^n} \\
&= \lim_{n\to\infty} (\exp(tL(X)/2^n)\exp(tL(Y)/2^n))^{2^n} \\
&= \exp(t(L(X)+L(Y)))
\end{aligned}
$$

and thus L is additive:

$$L(X+Y) = L(X) + L(Y).$$

We conclude that L is a linear transformation from the finite-dimensional vector space $\mathfrak{g}$ to the finite-dimensional vector space $\mathfrak{h}$. In particular, L is smooth. On the other hand, we have

$$\Phi(\exp(X)) = \exp(L(X)).$$

Since $\exp : \mathfrak{g} \to G$ and $\exp : \mathfrak{h} \to H$ are diffeomorphisms near the origin, we conclude that Φ is smooth in a neighbourhood of the identity. Using the homomorphism property (and the fact that the group operations are smooth for both G and H) we conclude that Φ is smooth everywhere, as required. $\qquad\square$

This fact has a pleasant corollary:

Corollary 2.4.7 (Uniqueness of Lie structure). *Any (global) topological group can be made into a Lie group in at most one manner. More precisely, given a topological group G, there is at most one smooth structure one can place on G that makes the group operations smooth.*

Proof. Suppose for the sake of contradiction that one could find two different smooth structures on G that make the group operations smooth, leading to two different Lie groups G', G'' based on G. The identity map from G' to G'' is a continuous homomorphism, and hence smooth by the preceding proposition; similarly for the inverse map from G'' to G'. This implies that the smooth structures coincide, and the claim follows. $\qquad\square$

Note that a general high-dimensional topological manifold may have more than one smooth structure, which may even be nondiffeomorphic to each other (as the example of *exotic spheres* [**Mi1956**] demonstrates), so this corollary is not entirely vacuous.

Exercise 2.4.5. Let G be a connected (global) Lie group, let H be another (global) Lie group, and let $\Phi : G \to H$ be a continuous homomorphism (which is thus smooth by Proposition 2.4.6). Show that Φ is uniquely determined by the derivative map $D\Phi(1) : \mathfrak{g} \to \mathfrak{h}$. In other words, if $\Phi' : G \to H$ is another continuous homomorphism with $D\Phi(1) = D\Phi'(1)$, then $\Phi = \Phi'$. (*Hint:* First prove this in a small neighbourhood of the origin. What group does this neighbourhood generate?) What happens if G is not connected?

Exercise 2.4.6 (Weak regularity implies strong regularity, local version). Let G, H be local Lie groups, and let $\Phi : G \to H$ be a continuous homomorphism. Show that Φ is smooth in a neighbourhood of the identity in G.

Now we can establish the final stage, at least, of the program outlined in Figure 1:

Exercise 2.4.7 (Local Lie implies Lie). Let G be a global topological group. Suppose that there is an open neighbourhood U of the identity such that the local group $G \downarrow_U$ can be given the structure of a local Lie group. Show that G can be given the structure of a global Lie group. (*Hint:* We already have at least one coordinate chart on G; translate it around to create an atlas of such charts. To show compatibility of the charts and global smoothness of the group, one needs to show that the conjugation maps $x \mapsto gxg^{-1}$ are smooth near the origin for any $g \in G$. To prove this, use Exercise 2.4.6.)

2.5. The Baker-Campbell-Hausdorff formula

We now study radially homogeneous $C^{1,1}$ local groups in more detail, in particular, filling in some of the last few steps in the program in Figure 1. We will show

Theorem 2.5.1 (Baker-Campbell-Hausdorff formula, qualitative version). *Let $V \subset \mathbf{R}^d$ be a radially homogeneous $C^{1,1}$ local group. Then the group operation $*$ is real analytic near the origin. In particular, after restricting V to a sufficiently small neighbourhood of the origin, one obtains a local Lie group.*

We will in fact give a more precise formula for $*$, known as the *Baker-Campbell-Haudorff-Dynkin formula*, in the course of proving Theorem 2.5.1. This formula is usually proven just for Lie groups, but it turns out that the proof of the formula extends without much difficulty to the $C^{1,1}$ local group setting (the main difference being that continuous operations, such as Riemann integrals, have to be replaced by discrete counterparts, such as Riemann sums).

Remark 2.5.2. In the case where V comes from viewing a general linear group $GL_n(\mathbf{C})$ in local exponential coordinates, the group operation $*$ is given by $x * y = \log(\exp(x)\exp(y))$ for sufficiently small $x, y \in M_n(\mathbf{C})$. Thus, a corollary of Theorem 2.5.1 is that this map is real analytic.

We begin the proof of Theorem 2.5.1. Throughout this section, $V \subset \mathbf{R}^d$ is a fixed radially homogeneous $C^{1,1}$ local group. We will need some variants of the basic bound (2.4).

Exercise 2.5.1 (Lipschitz bounds). If $x, y, z \in V$ are sufficiently small, establish the bounds

$$(2.15) \qquad x * y = x + y + O(|x + y||y|),$$

$$(2.16) \qquad x * y = x + y + O(|x + y||x|),$$

$$(2.17) \qquad x * y = x * z + O(|y - z|),$$

and

$$(2.18) \qquad y * x = z * x + O(|y - z|).$$

(*Hint:* To prove (2.15), start with the identity $(x * y) * (-y) = x$.)

Now we exploit the radial homogeneity to describe the conjugation operation $y \mapsto x * y * (-x)$ as a linear map:

Lemma 2.5.3 (Adjoint representation). *For all x sufficiently close to the origin, there exists a linear transformation $\mathrm{Ad}_x : \mathbf{R}^d \to \mathbf{R}^d$ such that $x * y * (-x) = \mathrm{Ad}_x(y)$ for all y sufficiently close to the origin.*

Remark 2.5.4. Using the matrix example from Remark 2.5.2, we are asserting here that

$$\exp(x)\exp(y)\exp(-x) = \exp(\mathrm{Ad}_x(y))$$

for some linear transform $\mathrm{Ad}_x(y)$ of y, and all sufficiently small x, y. Indeed, using the basic matrix identity $\exp(AxA^{-1}) = A\exp(x)A^{-1}$ for invertible A (coming from the fact that the conjugation map $x \mapsto AxA^{-1}$ is a continuous ring homomorphism) we see that we may take $\mathrm{Ad}(x) = \exp(x)y\exp(-x)$ here.

Proof. Fix x. The map $y \mapsto x * y * (-x)$ is continuous near the origin, so it will suffice to establish additivity, in the sense that

$$x * (y + z) * (-x) = (x * y * (-x)) + (x * z * (-x))$$

for y, z sufficiently close to the origin.

Let n be a large natural number. Then from (2.11) we have

$$(y+z) = \left(\frac{1}{n}y + \frac{1}{n}z\right)^{*n}.$$

Conjugating this by x, we see that

$$x * (y+z) * (-x) = \left(x * \left(\frac{1}{n}y + \frac{1}{n}z\right) * (-x)\right)^n$$

$$= n\left(x * \left(\frac{1}{n}y + \frac{1}{n}z\right) * (-x)\right).$$

But from (2.4) we have

$$\frac{1}{n}y + \frac{1}{n}z = \frac{1}{n}y * \frac{1}{n}z + O\left(\frac{1}{n^2}\right)$$

and thus (by Exercise 2.5.1)

$$x * \left(\frac{1}{n}y + \frac{1}{n}z\right) * (-x) = x * \frac{1}{n}y * \frac{1}{n}z * (-x) + O\left(\frac{1}{n^2}\right).$$

But if we split $x * \frac{1}{n}y * \frac{1}{n}z * (-x)$ as the product of $x * \frac{1}{n}y * (-x)$ and $x * \frac{1}{n}z * (-x)$ and use (2.4), we have

$$x * \frac{1}{n}y * \frac{1}{n}z * (-x) = x * \frac{1}{n}y * (-x) + x * \frac{1}{n}z * (-x) + O\left(\frac{1}{n^2}\right).$$

Putting all this together we see that

$$x * (y+z) * (-x) = n\left(x * \frac{1}{n}y * (-x) + x * \frac{1}{n}z * (-x) + O\left(\frac{1}{n^2}\right)\right)$$

$$= x * y * (-x) + x * z * (-x) + O\left(\frac{1}{n}\right);$$

sending $n \to \infty$ we obtain the claim. $\qquad\square$

From (2.4) we see that

$$\|\operatorname{Ad}_x - I\|_{\mathrm{op}} = O(|x|)$$

for x sufficiently small. Also from the associativity property we see that

(2.19) $$\qquad\qquad \operatorname{Ad}_{x*y} = \operatorname{Ad}_x \operatorname{Ad}_y$$

for all x, y sufficiently small. Combining these two properties (and using (2.15)) we conclude in particular that

(2.20) $$\qquad\qquad \|\operatorname{Ad}_x - \operatorname{Ad}_y\|_{\mathrm{op}} = O(|x - y|)$$

for x, y sufficiently small. Thus we see that Ad is a (locally) continuous linear representation. In particular, $t \mapsto \operatorname{Ad}_{tx}$ is a (locally) continuous homomorphism into a linear group, and so (by Proposition 1.0.1) we have the *Hadamard lemma*

$$\mathrm{Ad}_x = \exp(\mathrm{ad}_x)$$

for all sufficiently small x, where $\mathrm{ad}_x : \mathbf{R}^d \to \mathbf{R}^d$ is the linear transformation

$$\mathrm{ad}_x = \frac{d}{dt} \mathrm{Ad}_{tx} \mid_{t=0}.$$

From (2.19), (2.20), (2.4) we see that

$$\mathrm{Ad}_{tx} \mathrm{Ad}_{ty} = \mathrm{Ad}_{t(x+y)} + O(|t|^2)$$

for x, y, t sufficiently small, and so by the product rule we have

$$\mathrm{ad}_{x+y} = \mathrm{ad}_x + \mathrm{ad}_y .$$

Also, we clearly have $\mathrm{ad}_{tx} = t\, \mathrm{ad}_x$ for x, t small. Thus we see that ad_x is linear in x, and so we have

(2.21) $$\mathrm{ad}_x y = [x, y]$$

for some bilinear form $[,] : \mathbf{R}^d \to \mathbf{R}^d$.

One can show that this bilinear form in fact defines a Lie bracket (i.e., it is anti-symmetric and obeys the Jacobi identity), but for now, all we need is that it is manifestly real analytic (since all bilinear forms are polynomial and thus analytic). In particular, ad_x and Ad_x depend analytically on x.

We now give an important approximation to $x * y$ in the case when y is small:

Lemma 2.5.5. *For x, y sufficiently small, we have*

$$x * y = x + F(\mathrm{Ad}_x)y + O(|y|^2)$$

where

$$F(z) := \frac{z \log z}{z - 1}.$$

Proof. If we write $z := x * y - x$, then $z = O(|y|)$ (by (2.4)) and

$$(-x) * (x + z) = y.$$

We will shortly establish the approximation

(2.22) $$(-x) * (x + z) = \frac{1 - \exp(-\mathrm{ad}_x)}{\mathrm{ad}_x} z + O(|z|^2);$$

inverting

$$\frac{1 - \exp(-\mathrm{ad}_x)}{\mathrm{ad}_x} = \frac{\mathrm{Ad}_x - 1}{\mathrm{Ad}_x \log \mathrm{Ad}_x}$$

we obtain the claim.

It remains to verify (2.22). Let n be a large natural number. We can expand the left-hand side of (2.22) as a telescoping series

$$(2.23) \quad \sum_{j=0}^{n-1} \left(-\frac{j+1}{n}x\right) * \left(\frac{j+1}{n}x + \frac{j+1}{n}z\right) - \left(-\frac{j}{n}x\right) * \left(\frac{j}{n}x + \frac{j}{n}z\right).$$

Using (2.11), the first summand can be expanded as

$$\left(-\frac{j}{n}x\right) * \left(-\frac{x}{n}\right) * \left(\frac{x}{n} + \frac{z}{n}\right) * \left(\frac{j}{n}x + \frac{j}{n}z\right).$$

From (2.15) one has $\left(-\frac{x}{n}\right) * \left(\frac{x}{n} + \frac{z}{n}\right) = \frac{z}{n} + O(\frac{|z|}{n^2})$, so by (2.17), (2.18) we can write the preceding expression as

$$\left(-\frac{j}{n}x\right) * \frac{z}{n} * \left(\frac{j}{n}x + \frac{j}{n}z\right) + O\left(\frac{|z|}{n^2}\right)$$

which by definition of Ad can be rewritten as

$$(2.24) \quad \left(\mathrm{Ad}_{-\frac{j}{n}x}\frac{z}{n}\right) * \left(-\frac{j}{n}x\right) * \left(\frac{j}{n}x + \frac{j}{n}z\right) + O\left(\frac{|z|}{n^2}\right).$$

From (2.15) one has

$$\left(-\frac{j}{n}x\right) * \left(\frac{j}{n}x + \frac{j}{n}z\right) = O(|z|)$$

while from (2.20) one has $\mathrm{Ad}_{-\frac{j}{n}x}\frac{z}{n} = O(|z|/n)$, hence from (2.4) we can rewrite (2.24) as

$$\mathrm{Ad}_{-\frac{j}{n}x}\frac{z}{n} + \left(-\frac{j}{n}x\right) * \left(\frac{j}{n}x + \frac{j}{n}z\right) + O\left(\frac{|z|^2}{n}\right) + O\left(\frac{|z|}{n^2}\right).$$

Inserting this back into (2.23), we can thus write the left-hand side of (2.22) as

$$\left(\sum_{j=0}^{n-1} \mathrm{Ad}_{-\frac{j}{n}x}\frac{z}{n}\right) + O(|z|^2) + O\left(\frac{|z|}{n}\right).$$

Writing $\mathrm{Ad}_{-\frac{j}{n}x} = \exp\left(-\frac{j}{n}\mathrm{ad}_x\right)$, and then letting $n \to \infty$, we conclude (from the convergence of the Riemann sum to the Riemann integral) that

$$(-x) * (x + z) = \int_0^1 \exp(-t\,\mathrm{ad}_x)z \, dt + O(|z|^2)$$

and the claim follows. $\square$

Remark 2.5.6. In the matrix case, the key computation is to show that

$$\exp(-x)\exp(x+z) = 1 + \frac{1 - \exp(-\mathrm{ad}_x)}{\mathrm{ad}_x}z + O(|z|^2).$$

To see this, we can use the fundamental theorem of calculus to write the left-hand side as

$$1 + \int_0^1 \frac{d}{dt}(\exp(-tx)\exp(t(x+z)))\ dt.$$

Since $\frac{d}{dt}\exp(-tx) = \exp(-tx)(-x)$ and $\frac{d}{dt}\exp(t(x+z)) = (x+z)\exp(t(x+z))$, we can rewrite this as

$$1 + \int_0^1 \exp(-tx)z\exp(t(x+z))\ dt.$$

Since $\exp(t(x+z)) = \exp(tx) + O(|z|)$, this becomes

$$1 + \int_0^1 \exp(-tx)z\exp(tx)\ dt + O(|z|^2);$$

since $\exp(-tx)z\exp(tx) = \exp(-t\,\mathrm{ad}_x)z$, we obtain the desired claim.

We can integrate the above formula to obtain an exact formula for $*$:

Corollary 2.5.7 (Baker-Campbell-Hausdorff-Dynkin formula). *For x, y sufficiently small, one has*

$$x * y = x + \int_0^1 F(\mathrm{Ad}_x\,\mathrm{Ad}_{ty})y\ dt.$$

The right-hand side is clearly real analytic in x and y, and Theorem 2.5.1 follows.

Proof. Let n be a large natural number. We can express $x * y$ as the telescoping sum

$$x + \sum_{j=0}^{n-1} x * \left(\frac{j+1}{n}y\right) - x * \left(\frac{j}{n}y\right).$$

From (2.11) followed by Lemma 2.5.5 and (2.19), one has

$$x * \left(\frac{j+1}{n}y\right) = x * \left(\frac{j}{n}y\right) * \frac{y}{n}$$

$$= x * \left(\frac{j}{n}y\right) + F\left(\mathrm{Ad}_x\,\mathrm{Ad}_{\frac{j}{n}y}\right)\frac{y}{n} + O\left(\frac{1}{n^2}\right).$$

We conclude that

$$x * y = x + \frac{1}{n}\sum_{j=0}^{n-1} F\left(\mathrm{Ad}_x\,\mathrm{Ad}_{\frac{j}{n}y}\right)y + O\left(\frac{1}{n}\right).$$

Sending $n \to \infty$, so that the Riemann sum converges to a Riemann integral, we obtain the claim. $\qquad\square$

Remark 2.5.8. It is not immediately obvious from this formula alone why $*$ should be associative. A derivation of associativity from the Baker-Campbell-Hausdorff-Dynkin formula is given in Chapter 14.

Exercise 2.5.2. Use the Taylor-type expansion

$$F(z) = 1 - \frac{1/z - 1}{2} + \frac{(1/z - 1)^2}{3} - \frac{(1/z - 1)^3}{4} + \cdots$$

to obtain the explicit expansion

$$x * y = x + \sum_{n=0}^{\infty} \frac{(-1)^m}{n+1}$$

$$\sum_{\substack{r_i, s_i \geq 0 \\ (r_i, s_i) \neq (0,0)}} \frac{(\mathrm{ad}_y)^{r_1}(\mathrm{ad}_x)^{s_1} \ldots (\mathrm{ad}_y)^{r_n}(\mathrm{ad}_x)^{s_n}}{r_1! s_1! \ldots r_n! s_n! (r_1 + \cdots + r_n + 1)} y$$

where $m := n + r_1 + \cdots + r_n + s_1 + \cdots + s_n + 1$, and show that the series is absolutely convergent for x, y small enough. Invert this to obtain the alternate expansion

$$x * y = y + \sum_{n=0}^{\infty} \frac{(-1)^n}{n+1}$$

$$\sum_{\substack{r_i, s_i \geq 0 \\ (r_i, s_i) \neq (0,0)}} \frac{(\mathrm{ad}_x)^{r_1}(\mathrm{ad}_y)^{s_1} \ldots (\mathrm{ad}_x)^{r_n}(\mathrm{ad}_y)^{s_n}}{r_1! s_1! \ldots r_n! s_n! (r_1 + \cdots + r_n + 1)} x.$$

Exercise 2.5.3. Let V be a radially homogeneous $C^{1,1}$ local group. By Theorem 2.5.1, an open neighbourhood of the origin in V has the structure of a local Lie group, and thus by Exercise 2.3.4 is associated to a Lie algebra. Show that this Lie algebra is isomorphic to $\mathbf{R}^d$ and the Lie bracket $[,]$ is given by (2.19). Note that this establishes *a posteriori* the fact that the bracket $[,]$ occurring in (2.19) is anti-symmetric and obeys the Jacobi identity.

We now record some consequences of the Baker-Campbell-Hausdorff formula.

Exercise 2.5.4 (Lie groups are analytic). Let G be a global Lie group. Show that G is a real analytic manifold (i.e., one can find an atlas of smooth coordinate charts whose transition maps are all real analytic), and that the group operations are also real analytic (i.e., they are real analytic when viewed in the above-mentioned coordinate charts). Furthermore, show that any continuous homomorphism between Lie groups is also real analytic.

Exercise 2.5.5 (Lie's second theorem). Let G, H be global Lie groups, and let $\phi : \mathfrak{g} \to \mathfrak{h}$ be a Lie algebra homomorphism. Show that there exists an open neighbourhood U of the identity in G and a homomorphism $\Phi : U \to H$ from the local Lie group $G \restriction_U$ to H such that $D\Phi(1) = \phi$. If G is connected and simply connected, show that one can take U to be all of G.

Exercise 2.5.6 (Lie's third theorem). *Ado's theorem* asserts that every finite-dimensional Lie algebra is isomorphic to a subalgebra of $\mathfrak{gl}_n(\mathbf{R})$ for some n. This (somewhat difficult) theorem and its proof is discussed in Chapter 13. Assuming Ado's theorem as a "black box", conclude the following claims:

(i) (Lie's third theorem, local version) Every finite-dimensional Lie algebra is isomorphic to the Lie algebra of some local Lie group.

(ii) Every local or global Lie group has a neighbourhood of the identity that is isomorphic to a local *linear* Lie group (i.e., a local Lie group contained in $\mathrm{GL}_n(\mathbf{R})$ or $\mathrm{GL}_n(\mathbf{C})$ for some n).

(iii) (Lie's third theorem, global version) Every finite-dimensional Lie algebra $\mathfrak{g}$ is isomorphic to the Lie algebra of some global Lie group. (*Hint:* From (i) and (ii), one may identify $\mathfrak{g}$ with the Lie algebra of a local linear Lie group. Now consider the space of all smooth curves in the ambient linear group that are everywhere "tangent" to this local linear Lie group modulo "homotopy", and use this to build the global Lie group.)

(iv) (Lie's third theorem, simply connected version) Every finite-dimensional Lie algebra $\mathfrak{g}$ is isomorphic to the Lie algebra of some global connected, simply connected Lie group. Furthermore, this Lie group is unique up to isomorphism.

(v) Show that every local Lie group G has a neighbourhood of the identity that is isomorphic to a neighbourhood of the identity of a global connected, simply connected Lie group. Furthermore, this Lie group is unique up to isomorphism.

Remark 2.5.9. One does not need the full strength of Ado's theorem to establish conclusion (i) of the above exercise. Indeed, it suffices to show that the operation $*$ defined in Exercise 2.5.2 is associative near the origin. To do this, it suffices to verify associativity in the sense of formal power series; and then by abstract nonsense one can lift up to the free Lie algebra on d generators, and then down to the free *nilpotent* Lie algebra on d generators and of some arbitrary finite step s, which one can verify to be a finite-dimensional Lie algebra. Applying Ado's theorem for the special case of nilpotent Lie algebras (which is easier to establish than the general case of Ado's theorem, as discussed in Chapter 13), one can identify this nilpotent Lie algebra with a subalgebra of $\mathfrak{g}_n(\mathbf{R})$ for some n, and then one can argue as in the above exercise to conclude. See also Chapter 14 for an alternate way to establish associativity of $*$. However, I do not know how to establish conclusions (ii), (iii) or (iv) without using Ado's theorem in full generality (and (ii) is in fact *equivalent* to this theorem, at least in characteristic 0).

Remark 2.5.10. Lie's three theorems can be interpreted as establishing an *equivalence* between three different categories: the category of finite-dimensional Lie algebras; the category of local Lie groups (or more precisely, the category of local Lie group *germs*, formed by identifying local Lie groups that are identical near the origin); and the category of global connected, simply connected Lie groups. See Chapter 15 for further discussion.

The fact that we were able to establish the Baker-Campbell-Hausdorff formula at the $C^{1,1}$ regularity level will be useful for the purposes of proving results related to Hilbert's fifth problem. In particular, we have the following criterion for a group to be Lie (very much in accordance with the rigidity principle from the introduction):

Lemma 2.5.11 (Criterion for Lie structure). *Let G be a topological group. Then G is Lie if and only if there is a neighbourhood of the identity in G which is isomorphic (as a topological group) to a $C^{1,1}$ local group.*

This gives the last three steps of the program in Figure 1.

Proof. The "only if" direction is trivial. For the "if" direction, combine Corollary 2.4.4 with Theorem 2.5.1 and Exercise 2.4.7. $\qquad\square$

Remark 2.5.12. Informally, Lemma 2.5.11 asserts that $C^{1,1}$ regularity can automatically be upgraded to smooth (C^{∞}) or even real analytic (C^{ω}) regularity for topological groups. In contrast, note that a locally Euclidean group has neighbourhoods of the identity that are isomorphic to a "C^0 local group" (which is the same concept as a $C^{1,1}$ local group, but without the asymptotic (2.4)). Thus we have reduced Hilbert's fifth problem to the task of boosting C^0 regularity to $C^{1,1}$ regularity, rather than that of boosting C^0 regularity to C^{∞} regularity.

Exercise 2.5.7. Let G be a Lie group with Lie algebra $\mathfrak{g}$. For any $X, Y \in \mathfrak{g}$, show that

$$\exp([X,Y]) = \lim_{n\to\infty} \left(\exp(X/n)\exp(Y/n)\exp(-X/n)\exp(-Y/n)\right)^{n^2}.$$

Chapter 3

Building Lie structure from representations and metrics

Hilbert's fifth problem concerns the minimal hypotheses one needs to place on a topological group G to ensure that it is actually a Lie group. In Chapter 2, we saw that one could reduce the regularity hypothesis imposed on G to a "$C^{1,1}$" condition, namely that there was an open neighbourhood of G that was isomorphic (as a local group) to an open subset V of a Euclidean space $\mathbf{R}^d$ with identity element 0, and with group operation $*$ obeying the asymptotic

$$x * y = x + y + O(|x||y|)$$

for sufficiently small x, y. We will call such local groups $(V, *)$ $C^{1,1}$ *local groups*.

We now reduce the regularity hypothesis further, to one in which there is no explicit Euclidean space that is initially attached to G; this will flesh out another two steps of the diagram in Figure 1. Of course, Lie groups are still locally Euclidean, so if the hypotheses on G do not involve any explicit Euclidean spaces, then one must somehow build such spaces from other structures. One way to do so is to exploit an ambient space with Euclidean or Lie structure that G is embedded or immersed in. A trivial example of this is provided by the following basic fact from linear algebra:

Lemma 3.0.13. *If V is a finite-dimensional vector space (i.e., it is isomorphic to $\mathbf{R}^d$ for some d), and W is a linear subspace of V, then W is also a finite-dimensional vector space.*

We will establish a nonlinear version of this statement, known as *Cartan's theorem*. Recall that a subset S of a d-dimensional smooth manifold M is a *d'-dimensional smooth (embedded) submanifold* of M for some $0 \leq d' \leq d$ if for every point $x \in S$ there is a smooth coordinate chart $\phi : U \to V$ of a neighbourhood U of x in M that maps x to 0, such that $\phi(U \cap S) = V \cap \mathbf{R}^{d'}$, where we identify $\mathbf{R}^{d'} \equiv \mathbf{R}^{d'} \times \{0\}^{d-d'}$ with a subspace of $\mathbf{R}^d$. Informally, S locally sits inside M the same way that $\mathbf{R}^{d'}$ sits inside $\mathbf{R}^d$.

Theorem 3.0.14 (Cartan's theorem). *If H is a (topologically) closed subgroup of a Lie group G, then H is a smooth submanifold of G, and is thus also a Lie group.*

Note that the hypothesis that H is closed is essential; for instance, the rationals $\mathbf{Q}$ are a subgroup of the (additive) group of reals $\mathbf{R}$, but the former is not a Lie group even though the latter is.

Exercise 3.0.8. Let H be a subgroup of a locally compact group G. Show that H is closed in G if and only if it is locally compact.

A variant of the above results is provided by using (faithful) representations instead of embeddings. Again, the linear version is trivial:

Lemma 3.0.15. *If V is a finite-dimensional vector space, and W is another vector space with an injective linear transformation $\rho : W \to V$ from W to V, then W is also a finite-dimensional vector space.*

Here is the nonlinear version:

Theorem 3.0.16 (von Neumann's theorem). *If G is a Lie group, and H is a locally compact group with an injective continuous homomorphism $\rho : H \to G$, then H also has the structure of a Lie group.*

Actually, it will suffice for the homomorphism ρ to be locally injective rather than injective; related to this, von Neumann's theorem localises to the case when H is a local group rather than a Lie group. The requirement that H be locally compact is necessary, for much the same reason that the requirement that H be closed was necessary in Cartan's theorem.

Example 3.0.17. Let $G = (\mathbf{R}/\mathbf{Z})^2$ be the two-dimensional torus, let $H = \mathbf{R}$, and let $\rho : H \to G$ be the map $\rho(x) := (x, \alpha x)$, where $\alpha \in \mathbf{R}$ is a fixed real number. Then ρ is a continuous homomorphism which is locally injective, and is even globally injective if α is irrational, and so Theorem 3.0.16 is consistent with the fact that H is a Lie group. On the other hand, note that when α is irrational, then $\rho(H)$ is not closed; and so Theorem 3.0.16 does not follow immediately from Theorem 3.0.14 in this case. (We will see, though, that Theorem 3.0.16 follows from a local version of Theorem 3.0.14.)

As a corollary of Theorem 3.0.16, we observe that any locally compact Hausdorff group H with a faithful linear representation, i.e., a continuous injective homomorphism from H into a linear group such as $GL_n(\mathbf{R})$ or $GL_n(\mathbf{C})$, is necessarily a Lie group. This suggests a representation-theoretic approach to Hilbert's fifth problem. While this approach does not seem to readily solve the entire problem, it can be used to establish a number of important special cases with a well-understood representation theory, such as the compact case or the abelian case (for which the requisite representation theory is given by the *Peter-Weyl theorem* and *Pontryagin duality* respectively). We will discuss these cases further in later sections.

In all of these cases, one is not really building up Euclidean or Lie structure completely from scratch, because there is already a Euclidean or Lie structure present in another object in the hypotheses. Now we turn to results that can create such structure assuming only what is ostensibly a weaker amount of structure. In the linear case, one example of this is the following classical result in the theory of *topological vector spaces*.

Theorem 3.0.18. *Let V be a locally compact Hausdorff topological vector space. Then V is isomorphic (as a topological vector space) to $\mathbf{R}^d$ for some finite d.*

Remark 3.0.19. The *Banach-Alaoglu theorem* asserts that in a normed vector space V, the closed unit ball in the dual space V^* is always compact in the *weak-* topology*. Of course, this dual space V^* may be infinite-dimensional. This, however, does not contradict the above theorem, because the closed unit ball is *not* a neighbourhood of the origin in the weak-* topology (it is only a neighbourhood with respect to the strong topology).

The full nonlinear analogue of this theorem would be the Gleason-Yamabe theorem, which we are not yet ready to prove in this chapter. However, by using methods similar to that used to prove Cartan's theorem and von Neumann's theorem, one can obtain a partial nonlinear analogue which requires an additional hypothesis of a special type of metric, which we will call a *Gleason metric*:

Definition 3.0.20. Let G be a topological group. A *Gleason metric* on G is a left-invariant metric $d : G \times G \to \mathbf{R}^+$ which generates the topology on G and obeys the following properties for some constant $C > 0$, writing $\|g\|$ for $d(g, \mathrm{id})$:

(1) (Escape property) If $g \in G$ and $n \geq 1$ is such that $n\|g\| \leq \frac{1}{C}$, then $\|g^n\| \geq \frac{1}{C} n\|g\|$.

(2) (Commutator estimate) If $g, h \in G$ are such that $\|g\|, \|h\| \leq \frac{1}{C}$, then

(3.1)
$$\|[g, h]\| \leq C\|g\|\|h\|,$$

recalling that $[g, h] := g^{-1}h^{-1}gh$ is the *commutator* of g and h.

Exercise 3.0.9. Let G be a topological group that contains a neighbourhood of the identity isomorphic to a $C^{1,1}$ local group. Show that G admits at least one Gleason metric.

Theorem 3.0.21 (Building Lie structure from Gleason metrics). *Let G be a locally compact group that has a Gleason metric. Then G is isomorphic to a Lie group.*

We will rely on Theorem 3.0.21 (which represents the last five steps of Figure 1) to solve Hilbert's fifth problem; this theorem reduces the task of establishing Lie structure on a locally compact group to that of building a metric with suitable properties. Thus, much of the remainder of the solution of Hilbert's fifth problem will now be focused on the problem of how to construct good metrics on a locally compact group.

In all of the above results, a key idea is to use *one-parameter subgroups* to convert from the nonlinear setting to the linear setting. Recall from Chapter 2 that in a Lie group G, the one-parameter subgroups are in one-to-one correspondence with the elements of the Lie algebra $\mathfrak{g}$, which is a vector space. In a general topological group G, the concept of a one-parameter subgroup (i.e., a continuous homomorphism from $\mathbf{R}$ to G) still makes sense; the main difficulties are then to show that the space of such subgroups continues to form a vector space, and that the associated exponential map $\exp : \phi \mapsto \phi(1)$ is still a local homeomorphism near the origin.

Exercise 3.0.10. The purpose of this exercise is to illustrate the perspective that a topological group can be viewed as a nonlinear analogue of a vector space. Let G, H be locally compact groups. For technical reasons we assume that G, H are both *σ-compact* (i.e., the countable union of compact sets) and metrisable.

(i) (Open mapping theorem) Show that if $\phi : G \to H$ is a continuous homomorphism which is surjective, then it is *open* (i.e., the image of open sets is open). (*Hint:* Mimic the proof of the *open mapping theorem* for Banach spaces, as discussed for instance in [**Ta2010**, §1.7]. In particular, take advantage of the *Baire category theorem*.)

(ii) (Closed graph theorem) Show that if a homomorphism $\phi : G \to H$ is closed (i.e., its graph $\{(g, \phi(g)) : g \in G\}$ is a closed subset of $G \times H$), then it is continuous. (*Hint:* Mimic the derivation of the *closed graph theorem* from the open mapping theorem in the Banach space case, as again discussed in [**Ta2010**, §1.7].)

(iii) Let $\phi : G \to H$ be a homomorphism, and let $\rho : H \to K$ be a continuous injective homomorphism into another Hausdorff topological group K. Show that ϕ is continuous if and only if $\rho \circ \phi$ is continuous.

(iv) Relax the condition of metrisability to that of being Hausdorff. (*Hint:* Now one cannot use the Baire category theorem for metric spaces; but there is an analogue of this theorem for locally compact Hausdorff spaces.)

3.1. The theorems of Cartan and von Neumann

We now turn to the proof of Cartan's theorem. As indicated in the introduction, the fundamental concept here will be that of a one-parameter subgroup:

Definition 3.1.1 (One-parameter subgroups). Let G be a topological group. A *one-parameter subgroup* of G is a continuous homomorphism $\phi : \mathbf{R} \to G$. The space of all such one-parameter subgroups is denoted $L(G)$.

Remark 3.1.2. Strictly speaking, the terminology "one-parameter subgroup" is a misnomer, because it is the image $\phi(G)$ of ϕ which is a subgroup of G, rather than ϕ itself. Note that we consider reparameterisations $t \mapsto \phi(\lambda t)$ of a one-parameter subgroup $t \mapsto \phi(t)$, where λ is a nonzero real number, to be distinct from ϕ when $\lambda \neq 1$, even though both one-parameter subgroups have the same image.

We recall Exercise 2.4.4 from the preceding section, which we reformulate here as a lemma:

Lemma 3.1.3 (Classification of one-parameter subgroups). *Let G be a Lie group, with Lie algebra $\mathfrak{g}$. Then if X is an element of $\mathfrak{g}$, then $t \mapsto \exp(tX)$ is a one-parameter subgroup; conversely, if ϕ is a one-parameter subgroup, then there is a unique $X \in \mathfrak{g}$ such that $\phi(t) = \exp(tX)$ for all $t \in \mathbf{R}$. Thus we have a canonical one-to-one correspondence between $\mathfrak{g}$ and $L(G)$.*

Now let H be a closed subgroup of a Lie group G. Every one-parameter subgroup of H is clearly also a one-parameter subgroup of G, which by the above lemma can be viewed as an element of $\mathfrak{g}$:

$$L(H) \subset L(G) \equiv \mathfrak{g}.$$

Thus we can think of $L(H)$ as a subset $\mathfrak{h}$ of $\mathfrak{g}$:

$$\mathfrak{h} := \{X \in \mathfrak{g} : \exp(tX) \in H \text{ for all } t \in \mathbf{R}\}.$$

We claim that $\mathfrak{h}$ is in fact a linear subspace of $\mathfrak{g}$. Indeed, it contains the zero element of $\mathfrak{g}$ (which corresponds to the trivial one-parameter subgroup $t \mapsto 1$), and from reparameterisation we see that if $X \in \mathfrak{h}$, then $\lambda X \in \mathfrak{h}$ for all $\lambda \in \mathbf{R}$. Finally, if $X, Y \in \mathfrak{h}$, then by definition we have $\exp(tX), \exp(tY) \in H$ for all $t \in \mathbf{R}$. But recall from Exercise 2.4.3(ii) that

$$\exp(t(X + Y)) = \lim_{n \to \infty} (\exp(tX/2^n) \exp(tY/2^n))^{2^n}.$$

Since H is a group, we see that $(\exp(tX/2^n) \exp(tY/2^n))^{2^n}$ lies in H. Since H is closed, we conclude that $\exp(t(X + Y)) \in H$ for all $t \in \mathbf{R}$, which implies that $X + Y \in \mathfrak{h}$. Thus $\mathfrak{h}$ is closed under both addition and scalar multiplication, and so it is a vector space. (It turns out that $\mathfrak{h}$ is in fact a Lie algebra, but we will not need this fact yet.)

The next step is to show that $\mathfrak{h}$ is "large" enough to serve as the "Lie algebra" of H. To illustrate this type of fact, let us first establish a simple special case.

Lemma 3.1.4. *Suppose that the identity 1 is not an isolated point of H (i.e., H is not discrete). Then $\mathfrak{h}$ is nontrivial (i.e., it does not consist solely of 0).*

Proof. As 1 is not isolated, there exists a sequence $h_n \neq 1$ of elements of H that converge to 1. As $\exp : \mathfrak{g} \to G$ is a local homeomorphism near the identity, we may thus find a sequence $X_n \neq 0$ of elements of $\mathfrak{g}$ converging to zero such that $\exp(X_n) = h_n$ for all sufficiently large n.

Let us arbitrarily endow the finite-dimensional vector space $\mathfrak{g}$ with a norm (it will not matter which norm we select). Then the sequence $X_n/\|X_n\|$ lies on the unit sphere with respect to this norm, and thus by the Heine-Borel theorem (and passing to a subsequence) we may assume that $X_n/\|X_n\|$ converges to some element ω of norm 1.

Let t be any positive real number. Then $X_n \lfloor t/\|X_n\| \rfloor$ converges to $t\omega$, and so $\exp(X_n)^{\lfloor t/\|X_n\|\rfloor}$ converges to $\exp(t\omega)$. As $\exp(X_n) = h_n$ lies in H, so does $\exp(X_n)^{\lfloor t/\|X_n\|\rfloor}$; as H is closed, we conclude that $\exp(t\omega) \in H$ for all positive $t \in \mathbf{R}$, and hence for all $t \in \mathbf{R}$. We conclude that $\omega \in \mathfrak{h}$, and the claim follows. $\square$

Now we establish a stronger version of the above lemma:

Lemma 3.1.5. *There exists a neighbourhood U of the identity in H, and a neighbourhood V of the origin in $\mathfrak{h}$, such that $\exp : V \to U$ is a homeomorphism.*

Proof. Let V be a neighbourhood of the origin in $\mathfrak{h}$ such that $\exp : V \to \exp(V)$ is a homeomorphism (this exists since $\exp$ is a local homeomorphism in a neighbourhood of the origin in $\mathfrak{g}$. Clearly $\exp(V)$ lies in H and contains 1. If $\exp(V)$ contains a neighbourhood of 1 in H then we are done, so suppose that this is not the case. Then we can find a sequence $h_n \notin \exp(V)$ of elements in H that converge to 1. We may write $h_n = \exp(X_n)$ for some $X_n \notin V$ converging to zero in $\mathfrak{g}$.

As $\mathfrak{h}$ is a subspace of the finite-dimensional vector space $\mathfrak{g}$, we may write $\mathfrak{g} = \mathfrak{h} + \mathfrak{k}$ for some vector space $\mathfrak{k}$ transverse to $\mathfrak{h}$ (i.e., $\mathfrak{h} \cap \mathfrak{k} = 0$). (We do not require $\mathfrak{k}$ to be a Lie algebra.) From the inverse function theorem, the map $(Y, Z) \mapsto \exp(Y) \exp(Z)$ from $\mathfrak{h} \times \mathfrak{k}$ to G is a local homeomorphism near the identity. Thus we may write $\exp(X_n) = \exp(Y_n) \exp(Z_n)$ for sufficiently large n, where $Y_n \in \mathfrak{h}$ and $Z_n \in \mathfrak{k}$ both go to zero as $n \to \infty$. Since $X_n \notin V$, we see that Z_n is nonzero for n sufficiently large.

We arbitrarily place a norm on $\mathfrak{k}$. As before, we may pass to a subsequence and assume that $Z_n / \|Z_n\|$ converges to some limit ω in the unit sphere of $\mathfrak{k}$; in particular, $\omega \notin \mathfrak{h}$.

Since $\exp(X_n)$ and $\exp(Y_n)$ both lie in H, $\exp(Z_n)$ does also. By arguing as in the proof of Lemma 3.1.4 we conclude that $\exp(t\omega)$ lies in H for all $t \in \mathbf{R}$, and so $\omega \in \mathfrak{h}$, yielding the desired contradiction. $\square$

From the above lemma we see that H locally agrees with $\exp(V)$ near the identity, and thus locally agrees with $\exp(V)h$ near h for every $h \in H$. This implies that H is a smooth submanifold of G; since it is also a topological group, it is thus a Lie group. This establishes Cartan's theorem.

Remark 3.1.6. Observe *a posteriori* that $\mathfrak{h}$ is the Lie algebra of H and, in particular, is closed with respect to Lie brackets. This fact can also be established directly using Exercise 2.5.7.

There is a local version of Cartan's theorem, in which groups are replaced by local groups:

Theorem 3.1.7 (Local Cartan's theorem). *If H is a locally compact local subgroup of a local Lie group G, then there is an open neighbourhood H' of the identity in H that is a smooth submanifold of G, and is thus also a local Lie group.*

The proof of this theorem follows the lines of the global Cartan's theorem, with some minor technical changes, and we set this proof out in the following exercise.

Exercise 3.1.1. Define a *local one-parameter subgroup* of a local group H to be a continuous homomorphism $\phi : (-\varepsilon, \varepsilon) \to H$ from the (additive) local group $(-\varepsilon, \varepsilon)$ to H. Call two local one-parameter subgroups *equivalent* if they agree on a neighbourhood of the origin, and let $L(H)$ be the set of all equivalence classes of local one-parameter subgroups. Establish the following claims:

(i) If H is a global group, then there is a canonical one-to-one correspondence that identifies this definition of $L(H)$ with the definition of $L(H)$ given previously.

(ii) In the situation of Theorem 3.1.7, show that $L(H)$ can be identified with a linear subspace $\mathfrak{h}$ of $\mathfrak{g}$, namely

$$\mathfrak{h} := \{X \in \mathfrak{g} : \exp(tX) \in H \text{ for all sufficiently small } t\}.$$

(iii) Let the notation and assumptions be as in (ii). For any neighbourhood H' of the identity in H, there is a neighbourhood V of the origin in $\mathfrak{h}$ such that $\exp(V) \subset H'$.

(iv) Let the notation and assumptions be as in (ii). There exists a neighbourhood U of the identity in H, and a neighbourhood V of the origin in $\mathfrak{h}$, such that $\exp : V \to U$ is a homeomorphism.

(v) Prove Theorem 3.1.7.

One can then use Theorem 3.1.7 to establish von Neumann's theorem, as follows. Suppose that H is a locally compact group with an injective continuous homomorphism $\rho : H \to G$ into a Lie group G. As H is locally compact, there is an open neighbourhood U of the origin in H whose closure $\overline{U}$ is compact. The map ρ from $\overline{U}$ to $\rho(\overline{U})$ is a continuous bijection from a compact set to a Hausdorff set, and is therefore a homeomorphism (since it maps closed (and hence compact) subsets of $\overline{U}$ to compact (and hence closed) subsets of $\rho(\overline{U})$). The set $\rho(U)$ is then a locally compact local subgroup of G and thus has a neighbourhood of the identity which is a local Lie group, by Theorem 3.1.7. Pulling this back by ρ, we see that some neighbourhood of the identity in H is a local Lie group, and thus H is a global Lie group by Exercise 2.4.7.

Exercise 3.1.2. State and prove a local version of von Neumann's theorem, in which G and H are local groups rather than global groups, and the global injectivity condition is similarly replaced by local injectivity.

3.2. Locally compact vector spaces

We will now turn to the study of *topological vector spaces*, which we will need to establish Theorem 3.0.21. We begin by recalling the definition of a topological vector space.

Definition 3.2.1 (Topological vector space). A *topological vector space* is a (real) vector space V equipped with a topology that makes the vector space operations $+ : V \times V \to V$ and $\cdot : \mathbf{R} \times V \to \mathbf{R}$ (jointly) continuous. (In particular, $(V, +)$ is necessarily a topological group.)

One can also consider complex topological vector spaces, but the theory for such spaces is almost identical to the real case, and we will only need the real case for what follows. In the literature, it is often common to restrict attention to Hausdorff topological vector spaces, although this is not a severe restriction in practice, as the following exercise shows:

Exercise 3.2.1. Let V be a topological vector space. Show that the closure $W := \overline{\{0\}}$ of the origin is a closed subspace of V, and the quotient space V/W is a Hausdorff topological vector space. Furthermore, show that a set is open in V if and only if it is the preimage of an open set in V/W under the quotient map $\pi : V \to V/W$.

An important class of topological vector spaces are the *normed vector spaces*, in which the topology is generated by a norm $\|\|$ on the vector space. However, not every topological vector space is generated by a norm. See [**Ta2010**, §1.9] for some further discussion.

We emphasise that in order to be a topological vector space, the vector space operations $+, \cdot$ need to be *jointly continuous*; merely being continuous in the individual variables is not sufficient to qualify for being a topological vector space. We illustrate this with some nonexamples of topological vector spaces:

Example 3.2.2. Consider the one-dimensional vector space $\mathbf{R}$ with the *cocompact* topology (a nonempty set is open iff its complement is compact in the usual topology). In this topology, the space is a T_1 *space*[1] (though not Hausdorff), the scalar multiplication map $\cdot : \mathbf{R} \times \mathbf{R} \to \mathbf{R}$ is jointly continuous as long as one excludes the scalar zero, and the addition map $+ : \mathbf{R} \times \mathbf{R} \to \mathbf{R}$ is continuous in each coordinate (i.e., translations are continuous), but not jointly continuous; for instance, the set $\{(x, y) \in \mathbf{R} : x + y \notin [0, 1]\}$ does not contain a nontrivial Cartesian product of two sets that are open in the cocompact topology. So this is not a topological vector space. Similarly for the cocountable or cofinite topologies on $\mathbf{R}$ (the latter topology, incidentally, is the same as the *Zariski topology* on $\mathbf{R}$).

Example 3.2.3. Consider the topology of $\mathbf{R}$ inherited by pulling back the usual topology on the unit circle $\mathbf{R}/\mathbf{Z}$. This pullback topology is not quite Hausdorff, but the addition map $+ : \mathbf{R} \times \mathbf{R} \to \mathbf{R}$ is jointly continuous (so that this gives $\mathbf{R}$ the structure of a topological group). On the other hand, the scalar multiplication map $\cdot : \mathbf{R} \times \mathbf{R} \to \mathbf{R}$ is not continuous at all. A

---

[1] A T_1 space is a topological space in which all points are closed.

slight variant of this topology comes from pulling back the usual topology on the torus $(\mathbf{R}/\mathbf{Z})^2$ under the map $x \mapsto (x, \alpha x)$ for some irrational α; this restores the Hausdorff property, and addition is still jointly continuous, but multiplication remains discontinuous.

Example 3.2.4. Consider $\mathbf{R}$ with the discrete topology; here, the topology is Hausdorff, addition is jointly continuous, and every dilation is continuous, but multiplication is not jointly continuous. If one instead gives $\mathbf{R}$ the *half-open topology*, then again the topology is Hausdorff and addition is jointly continuous, but scalar multiplication is only jointly continuous once one restricts the scalar to be nonnegative.

These examples illustrate that a vector space such as $\mathbf{R}$ can have many topologies on it (and many topological group structures), but only one topological vector space structure. More precisely, we have

Theorem 3.2.5. *Every finite-dimensional Hausdorff topological vector space has the usual topology.*

Proof. Let V be a finite-dimensional Hausdorff topological vector space, with topology $\mathcal{F}$. We need to show that every set which is open in the usual topology, is open in $\mathcal{F}$, and conversely.

Let $v_1, \ldots, v_n$ be a basis for the finite-dimensional space V. From the continuity of the vector space operations, we easily verify that the linear map $T : \mathbf{R}^n \to V$ given by

$$T(x_1, \ldots, x_n) := x_1 v_1 + \cdots + x_n v_n$$

is continuous. From this, we see that any set which is open in $\mathcal{F}$, is also open in the usual topology.

Now we show conversely that every set which is open in the usual topology, is open in $\mathcal{F}$. It suffices to show that there is a bounded open neighbourhood of the origin in $\mathcal{F}$, since one can then translate and dilate this open neighbourhood to obtain a (sub-)base for the usual topology. (Here, "bounded" refers to the usual sense of the term, for instance, with respect to an arbitrarily selected norm on V (note that on a finite-dimensional space, all norms are equivalent).)

We use T to identify V (as a vector space) with $\mathbf{R}^n$. As T is continuous, every set which is compact in the usual topology, is compact in $\mathcal{F}$. In particular, the unit sphere $S^{n-1} := \{x \in \mathbf{R}^n : \|x\| = 1\}$ (in, say, the Euclidean norm $\|\|$ on $\mathbf{R}^n$) is compact in $\mathcal{F}$. Using this and the Hausdorff assumption on $\mathcal{F}$, we can find an open neighbourhood U of the origin in F which is disjoint from S^{n-1}.

At present, U need not be bounded (note that we are not assuming V to be locally connected *a priori*). However, we can fix this as follows. Using the joint continuity of the scalar multiplication map, one can find another open neighbourhood U' of the origin and an open interval $(-\varepsilon, \varepsilon)$ around 0 such that the product set $(-\varepsilon, \varepsilon) \cdot U' := \{tx : t \in (-\varepsilon, \varepsilon); x \in U'\}$ is contained in U. In particular, since U avoids the unit sphere S^{n-1}, U'' must avoid the region $\{x \in \mathbf{R}^n : \|x\| > 1/\varepsilon\}$ and is thus bounded, as required. $\square$

We isolate one important consequence of the above theorem:

Corollary 3.2.6. *In a Hausdorff topological vector space V, every finite-dimensional subspace W is closed.*

Proof. It suffices to show that every vector $x \in V \backslash W$ is in the exterior of W. But this follows from Theorem 3.2.5 after restricting to the finite-dimensional space spanned by W and x. $\square$

We can now prove Theorem 3.0.18. Let V be a locally compact Hausdorff space, thus there exists a compact neighbourhood K of the origin. Then the dilate $\frac{1}{2}K$ is also a neighbourhood of the origin, and so by compactness K can be covered by finitely many translates of $\frac{1}{2}K$, thus

$$K \subset S + \frac{1}{2}K$$

for some finite set S. If we let W be the finite-dimensional vector space generated by S, we conclude that

$$K \subset W + \frac{1}{2}K.$$

Iterating this we have

$$K \subset W + 2^{-n}K$$

for any $n \geq 1$. On the other hand, if U is a neighbourhood of the origin, then for every $x \in V$ we see that $2^{-n}x \in U$ for sufficiently large n. By compactness of K (and continuity of the scalar multiplication map at zero), we conclude that $2^{-n}K \subset U$ for some sufficiently large n, and thus

$$K \subset W + U$$

for any neighbourhood U of the origin; thus K is in the closure of W. By Corollary 3.2.6, we conclude that

$$K \subset W.$$

But K is a neighbourhood of the origin, thus for every $x \in V$ we have $2^{-n}x \in K$ for all sufficiently large n, and thus $x \in 2^n W = W$. Thus $V = W$, and the claim follows.

Exercise 3.2.2. Establish the *Riesz lemma*: If $V = (V, \|\|)$ is a *normed vector space*, W is a proper closed subspace of V, and $\varepsilon > 0$, then there exists a vector x in V with $\|x\| = 1$ and $\text{dist}(x, V) \geq 1 - \varepsilon$. (*Hint:* Pick an element y of V not in W, and then pick $z \in W$ that nearly minimises $\|y - z\|$. Use these two vectors to construct a suitable x.) Using this lemma and the Heine-Borel theorem, give an alternate proof of Theorem 3.0.18 in the case when V is a normed vector space.

3.3. From Gleason metrics to Lie groups

Now we prove Theorem 3.0.21. The argument will broadly follow the lines of Cartan's theorem, but we will have to work harder in many stages of the argument in order to compensate for the lack of an obvious ambient Lie structure in the initial hypotheses. In particular, the Gleason metric hypothesis will substitute for the $C^{1,1}$ type structure enjoyed by Lie groups, which as we saw in Chapter 2 was needed to obtain good control on the exponential map.

Henceforth, G is a locally compact group with a Gleason metric d (and an associated "norm" $\|g\| = d(g, \text{id})$). In particular, by the Heine-Borel theorem, G is complete with this metric.

We use the asymptotic notation $X \ll Y$ in place of $X \leq CY$ for some constant C that can vary from line to line (in particular, C need not be the constant appearing in the definition of a Gleason metric), and write $X \sim Y$ for $X \ll Y \ll X$. We also let $\varepsilon > 0$ be a sufficiently small constant (depending only on the constant in the definition of a Gleason metric) to be chosen later.

Note that the left-invariant metric properties of d give the symmetry property

$$\|g^{-1}\| = \|g\|$$

and the triangle inequality

$$\|g_1 \cdots g_n\| \leq \sum_{i=1}^{n} \|g_i\|.$$

From the commutator estimate (3.1) and the triangle inequality we also obtain a conjugation estimate

$$\|ghg^{-1}\| \sim \|h\|$$

whenever $\|g\|, \|h\| \leq \varepsilon$. Since left-invariance gives

$$d(g, h) = \|g^{-1}h\|$$

we then conclude an approximate right invariance

$$d(gk, hk) \sim d(g, h)$$

whenever $\|g\|, \|h\|, \|k\| \leq \varepsilon$. In a similar spirit, the commutator estimate (3.1) also gives

(3.2) $$d(gh, hg) \ll \|g\| \|h\|$$

whenever $\|g\|, \|h\| \leq \varepsilon$.

This has the following useful consequence, which asserts that the power maps $g \mapsto g^n$ behave like dilations:

Lemma 3.3.1. *If $n \geq 1$ and $\|g\|, \|h\| \leq \varepsilon/n$, then*

$$d(g^n h^n, (gh)^n) \ll n^2 \|g\| \|h\|$$

and

$$d(g^n, h^n) \sim n d(g, h).$$

Proof. We begin with the first inequality. By the triangle inequality, it suffices to show that

(3.3) $$d((gh)^i g^{n-i} h^{n-i}, (gh)^{i+1} g^{n-i-1} h^{n-i-1}) \ll n \|g\| \|h\|$$

uniformly for all $0 \leq i < n$. By left-invariance and approximate right-invariance, the left-hand side is comparable to

$$d(g^{n-i-1} h, h g^{n-i-1}),$$

which by (3.2) is bounded above by

$$\ll \|g^{n-i-1}\| \|h\| \ll n \|g\| \|h\|$$

as required.

Now we prove the second estimate. Write $g = hk$, then $\|k\| = d(g, h) \leq 2\varepsilon/n$. We have

$$d(h^n k^n, h^n) = \|k^n\| \sim n \|k\|$$

thanks to the escape property (shrinking ε if necessary). On the other hand, from the first inequality, we have

$$d(g^n, h^n k^n) \ll n^2 \|h\| \|k\|.$$

If ε is small enough, the claim now follows from the triangle inequality. $\square$

Remark 3.3.2. Lemma 3.3.1 implies (by a standard covering argument) that the group G is locally of bounded doubling, though we will not use this fact here. The bounds above should be compared with the bounds in Exercise 2.4.1. Indeed, just as the bounds in that exercise were used in the previous sections to build the exponential map for Lie groups, the bounds in Lemma 3.3.1 are crucial for controlling the exponential function on the locally compact group G equipped with the Gleason metric d.

Now we bring in the space $L(G)$ of one-parameter subgroups. We give this space the *compact-open topology*, thus the topology is generated by balls of the form

$$\{\phi \in L(G) : \sup_{t \in I} d(\phi(t), \phi_0(t)) < r\}$$

for $\phi_0 \in L(G)$, $r > 0$, and compact I. Actually, using the homomorphism property, one can use a single compact interval I, such as $[-1, 1]$, to generate the topology if desired, thus making $L(G)$ a metric space.

Given that G is eventually going to be shown to be a Lie group, $L(G)$ must be isomorphic to a Euclidean space. We now move towards this goal by establishing various properties of $L(G)$ that Euclidean spaces enjoy.

Lemma 3.3.3. $L(G)$ *is locally compact.*

Proof. It is easy to see that $L(G)$ is complete. Let $\phi_0 \in L(G)$. As ϕ_0 is continuous, we can find an interval $I = [-T, T]$ small enough that $\|\phi_0(t)\| \leq \varepsilon$ for all $t \in [-T, T]$. By the Heine-Borel theorem, it will suffice to show that the set

$$B := \{\phi \in L(G) : \sup_{t \in [-T,T]} d(\phi(t), \phi_0(t)) < \varepsilon\}$$

is totally bounded. By the *Arzelá-Ascoli theorem*, it suffices to show that the family of functions in B is *equicontinuous*.

By construction, we have $\|\phi(t)\| \leq 2\varepsilon$ whenever $|t| \leq T$. By the escape property, this implies (for ε small enough, of course) that $\|\phi(t/n)\| \ll \varepsilon/n$ for all $|t| \leq T$ and $n \geq 1$, thus $\|\phi(t)\| \ll \varepsilon|t|/T$ whenever $|t| \leq T$. From the homomorphism property, we conclude that $d(\phi(t), \phi(t')) \ll \varepsilon|t-t'|/T$ whenever $|t|, |t'| \leq T$, which gives uniform Lipschitz control and hence equicontinuity as desired. $\square$

We observe for future reference that the proof of the above lemma also shows that all one-parameter subgroups are locally Lipschitz.

Now we put a vector space structure on $L(G)$, which we define by analogy with the Lie group case, in which each tangent vector X generates a one-parameter subgroup $t \mapsto \exp(tX)$. From this analogy, the scalar multiplication operation has an obvious definition: If $\phi \in L(G)$ and $c \in \mathbf{R}$, we define $c\phi \in L(G)$ to be the one-parameter subgroup

$$(3.4) \qquad\qquad c\phi(t) := \phi(ct)$$

which is easily seen to actually be a one-parameter subgroup.

Now we turn to the addition operation. In the Lie group case, one can express the one-parameter subgroup $t \mapsto \exp(t(X + Y))$ in terms of the one-parameter subgroups $t \mapsto \exp(tX)$, $t \mapsto \exp(tY)$ by the limiting formula

$$\exp(t(X + Y)) = \lim_{n \to \infty} (\exp(tX/n) \exp(tY/n))^n ;$$

cf. Exercise 2.4.3. In view of this, we would like to define the sum $\phi + \psi$ of two one-parameter subgroups $\phi, \psi \in L(G)$ by the formula

$$(\phi + \psi)(t) := \lim_{n \to \infty} (\phi(t/n)\psi(t/n))^n.$$

Lemma 3.3.4. *If $\phi, \psi \in L(G)$, then $\phi + \psi$ is well-defined and also lies in $L(G)$.*

Proof. To show well-definedness, it suffices to show that for each t, the sequence $(\phi(t/n)\psi(t/n))^n$ is a Cauchy sequence. It suffices to show that

$$\sup_{m \geq 1} d\left((\phi(t/n)\psi(t/n))^n, (\phi(t/nm)\psi(t/nm))^{nm}\right) \to 0$$

as $n \to \infty$. We will in fact prove the slightly stronger claim

$$\sup_{m \geq 1} \sup_{1 \leq n' \leq n} d\left((\phi(t/n)\psi(t/n))^{n'}, (\phi(t/nm)\psi(t/nm))^{n'm}\right) \to 0.$$

Observe from continuity of multiplication that to prove this claim for a given t, it suffices to do so for $t/2$; thus we may assume without loss of generality that t is small.

Let $\varepsilon > 0$ be a small number to be chosen later. Since ϕ, ψ are locally Lipschitz, we see (if t is sufficiently small depending on ε) that

$$\|\phi(t/n)\|, \|\psi(t/n)\| \ll \varepsilon/n$$

for all n. From Lemma 3.3.1, we conclude that

$$d\left(\phi(t/n)\psi(t/n), (\phi(t/nm)\psi(t/nm)^m)\right) \ll m^2(\varepsilon/nm)(\varepsilon/nm) = \varepsilon^2/n^2$$

if $m \geq 1$ and n is sufficiently large. Another application of Lemma 3.3.1 then gives

$$d\left((\phi(t/n)\psi(t/n))^{n'}, \left(\phi(t/nm)\psi(t/nm)^{n'm}\right)\right) \ll n'\varepsilon^2/n^2 \ll \varepsilon^2/n$$

if $m \geq 1$, n is sufficiently large, and $1 \leq n' \leq \varepsilon n$. The claim follows.

The above argument in fact shows that $(\phi(t/n)\psi(t/n))^n$ is uniformly Cauchy for t in a compact interval, and so the pointwise limit $\phi + \psi$ is in fact a uniform limit of continuous functions and is thus continuous. To prove that $\phi + \psi$ is a homomorphism, it suffices by density of the rationals to show that

$$(\phi + \psi)(at)(\phi + \psi)(bt) = (\phi + \psi)((a + b)t)$$

and

$$(\phi + \psi)(-t) = (\phi + \psi)(t)^{-1}$$

for all $t \in \mathbf{R}$ and all positive integers a, b. To prove the first claim, we observe that

$$
\begin{aligned}
(\phi + \psi)(at) &= \lim_{n \to \infty} (\phi(at/n)\psi(at/n))^n \\
&= \lim_{n \to \infty} (\phi(t/n)\psi(t/n))^{an}
\end{aligned}
$$

and similarly for $(\phi + \psi)(bt)$ and $(\phi + \psi)((a + b)t)$, whence the claim. To prove the second claim, we see that

$$
\begin{aligned}
(\phi + \psi)(-t)^{-1} &= \lim_{n \to \infty} (\phi(-t/n)\psi(-t/n))^{-n} \\
&= \lim_{n \to \infty} (\psi(t/n)\phi(t/n))^n,
\end{aligned}
$$

but $(\psi(t/n)\phi(t/n))^n$ is $(\phi(t/n)\psi(t/n))^n$ conjugated by $\psi(t/n)$, which goes to the identity; and the claim follows. $\qquad \square$

$L(G)$ also has an obvious zero element, namely the trivial one-parameter subgroup $t \mapsto \mathrm{id}$.

Lemma 3.3.5. $L(G)$ *is a topological vector space.*

Proof. We first show that $L(G)$ is a vector space. It is clear that the zero element 0 of $L(G)$ is an additive and scalar multiplication identity, and that scalar multiplication is associative. To show that addition is commutative, we again use the observation that $(\psi(t/n)\phi(t/n))^n$ is $(\phi(t/n)\psi(t/n))^n$ conjugated by an element that goes to the identity. A similar argument shows that $(-\phi) + (-\psi) = -(\phi + \psi)$, and a change of variables argument shows that $(a\phi) + (a\psi) = a(\phi + \psi)$ for all positive integers a, hence for all rational a, and hence by continuity for all real a. The only remaining thing to show is that addition is associative, thus if $\phi, \psi, \eta \in L(G)$, that $((\phi + \psi) + \eta)(t) = (\phi + (\psi + \eta))(t)$ for all $t \in \mathbf{R}$. By the homomorphism property, it suffices to show this for all sufficiently small t.

An inspection of the argument used to establish (3.3.4) reveals that there is a constant $\varepsilon > 0$ such that

$$
d\left((\phi + \psi)(t), (\phi(t/n)\psi(t/n))^n\right) \ll \varepsilon^2/n
$$

for all small t and all large n, and hence also that

$$
d\left((\phi + \psi)(t/n), \phi(t/n)\psi(t/n)\right) \ll \varepsilon^2/n^2
$$

(thanks to Lemma 3.3.1). Similarly we have (after adjusting ε if necessary)

$$
d\left(((\phi + \psi) + \eta)(t), ((\phi + \psi)(t/n)\eta(t/n))^n\right) \ll \varepsilon^2/n.
$$

From Lemma 3.3.1 we have

$$
d\left(((\phi + \psi)(t/n)\eta(t/n))^n, (\phi(t/n)\psi(t/n)\eta(t/n))^n\right) \ll \varepsilon^2/n
$$

and thus
$$d\left(((\phi + \psi) + \eta)(t), (\phi(t/n)\psi(t/n)\eta(t/n))^n\right) \ll \varepsilon^2/n.$$
Similarly for $\phi + (\psi + \eta)$. By the triangle inequality we conclude that
$$d\left(((\phi + \psi) + \eta)(t), (\phi + (\psi + \eta))(t)\right) \ll \varepsilon^2/n;$$
sending t to zero, the claim follows.

Finally, we need to show that the vector space operations are continuous. It is easy to see that scalar multiplication is continuous, as are the translation operations; the only remaining thing to verify is that addition is continuous at the origin. Thus, for every $\epsilon > 0$ we need to find a $\delta > 0$ such that $\sup_{t \in [-1,1]} \|(\phi + \psi)(t)\| \leq \epsilon$ whenever $\sup_{t \in [-1,1]} \|\phi(t)\| \leq \delta$ and $\sup_{t \in [-1,1]} \|\psi(t)\| \leq \delta$. But if ϕ, ψ are as above, then by the escape property (assuming δ small enough) we conclude that $\|\phi(t)\|, \|\psi(t)\| \ll \delta|t|$ for $t \in [-1, 1]$, and then from the triangle inequality we conclude that $\|(\phi + \psi)(t)\| \ll \delta$ for $t \in [-1, 1]$, giving the claim. $\square$

Exercise 3.3.1. Show that for any $\phi \in L(G)$, the quantity
$$\|\phi\| := \lim_{n \to \infty} n\|\phi(1/n)\|$$
exists and defines a norm on $L(G)$ that generates the topology on $L(G)$.

As $L(G)$ is both locally compact, metrisable, and a topological vector space, it must be isomorphic to a finite-dimensional vector space $\mathbf{R}^n$ with the usual topology, thanks to Theorem 3.2.5.

In analogy with the Lie algebra setting, we define the *exponential map* $\exp : L(G) \to G$ by setting $\exp(\phi) := \phi(1)$. Given the topology on $L(G)$, it is clear that this is a continuous map.

Exercise 3.3.2. Show that the exponential map is locally injective near the origin. (*Hint:* From Lemma 3.3.1, obtain the unique square roots property: if $g, h \in G$ are sufficiently close to the identity and $g^2 = h^2$, then $g = h$.)

We have proved a number of useful things about $L(G)$, but at present we have not established that $L(G)$ is *large* in any substantial sense; indeed, at present, $L(G)$ could be completely trivial even if G was large. In particular, the image of the exponential map exp could conceivably be quite small. We now address this issue. As a warmup, we show that $L(G)$ is at least nontrivial if G is nondiscrete (cf. Lemma 3.1.4):

Proposition 3.3.6. *Suppose that G is not a discrete group. Then $L(G)$ is nontrivial.*

Of course, the converse is obvious; discrete groups do not admit any nontrivial one-parameter subgroups.

Proof. As G is not discrete, there is a sequence g_n of nonidentity elements of G such that $\|g_n\| \to 0$ as $n \to \infty$. Writing N_n for the integer part of $\varepsilon/\|g_n\|$, then $N_n \to \infty$ as $n \to \infty$, and we conclude from the escape property that $\|g_n^{N_n}\| \sim \varepsilon$ for all n.

We define the approximate one-parameter subgroups $\phi_n : [-1, 1] \to G$ by setting

$$\phi_n(t) := g_n^{\lfloor tN_n \rfloor}.$$

Then we have $\|\phi_n(t)\| \ll \varepsilon|t| + \frac{\varepsilon}{N_n}$ for $|t| \leq 1$, and we have the approximate homomorphism property

$$d(\phi_n(t+s), \phi_n(t)\phi_n(s)) \to 0$$

uniformly whenever $|t|, |s|, |t+s| \leq 1$. As a consequence, ϕ_n is asymptotically equicontinuous on $[-1, 1]$, and so by (a slight generalisation of) the Arzéla-Ascoli theorem, we may pass to a subsequence in which ϕ_n converges uniformly to a limit $\phi : [-1, 1] \to G$, which is a genuine homomorphism that is genuinely continuous, and thus can be extended to a one-parameter subgroup. Also, $\|\phi_n(1)\| = \|g_n^{N_n}\| \sim \varepsilon$ for all n, and thus $\|\phi(1)\| \sim \varepsilon$; in particular, ϕ is nontrivial, and the claim follows. $\square$

We now generalise the above proposition to a more useful result (cf. Lemma 3.1.5).

Proposition 3.3.7. *For any neighbourhood K of the origin in $L(G)$, $\exp(K)$ is a neighbourhood of the identity in G.*

Proof. We use an argument[2] of Hirschfeld [**Hi1990**]. By shrinking K if necessary, we may assume that K is a compact star-shaped neighbourhood, with $\exp(K)$ contained in the ball of radius ε around the origin. As K is compact, $\exp(K)$ is compact also.

Suppose for contradiction that $\exp(K)$ is not a neighbourhood of the identity, then there is a sequence g_n of elements of $G \backslash K$ such that $\|g_n\| \to 0$ as $n \to \infty$. By the compactness of K, we can find an element h_n of K that minimises the distance $d(g_n, h_n)$. If we then write $g_n = h_n k_n$, then

$$\|k_n\| = d(g_n, h_n) \leq d(g_n, \mathrm{id}) = \|g_n\|$$

and hence $\|h_n\|, \|k_n\| \to 0$ as $n \to \infty$.

Let N_n be the integer part of $\varepsilon_n/\|k_n\|$, then $N_n \to \infty$ as $n \to \infty$, and $\|k_n^{N_n}\| \sim \varepsilon$ for all n.

Let $\phi_n : [-1, 1] \to G$ be the approximate one-parameter subgroups defined as

$$\phi_n(t) := k_n^{\lfloor tN_n \rfloor}.$$

---

[2]The author thanks Lou van den Dries and Isaac Goldbring for bringing this argument to his attention.

As before, we may pass to a subsequence such that ϕ_n converges uniformly to a limit $\phi : [-1, 1] \to G$, which extends to a one-parameter subgroup $\phi \in L(G)$.

In a similar vein, since $h_n \in \exp(K)$, we can find $\psi_n \in K$ such that $\psi_n(1) = h_n$, which by the escape property (and the smallness of K implies that $\|\psi_n(t)\| \ll t\|h_n\|$ for $|t| \leq 1$. In particular, ψ_n goes to zero in $L(G)$.

We now claim that $\exp(\psi_n + \frac{1}{N_n}\phi)$ is close to g_n. Indeed, from Lemma 3.3.1 we see that

$$d\left(\exp\left(\psi_n + \frac{1}{N_n}\phi\right), \exp(\psi_n)\exp\left(\frac{1}{N_n}\phi\right)\right) \ll \frac{1}{N_n}\|h_n\|.$$

Since $\exp(\psi_n) = h_n$, we conclude from the triangle inequality and left-invariance that

$$d\left(\exp\left(\psi_n + \frac{1}{N_n}\phi\right), g_n\right) \ll \frac{1}{N_n}\|h_n\| + d\left(k_n, \exp\left(\frac{1}{N_n}\phi\right)\right).$$

But from Lemma 3.3.1 again, one has

$$d\left(k_n, \exp\left(\frac{1}{N_n}\phi\right)\right) \ll \frac{1}{N_n}d\left(k_n^{N_n}, \exp(\phi)\right) = o(1/N_n)$$

and thus

$$d\left(\exp\left(\psi_n + \frac{1}{N_n}\phi\right), g_n\right) = o(1/N_n).$$

But for n large enough, $\psi_n + \frac{1}{N_n}\phi$ lies in K, and so the distance from g_n to K is $o(1/N_n) = o(d(g_n, h_n))$. But this contradicts the minimality of h_n for n large enough, and the claim follows. $\square$

If K is a sufficiently small compact neighbourhood of the identity in $L(G)$, then $\exp : K \to \exp(K)$ is bijective by Lemma 3.3.2; since it is also continuous, K is compact, and $\exp(K)$ is Hausdorff, we conclude that $\exp : K \to \exp(K)$ is a homeomorphism. The local group structure $G\mid_{\exp(K)}$ on $\exp(K)$ then pulls back to a local group structure on K.

Exercise 3.3.3. If we identify $L(G)$ with $\mathbf{R}^d$ for some d, show that the exponential map $\exp : K \to \exp(K)$ is bilipschitz.

Proposition 3.3.8. *K is a radially homogeneous $C^{1,1}$ local group (as defined in Definition 2.4.1 and (2.11)), after identifying $L(G)$ with $\mathbf{R}^d$ for some finite d.*

Proof. The radial homogeneity is clear from (3.4) and the homomorphism property, so the main task is to establish the $C^{1,1}$ property

$$x * y = x + y + O(|x||y|)$$

for the local group law $*$ on K. By Exercise 3.3.3, this is equivalent to the assertion that

$$d(\phi(1)\psi(1), (\phi + \psi)(1)) \ll \|\phi(1)\| \|\psi(1)\|$$

for ϕ, ψ sufficiently close to the identity in $L(G)$. By definition of $\phi + \psi$, it suffices to show that

$$d(\phi(1)\psi(1), (\phi(1/n)\psi(1/n))^n) \ll \|\phi(1)\| \|\psi(1)\|$$

for all n; but this follows from Lemma 3.3.1 (and the observation, from the escape property, that $\|\phi(1/n)\| \ll \|\phi(1)\|/n$ and $\|\psi(1/n)\| \ll \|\psi(1)\|/n$).

$\square$

Combining this proposition with Lemma 2.5.11, we obtain Theorem 3.0.21.

Exercise 3.3.4. State and prove a version of Theorem 3.0.21 for local groups. (In order to do this, you must first decide how to define an analogue of a Gleason metric on a local group.)

Chapter 4

Haar measure, the Peter-Weyl theorem, and compact or abelian groups

In Chapters 2, 3, we have been steadily reducing the amount of regularity needed on a topological group in order to be able to show that it is in fact a Lie group, in the spirit of Hilbert's fifth problem. Now, we will work on Hilbert's fifth problem from the other end, starting with the minimal assumption of *local compactness* on a topological group G, and seeing what kind of structures one can build using this assumption. (For simplicity we shall mostly confine our discussion to global groups rather than local groups for now.) In view of the preceding sections, we would like to see two types of structures emerge, in particular:

(1) *representations* of G into some more structured group, such as a matrix group $GL_n(\mathbf{C})$; and

(2) *metrics* on G that capture the escape and commutator structure of G (i.e., Gleason metrics).

To build either of these structures, a fundamentally useful tool is that of (left-) *Haar measure* — a left-invariant *Radon measure* μ on G. (One can of course also consider right-Haar measures; in many cases (such as for compact or abelian groups), the two concepts are the same, but this is not always the case.) This concept generalises the concept of *Lebesgue measure* on Euclidean spaces $\mathbf{R}^d$, which is of course fundamental in analysis on those spaces.

Haar measures will help us build useful representations and useful metrics on locally compact groups G. For instance, a Haar measure μ gives rise to the *regular representation* $\tau : G \to U(L^2(G, d\mu))$ that maps each element $g \in G$ of G to the unitary translation operator $\rho(g) : L^2(G, d\mu) \to L^2(G, d\mu)$ on the Hilbert space $L^2(G, d\mu)$ of square-integrable measurable functions on G with respect to this Haar measure by the formula

$$\tau(g)f(x) := f(g^{-1}x).$$

(The presence of the inverse g^{-1} is convenient in order to obtain the homomorphism property $\tau(gh) = \tau(g)\tau(h)$ without a reversal in the group multiplication.) In general, this is an infinite-dimensional representation; but in many cases (and, in particular, in the case when G is compact) we can decompose this representation into a useful collection of finite-dimensional representations, leading to the *Peter-Weyl theorem*, which is a fundamental tool for understanding the structure of compact groups. This theorem is particularly simple in the compact abelian case, where it turns out that the representations can be decomposed into one-dimensional representations $\chi : G \to U(\mathbf{C}) \equiv S^1$, better known as *characters*, leading to the theory of Fourier analysis on general compact abelian groups. With this and some additional (largely combinatorial) arguments, we will also be able to obtain satisfactory structural control on locally compact abelian groups as well.

The link between Haar measure and useful metrics on G is a little more complicated. First, once one has the regular representation $\tau : G \to U(L^2(G, d\mu))$, and given a suitable "test" function $\psi : G \to \mathbf{C}$, one can then embed G into $L^2(G, d\mu)$ (or into other function spaces on G, such as $C_c(G)$ or $L^\infty(G)$) by mapping a group element $g \in G$ to the translate $\tau(g)\psi$ of ψ in that function space. (This map might not actually be an embedding if ψ enjoys a nontrivial translation symmetry $\tau(g)\psi = \psi$, but let us ignore this possibility for now.) One can then pull the metric structure on the function space back to a metric on G, for instance, defining an $L^2(G, d\mu)$-based metric

$$d(g, h) := \|\tau(g)\psi - \tau(h)\psi\|_{L^2(G, d\mu)}$$

if ψ is square-integrable, or perhaps a $C_c(G)$-based metric

(4.1) $$d(g, h) := \|\tau(g)\psi - \tau(h)\psi\|_{C_c(G)}$$

if ψ is continuous and compactly supported (with $\|f\|_{C_c(G)} := \sup_{x \in G} |f(x)|$ denoting the supremum norm). These metrics tend to have several nice properties (for instance, they are automatically left-invariant), particularly if the test function is chosen to be sufficiently "smooth". For instance, if we introduce the differentiation (or more precisely, finite difference) operators

$$\partial_g := 1 - \tau(g)$$

(so that $\partial_g f(x) = f(x) - f(g^{-1}x)$) and use the metric (4.1), then a short computation (relying on the translation-invariance of the $C_c(G)$ norm) shows that

$$d([g,h], \mathrm{id}) = \|\partial_g \partial_h \psi - \partial_h \partial_g \psi\|_{C_c(G)}$$

for all $g, h \in G$. This suggests that commutator estimates, such as those appearing in Definition 3.0.20, might be available if one can control "second derivatives" of ψ; informally, we would like our test functions ψ to have a "$C^{1,1}$" type regularity.

If G was already a Lie group (or something similar, such as a $C^{1,1}$ local group), then it would not be too difficult to concoct such a function ψ by using local coordinates. But of course the whole point of Hilbert's fifth problem is to do without such regularity hypotheses, and so we need to build $C^{1,1}$ test functions ψ by other means, and here is where the Haar measure comes in: It provides the fundamental tool of *convolution*

$$\phi * \psi(x) := \int_G \phi(xy^{-1}) \psi(y) d\mu(y)$$

between two suitable functions $\phi, \psi : G \to \mathbf{C}$, which can be used to build smoother functions out of rougher ones. For instance:

Exercise 4.0.5. Let $\phi, \psi : \mathbf{R}^d \to \mathbf{C}$ be continuous, compactly supported functions which are Lipschitz continuous. Show that the convolution $\phi * \psi$ using Lebesgue measure on $\mathbf{R}^d$ obeys the $C^{1,1}$-type commutator estimate

$$\|\partial_g \partial_h (\phi * \psi)\|_{C_c(\mathbf{R}^d)} \le C \|g\| \|h\|$$

for all $g, h \in \mathbf{R}^d$ and some finite quantity C depending only on ϕ, ψ.

This exercise suggests a strategy to build Gleason metrics by convolving together some "Lipschitz" test functions and then using the resulting convolution as a test function to define a metric. This strategy may seem somewhat circular because one needs a notion of metric in order to define Lipschitz continuity in the first place, but it turns out that the properties required on that metric are weaker than those that the Gleason metric will satisfy, and so one will be able to break the circularity by using a "bootstrap" or "induction" argument.

We will discuss this strategy — which is due to Gleason, and is fundamental to all currently known solutions to Hilbert's fifth problem — in later sections. In this section, we will construct Haar measure on general locally compact groups, and then establish the Peter-Weyl theorem, which in turn can be used to obtain a reasonably satisfactory structural classification of both compact groups and locally compact abelian groups.

4.1. Haar measure

For technical reasons, it is convenient to not work with an absolutely general locally compact group, but to instead restrict attention to those groups that are both *σ-compact* and *Hausdorff*, in order to access measure-theoretic tools such as the *Fubini-Tonelli theorem* and the *Riesz representation theorem* without bumping into unwanted technical difficulties. Intuitively, σ-compact groups are those groups that do not have enormously "large" scales — scales are too coarse to be "seen" by any compact set. Similarly, Hausdorff groups are those groups that do not have enormously "small" scales — scales that are too small to be "seen" by any open set. A simple example of a locally compact group that fails to be σ-compact is the real line $\mathbf{R} = (\mathbf{R}, +)$ with the discrete topology; conversely, a simple example of a locally compact group that fails to be Hausdorff is the real line $\mathbf{R}$ with the trivial topology.

As the two exercises below show, one can reduce to the σ-compact Hausdorff case without much difficulty, either by restricting to an open subgroup to eliminate the largest scales and recover σ-compactness, or to quotient out by a compact normal subgroup to eliminate the smallest scales and recover the Hausdorff property. (This gives the very first step in Figure 1.)

Exercise 4.1.1. Let G be a locally compact group. Show that there exists an open subgroup G_0 which is locally compact and σ-compact. (*Hint:* Take the group generated by a compact neighbourhood of the identity.)

Exercise 4.1.2. Let G be a locally compact group. Let $H = \overline{\{\mathrm{id}\}}$ be the topological closure of the identity element.

(i) Show that given any open neighbourhood U of a point x in G, there exists a neighbourhood V of x whose closure lies in U. (*Hint:* Translate x to the identity and select V so that $V^2 \subset U$.) In other words, G is a *regular space*.

(ii) Show that for any group element $g \in G$, that the sets gH and H are either equal or disjoint.

(iii) Show that H is a compact normal subgroup of G.

(iv) Show that the quotient group G/H (equipped with the quotient topology) is a locally compact Hausdorff group.

(v) Show that a subset of G is open if and only if it is the preimage of an open set in G/H.

Now that we have restricted attention to the σ-compact Hausdorff case, we can now define the notion of a Haar measure.

Definition 4.1.1 (Radon measure). Let X be a σ-compact locally compact Hausdorff topological space. The *Borel σ-algebra* $\mathcal{B}[X]$ on X is the σ-algebra generated by the open subsets of X. A *Borel measure* is a countably additive nonnegative measure $\mu : \mathcal{B}[X] \to [0, +\infty]$ on the Borel σ-algebra. A *Radon measure* is a Borel measure obeying three additional axioms:

(i) (Local finiteness) One has $\mu(K) < \infty$ for every compact set K.

(ii) (Inner regularity) One has $\mu(E) = \sup_{K \subset E, K \text{ compact}} \mu(K)$ for every Borel measurable set E.

(iii) (Outer regularity) One has $\mu(E) = \inf_{U \supset E, U \text{ open}} \mu(U)$ for every Borel measurable set E.

Definition 4.1.2 (Haar measure). Let $G = (G, \cdot)$ be a σ-compact locally compact Hausdorff group. A Radon measure μ is *left-invariant* (resp. *right-invariant*) if one has $\mu(gE) = \mu(E)$ (resp. $\mu(Eg) = \mu(E)$) for all $g \in G$ and Borel measurable sets E. A *left-invariant Haar measure* is a nonzero Radon measure which is left-invariant; a right-invariant Haar measure is defined similarly. A *bi-invariant Haar measure* is a Haar measure which is both left-invariant and right-invariant.

Note that we do not consider the zero measure to be a Haar measure.

Example 4.1.3. A large part of the foundations of Lebesgue measure theory (e.g., most of [**Ta2011**, §1.2]) can be summed up in the single statement that Lebesgue measure is a (bi-invariant) Haar measure on Euclidean spaces $\mathbf{R}^d = (\mathbf{R}^d, +)$.

Example 4.1.4. If G is a countable discrete group, then *counting measure* is a bi-invariant Haar measure.

Example 4.1.5. If μ is a left-invariant Haar measure on a σ-compact locally compact Hausdorff group G, then the reflection $\tilde{\mu}$ defined by $\tilde{\mu}(E) := \mu(E^{-1})$ is a right-invariant Haar measure on G, and the scalar multiple $\lambda\mu$ is a left-invariant Haar measure on G for any $0 < \lambda < \infty$.

Exercise 4.1.3. If μ is a left-invariant Haar measure on a σ-compact locally compact Hausdorff group G, show that $\mu(U) > 0$ for any nonempty open set U.

Let μ be a left-invariant Haar measure on a σ-compact locally compact Hausdorff group. Let $C_c(G)$ be the space of all continuous, compactly supported complex-valued functions $f : G \to \mathbf{C}$; then f is absolutely integrable with respect to μ (thanks to local finiteness), and one has

$$\int_G f(gx) \, d\mu(x) = \int_G f(x) \, dx$$

for all $g \in G$ (thanks to left-invariance). Similarly for right-invariant Haar measures (but now replacing gx by xg).

The fundamental theorem regarding Haar measures is:

Theorem 4.1.6 (Existence and uniqueness of Haar measure). *Let G be a σ-compact locally compact Hausdorff group. Then there exists a left-invariant Haar measure μ on G. Furthermore, this measure is unique up to scalars: if μ, ν are two left-invariant Haar measures on G, then $\nu = \lambda\mu$ for some scalar $\lambda > 0$.*

Similarly if "left-invariant" is replaced by "right-invariant" throughout. (However, we do not claim that every left-invariant Haar measure is automatically right-invariant, or vice versa.)

This theorem gives half of the second step of Figure 1. (The other half is the *Birkhoff-Kakutani theorem*, see Theorem 5.1.1 and Exercise 5.1.3, which also requires passage to a subquotient.)

To prove Theorem 4.1.6, we will rely on the *Riesz representation theorem* (see, e.g., [**Ta2010**, §1.10] for a proof):

Theorem 4.1.7 (Riesz representation theorem). *Let X be a σ-compact locally compact Hausdorff space. Then to every linear functional $I : C_c(X) \to \mathbf{R}$ which is nonnegative (thus $I(f) \geq 0$ whenever $f \geq 0$), one can associate a unique Radon measure μ such that $I(f) = \int_X f \, d\mu$ for all $f \in C_c(X)$. Conversely, for each Radon measure μ, the functional $I_\mu : f \mapsto \int_X f \, d\mu$ is a nonnegative linear functional on $C_c(X)$.*

We now establish the uniqueness component of Theorem 4.1.6. We shall just prove the uniqueness of left-invariant Haar measure, as the right-invariant case is similar (and also follows from the left-invariant case by Example 4.1.5). Let μ, ν be two left-invariant Haar measures on G. We need to prove that ν is a scalar multiple of μ. From the Riesz representation theorem, it suffices to show that I_ν is a scalar multiple of I_μ. Equivalently, it suffices to show that

$$I_\nu(f)I_\mu(g) = I_\mu(f)I_\nu(g)$$

for all $f, g \in C_c(G)$.

To show this, the idea is to approximate both f and g by superpositions of translates of the same function ψ_ε. More precisely, fix $f, g \in C_c(G)$, and let $\varepsilon > 0$. As the functions f and g are continuous and compactly supported, they are uniformly continuous, in the sense that we can find an open neighbourhood U_ε of the identity such that $|f(xy) - f(x)| \leq \varepsilon$ and $|g(xy) - g(x)| \leq \varepsilon$ for all $x \in G$ and $y \in U_\varepsilon$; we may also assume that the U_ε are contained in a compact set that is uniform in ε. By Exercise 4.1.3 and Urysohn's lemma, we can then find an "approximation to the identity" $\psi_\varepsilon \in C_c(U)$ supported in U such that $\int_G \psi_\varepsilon(y) \, d\mu(y) = 1$. Since

$$f(xy) = f(x) + O(\varepsilon)$$

for all y in the support of ψ, we conclude that

$$\int_G f(xy)\psi_\varepsilon(y) \, d\mu(y) = f(x) + O(\varepsilon)$$

uniformly in $x \in G$; also, the left-hand side has uniformly compact support in ε. If we integrate against ν, we conclude that

$$\int_G \int_G f(xy)\psi_\varepsilon(y) \, d\mu(y) d\nu(x) = I_\nu(f) + O(\varepsilon)$$

where the implied constant in the $O()$ notation can depend on μ, ν, f, g but not on ε. But by the left-invariance of μ, the left-hand side is also

$$\int_G \int_G f(y)\psi_\varepsilon(x^{-1}y) \, d\mu(y)d\nu(x),$$

which by the Fubini-Tonelli theorem is

$$\int_G f(y) \left(\int_G \psi_\varepsilon(x^{-1}y) \, d\nu(x) \right) \, d\mu(y)$$

which by the left-invariance of ν is

$$\int_G f(y) \left(\int_G \psi_\varepsilon(x^{-1}) \, d\nu(x) \right) \, d\mu(y)$$

which simplifies to $I_\mu(f) \int_G \psi_\varepsilon(x^{-1}) \, d\nu(x)$. We conclude that

$$I_\nu(f) = I_\mu(f) \int_G \psi_\varepsilon(x^{-1}) \, d\nu(x) + O(\varepsilon)$$

and similarly

$$I_\nu(g) = I_\mu(g) \int_G \psi_\varepsilon(x^{-1}) \, d\nu(x) + O(\varepsilon)$$

which implies that

$$I_\nu(f)I_\mu(g) - I_\mu(f)I_\nu(g) = O(\varepsilon).$$

Sending $\varepsilon \to 0$ we obtain the claim.

Exercise 4.1.4. Obtain another proof of uniqueness of Haar measure by investigating the translation-invariance properties of the Radon-Nikodym derivative $\frac{d\mu}{d(\mu+\nu)}$ of μ with respect to $\mu + \nu$.

Now we show existence of Haar measure. Again, we restrict attention to the left-invariant case (using Example 4.1.5 if desired). By the Riesz representation theorem, it suffices to find a functional $I : C_c(G)^+ \to \mathbf{R}^+$ from the space $C_c(G)^+$ of nonnegative continuous compactly supported functions to the nonnegative reals obeying the following axioms:

(1) (Homogeneity) $I(\lambda f) = \lambda I(f)$ for all $\lambda > 0$ and $f \in C_c(G)^+$.
(2) (Additivity) $I(f + g) = I(f) + I(g)$ for all $f, g \in C_c(G)^+$.

(3) (Left-invariance) $I(\tau(x)f) = I(f)$ for all $f \in C_c(G)^+$ and $x \in G$.

(4) (Nondegeneracy) $I(f_0) > 0$ for at least one $f_0 \in C_c(G)^+$.

Here, $\tau(x)$ is the translation operation $\tau(x)f(y) := f(x^{-1}y)$ as discussed in the introduction.

We will construct this functional by an approximation argument. Specifically, we fix a nonzero $f_0 \in C_c(G)^+$. We will show that given any finite number of functions $f_1, \ldots, f_n \in C_c(G)^+$ and any $\varepsilon > 0$, one can find a functional $I = I_{f_1,\ldots,f_n,\varepsilon} : C_c(G)^+ \to \mathbf{R}^+$ that obeys the following axioms:

(1) (Homogeneity) $I(\lambda f) = \lambda I(f)$ for all $\lambda > 0$ and $f \in C_c(G)^+$.

(2) (Approximate additivity) $|I(f_i + f_j) - I(f_i) - I(f_j)| \le \varepsilon$ for all $1 \le i, j \le n$.

(3) (Left-invariance) $I(\tau(x)f) = I(f)$ for all $f \in C_c(G)^+$ and $x \in G$.

(4) (Uniform bound) For each $f \in C_c(G)^+$, we have $I(f) \le K(f)$, where $K(f)$ does not depend on $f_1, \ldots, f_n$ or ε.

(5) (Normalisation) $I(f_0) = 1$.

Once one has established the existence of these approximately additive functionals $I_{f_1,\ldots,f_n,\varepsilon}$, one can then construct the genuinely additive functional I (and thus a left-invariant Haar measure) by a number of standard compactness arguments. For instance:

(1) One can observe (from *Tychonoff's theorem*) that the space of all functionals $I : C_c(G)^+ \to \mathbf{R}^+$ obeying the uniform bound $I(f) \le K(f)$ is a compact subset of the product space $(\mathbf{R}^+)^{C_c(G)^+}$; in particular, any collection of closed sets in this space obeying the *finite intersection property* has nonempty intersection. Applying this fact to the closed sets $F_{f_1,\ldots,f_n,\varepsilon}$ of functionals obeying the homogeneity, approximate additivity, left-invariance, uniform bound, and normalisation axioms for various $f_1, \ldots, f_n, \varepsilon$, we conclude that there is a functional I that lies in all such sets, giving the claim.

(2) If one lets $\mathcal{C}$ be the space of all tuples $(f_1, \ldots, f_n, \varepsilon)$, one can use the *Hahn-Banach theorem* to construct a bounded real linear functional $\lambda : \ell^\infty(\mathcal{C}) \to \mathbf{R}$ that maps the constant sequence 1 to 1. If one then applies this functional to the $I_{f_1,\ldots,f_n,\varepsilon}$ one can obtain a functional I with the required properties.

(3) One can also adopt a *nonstandard analysis* approach, taking an ultralimit of all the $I_{f_1,\ldots,f_n,\varepsilon}$ and then taking a standard part to recover I.

(4) A closely related method is to obtain I from the $I_{f_1,\ldots,f_n,\varepsilon}$ by using the *compactness theorem* in logic.

(5) In the case when G is metrisable (and hence *separable*, by σ-compactness), then $C_c(G)$ becomes separable, and one can also use the *Arzelá-Ascoli theorem* in this case. (One can also try in this case to directly ensure that the $I_{f_1,\ldots,f_n,\varepsilon}$ converge pointwise, without needing to pass to a further subsequence, although this requires more effort than the compactness-based methods.)

These approaches are more or less equivalent to each other, and the choice of which approach to use is largely a matter of personal taste.

It remains to obtain the approximate functionals $I_{f_1,\ldots,f_n,\varepsilon}$ for a given $f_0, f_1, \ldots, f_n$ and ε. As with the uniqueness claim, the basic idea is to approximate all the functions $f_0, f_1, \ldots, f_n$ by translates $\tau(y)\psi$ of a given function ψ. More precisely, let $\delta > 0$ be a small quantity (depending on $f_0, f_1, \ldots, f_n$ and ε) to be chosen later. By uniform continuity, we may find a neighbourhood U of the identity such that $f_i(xy) = f_i(x) + O(\delta)$ for all $x \in G$ and $y \in U$. Let $\psi \in C_c(G)^+$ be a function, not identically zero, which is supported in U.

To motivate the argument that follows, pretend temporarily that we have a left-invariant Haar measure μ available, and let $\kappa := \int_G \psi \, d\mu$ be the integral of ψ with respect to this measure. Then $0 < \kappa < \infty$, and by left-invariance one has

$$\int_G \tau(y)\psi(x) \, d\mu(x) = \kappa,$$

and thus

$$\int_G \sum_{k=1}^{K} c_k \tau(y_k)\psi(x) \, d\mu = \kappa \sum_{k=1}^{K} c_k$$

for any scalars $c_1, \ldots, c_K \in \mathbf{R}^+$ and $y_1, \ldots, y_K \in G$. In particular, if we introduce the *covering number*

$$[f : \psi] := \inf \left\{ \sum_{k=1}^{K} c_k : c_1, \ldots, c_K \in \mathbf{R}^+; f(x) \leq \sum_{k=1}^{K} c_k \tau(y_k)\psi(x) \text{ for all } x \in G \right\}$$

of a given function $f \in C_c(G)^+$ by ψ, we have

$$\int_G f \, d\mu \leq \kappa[f : \psi].$$

This suggests using a scalar multiple of $f \mapsto [f : \psi]$ as the approximate linear functional (noting that $[f : \psi]$ can be defined without reference to any existing Haar measure); in view of the normalisation $I(f_0) = 1$, it is then natural to introduce the functional

$$I(f) := \frac{[f : \psi]}{[f_0 : \psi]}.$$

(This functional is analogous in some ways to the concept of *outer measure* or the *upper Darboux integral* in measure theory.) Note from compactness that $[f : \psi]$ is finite for every $f \in C_c(G)^+$, and from the nontriviality of f_0 we see that $[f_0 : \psi] > 0$, so I is well-defined as a map from $C_c(G)^+$ to $\mathbf{R}$. It is also easy to verify that I obeys the homogeneity, left-invariance, and normalisation axioms. From the easy inequality

$$(4.2) \qquad\qquad [f : \psi] \leq [f : f_0][f_0 : \psi]$$

we also obtain the uniform bound axiom, and from the infimal nature of $[f : \psi]$ we also easily obtain the subadditivity property

$$I(f + g) \leq I(f) + I(g).$$

To finish the construction, it thus suffices to show that

$$I(f_i + f_j) \geq I(f_i) + I(f_j) - \varepsilon$$

for each $1 \leq i, j \leq n$, if $\delta > 0$ is chosen sufficiently small depending on $\varepsilon, f_0, f_1, \ldots, f_n$.

Fix f_i, f_j. By definition, we have the pointwise bound

$$(4.3) \qquad\qquad f_i(x) + f_j(x) \leq \sum_{k=1}^{K} c_k \tau(y_k) \psi(x)$$

for some $c_1, \ldots, c_K$ with

$$(4.4) \qquad\qquad \sum_{k=1}^{K} c_k \leq \left(I(f_i + f_j) + \frac{\varepsilon}{2} \right) [f_0 : \psi].$$

If we then write $c_k = c_k' + c_k''$ where

$$c_k' := c_k \frac{f_i(y_k) + \delta}{f_i(y_k) + f_j(y_k) + 2\delta}$$

and

$$c_k'' := c_k \frac{f_j(y_k) + \delta}{f_i(y_k) + f_j(y_k) + 2\delta},$$

then we claim that

$$(4.5) \qquad\qquad f_i(x) \leq \sum_{k=1}^{K} c_k' \tau(y_k) \psi(x) + 4\delta$$

and

$$(4.6) \qquad\qquad f_j(x) \leq \sum_{k=1}^{K} c_k'' \tau(y_k) \psi(x) + 4\delta$$

if δ is small enough. Indeed, we have

$$\sum_{k=1}^{K} c_k' \tau(y_k)\psi(x) = \sum_{k=1}^{K} c_k \psi(y_k^{-1}x) \frac{f_i(y_k) + \delta}{f_i(y_k) + f_j(y_k) + 2\delta}.$$

If $\psi(y_k^{-1}x)$ is nonzero, then by the construction of ψ and U, one has $|f_i(y_k) - f_i(x)| \leq \delta$ and $|f_j(y_k) - f_j(x)| \leq \delta$, which implies that

$$\frac{f_i(y_k) + \delta}{f_i(y_k) + f_j(y_k) + 2\delta} = \frac{f_i(x)}{f_i(x) + f_j(x) + 4\delta}.$$

Using (4.3) we thus have

$$\sum_{k=1}^{K} c_k' \tau(y_k)\psi(x) + 4\delta \geq \frac{f_i(x)}{f_i(x) + f_j(x) + 4\delta}(f_i(x) + f_j(x)) + 4\delta$$

which gives (4.5); a similar argument gives (4.6). From the subadditivity (and monotonicity) of I, we conclude that

$$I(f_i) \leq \frac{\sum_{k=1}^{K} c_k'}{[f_0 : \psi]} + 4\delta I(g)$$

and

$$I(f_j) \leq \frac{\sum_{k=1}^{K} c_k''}{[f_0 : \psi]} + 4\delta I(g)$$

where $g \in C_c(G)$ equals 1 on the support of f_i, f_j. Summing and using (4.4), we conclude that

$$I(f_i) + I(f_j) \leq I(f_i + f_j) + \frac{\varepsilon}{2} + 8\delta I(g)$$

and the claim follows by taking δ small enough. This concludes the proof of Theorem 4.1.6.

Exercise 4.1.5. State and prove a generalisation of Theorem 4.1.6 in which the hypothesis that G is Hausdorff and σ-compact are dropped. (This requires extending concepts such as "Borel σ-algebra", "Radon measure", and "Haar measure" to the non-Hausdorff or non-σ-compact setting. Note that different texts sometimes have inequivalent definitions of these concepts in such settings; because of this (and also because of the potential breakdown of some basic measure-theoretic tools such as the Fubini-Tonelli theorem), it is usually best to avoid working with Haar measure in the non-Hausdorff or non-σ-compact case unless one is very careful.)

Remark 4.1.8. An important special case of the Haar measure construction arises for *compact* groups G. Here, we can normalise the Haar measure by requiring that $\mu(G) = 1$ (i.e., μ is a probability measure), and so there is now a unique (left-invariant) Haar probability measure on such a group. In Exercise 4.1.7 we will see that this measure is in fact bi-invariant.

Remark 4.1.9. The above construction, based on the Riesz representation theorem, is not the only way to construct Haar measure. Another approach that is common in the literature is to first build a left-invariant outer measure and then use the *Carathéodory extension theorem*. Roughly speaking, the main difference between that approach and the one given here is that it is based on covering compact or open sets by other compact or open sets, rather than covering continuous, compactly supported functions by other continuous, compactly supported functions. In the compact case, one can also construct Haar probability measure by defining $\int_G f \, d\mu$ to be the mean of f, or more precisely, the unique constant function that is an average of translates of f. See [**Ta2010**, Exercise 1.12.6] for further discussion (the text there focuses on the abelian case, but the argument extends to the nonabelian setting).

Exercise 4.1.6 (Alternate proof of uniqueness of Haar measure). Let μ, ν be two left-invariant Haar measures on a σ-compact locally compact Hausdorff group G.

(i) For any functions $f, g \in C_c(G)$, establish the identity

$$\int_G f(x) \left(\int_G g(yx) \, d\nu(y) \right) d\mu(x) = \left(\int_G f(y^{-1}) d\nu(y) \right) \left(\int_G g(x) \, d\mu(x) \right).$$

   (*Hint:* Use the Fubini-Tonelli theorem twice and left-invariance twice.)

(ii) Show that for any $g \in C_c(G)$, $\int_G g(x) \, d\mu(x) = 0$ implies

$$\int_G g(y) \, d\nu(y) = 0$$

   and vice versa.

(iii) Conclude that one has $\mu = c\nu$ for some scalar $c \in \mathbf{R}^+$.

The following exercise explores the distinction between left-invariance and right-invariance.

Exercise 4.1.7. Let G be a σ-compact locally compact Hausdorff group, and let μ be a left-invariant Haar measure on G.

(i) Show that for each $y \in G$, there exists a unique positive real $c(y)$ (independent of the choice of μ) such that $\mu(Ey) = c(y)\mu(E)$ for all Borel measurable sets E and $\int_G f(xy^{-1}) \, d\mu(x) = c(y) \int_G f(x) \, d\mu(x)$ for all absolutely integrable f. In particular, a left-invariant Haar measure is right-invariant if and only if $c(y) = 1$ for all $y \in G$.

(ii) Show that the map $y \mapsto c(y)$ is a continuous homomorphism from G to the multiplicative group $\mathbf{R}^+ = (\mathbf{R}^+, \cdot)$. (This homomorphism is known as the *modular function*, and G is said to be *unimodular* if c is identically equal to 1.)

- Show that for any $f \in C_c(G)$, one has that $\int_G f(x^{-1}) \, d\mu(x) = \int_G c(x)^{-1} f(x) \, d\mu(x)$. (*Hint:* Take another function $g \in C_c(G)$ and evaluate $\int_G \int_G g(yx) c(x)^{-1} f(x^{-1}) \, d\mu(x) d\mu(y)$ in two different ways, one of which involves replacing x by $y^{-1}x$.) In particular, in a unimodular group one has $\mu(E^{-1}) = \mu(E)$ and $\int_G f(x^{-1}) \, dx = \int_G f(x) \, dx$ for any Borel set E and any $f \in C_c(G)$.

(iii) Show that G is unimodular if it is compact.

(iv) If G is a Lie group with Lie algebra $\mathfrak{g}$, show that $c(g) = |\det \mathrm{Ad}_g|$, where $\mathrm{Ad}_g : \mathfrak{g} \to \mathfrak{g}$ is the *adjoint representation* of g, defined by requiring $\exp(t \, \mathrm{Ad}_g X) = g \exp(tX) g^{-1}$ for all $X \in \mathfrak{g}$ (cf. Lemma 2.5.3).

(v) If G is a connected Lie group with Lie algebra $\mathfrak{g}$, show that G is unimodular if and only if $\mathrm{tr} \, \mathrm{ad}_X = 0$ for all $X \in \mathfrak{g}$, where $\mathrm{ad}_X : Y \mapsto [X, Y]$ is the *adjoint representation* of X.

(vi) Show that G is unimodular if it is a connected *nilpotent* Lie group.

(vii) Let G be a connected Lie group whose Lie algebra $\mathfrak{g}$ is such that $[\mathfrak{g}, \mathfrak{g}] = \mathfrak{g}$ (where $[\mathfrak{g}, \mathfrak{g}]$ is the linear span of the commutators $[X, Y]$ with $X, Y \in \mathfrak{g}$). (This condition is, in particular, obeyed when the Lie algebra $\mathfrak{g}$ is *semisimple*.) Show that G is unimodular.

(viii) Let G be the group of pairs $(a, b) \in \mathbf{R}^+ \times \mathbf{R}$ with the composition law $(a, b)(c, d) := (ac, ad + b)$. (One can interpret G as the group of orientation-preserving affine transformations $x \mapsto ax + b$ on the real line.) Show that G is a connected Lie group that is not unimodular.

In the case of a Lie group, one can also build Haar measures by starting with a noninvariant smooth measure, and then correcting it. Given a smooth manifold M, define a *smooth measure* μ on M to be a Radon measure which is a smooth multiple of Lebesgue measure when viewed in coordinates, thus for any smooth coordinate chart $\phi : U \to V$, the pushforward measure $\phi_*(\mu \lfloor_U)$ takes the form $f(x) \, dx \lfloor_V$ for some smooth function $f : V \to \mathbf{R}^+$, thus

$$\mu(E) = \int_{\phi(E)} f(x) \, dx$$

for all $E \subset U$. We say that the smooth measure is *nonvanishing* if f is nonzero on V for every coordinate chart $\phi : U \to V$.

Exercise 4.1.8. Let G be a Lie group, and let μ be a nonvanishing smooth measure on G.

(i) Show that for every $g \in G$, there exists a unique smooth function $\rho_g : G \to \mathbf{R}^+$ such that

$$\int_G f(g^{-1}x) \, d\mu(x) = \int_G f(x)\rho_g(x) \, d\mu(x).$$

(ii) Verify the *cocycle equation* $\rho_{gh}(x) = \rho_g(x)\rho_h(gx)$ for all $g, h, x \in G$.

(iii) Show that the measure ν defined by

$$\nu(E) := \int_E \rho_x(\mathrm{id})^{-1} \, d\mu(x)$$

is a left-invariant Haar measure on G.

There are a number of ways to generalise the Haar measure construction. For instance, one can define a local Haar measure on a local group G. If U is a neighbourhood of the identity in a σ-compact locally compact Hausdorff local group G, we define a *local left-invariant Haar measure* on U to be a nonzero Radon measure on U with the property that $\mu(gE) = \mu(E)$ whenever $g \in G$ and $E \subset U$ is a Borel set such that gE is well-defined and also in U.

Exercise 4.1.9 (Local Haar measure). Let G be a σ-compact locally compact Hausdorff local group, and let U be an open neighbourhood of the identity in G such that U is symmetric (i.e., U^{-1} is well-defined and equal to U) and U^{10} is well-defined in G. By adapting the arguments above, show that there is a local left-invariant Haar measure on U, and that it is unique up to scalar multiplication. (*Hint:* A new technical difficulty is that there are now multiple covering numbers of interest, namely the covering numbers $[f, g]_{U^m}$ associated to various small powers U^m of m. However, as long as one keeps track of which covering number to use at various junctures, this will not cause difficulty.)

One can also sometimes generalise the Haar measure construction from groups G to spaces X that G acts transitively on.

Definition 4.1.10 (Group actions). Given a topological group G and a topological space X, define a (left) *continuous action* of G on X to be a continuous map $(g, x) \mapsto gx$ from $G \times X$ to X such that $g(hx) = (gh)x$ and $\mathrm{id}\, x = x$ for all $g, h \in G$ and $x \in X$.

This action is said to be *transitive* if for any $x, y \in X$, there exists $g \in G$ such that $gx = y$, and in this case X is called a *homogeneous space* with structure group G, or *homogenous G-space* for short.

For any $x_0 \in X$, we call $\mathrm{Stab}(x_0) := \{g \in G : gx_0 = x_0\}$ the *stabiliser* of x_0; this is a closed subgroup of G.

If G, X are smooth manifolds (so that G is a Lie group) and the action $(g, x) \mapsto gx$ is a smooth map, then we say that we have a *smooth action* of G on X.

Exercise 4.1.10. If G acts transitively on a space X, show that all the stabilisers $\mathrm{Stab}(x_0)$ are conjugate to each other, and X is homeomorphic to the quotient spaces $G/\mathrm{Stab}(x_0)$ after weakening the topology of the quotient space (or strengthening the topology of the space X).

If G and X are σ-compact, locally compact, and Hausdorff, a (left) *Haar measure* is a nonzero Radon measure on X such that $\mu(gE) = \mu(E)$ for all Borel $E \subset X$ and $g \in G$.

Exercise 4.1.11. Let G be a σ-compact, locally compact, and Hausdorff group (left) acting continuously and transitively on a σ-compact, locally compact, and Hausdorff space X.

   (i) (Uniqueness up to scalars) Show that if μ, ν are (left) Haar measures on X, then $\mu = \lambda \nu$ for some $\nu > 0$.

  (ii) (Compact case) Show that if G is compact, then X is compact too, and a Haar measure on X exists.

 (iii) (Smooth unipotent case) Suppose that the action is smooth (so that G is a Lie group and X is a smooth manifold). Let x_0 be a point of X. Suppose that for each $g \in \mathrm{Stab}(x_0)$, the derivative map $Dg(x_0) : T_{x_0} X \to T_{x_0} X$ of the map $g : x \mapsto gx$ at x_0 is unimodular (i.e., it has determinant ± 1). Show that a Haar measure on X exists.

 (iv) (Smooth case) Suppose that the action is smooth. Show that any Haar measure on X is necessarily smooth. Conclude that a Haar measure exists if and only if the derivative maps Dg are unimodular.

  (v) (Counterexample) Let G be the $ax+b$ group from Exercise 4.1.7(viii), acting on $\mathbf{R}$ by the action $(a, b)x := ax + b$. Show that there is no Haar measure on $\mathbf{R}$. (This can be done either through (iv), or by an elementary direct argument.)

4.2. The Peter-Weyl theorem

We now restrict attention to compact groups G, which we will take to be Hausdorff for simplicity (although the results in this section will easily extend to the non-Hausdorff case using Exercise 4.1.2). By the previous discussion, there is a unique bi-invariant Haar probability measure μ on G, which gives rise in particular to the Hilbert space $L^2(G) = L^2(G, d\mu)$ of square-integrable functions $f : G \to \mathbf{C}$ on G (quotiented out by almost everywhere equivalence, as usual), with norm

$$\|f\|_{L^2(G)} := \left(\int_G |f(x)|^2 \, d\mu(x) \right)^{1/2}$$

and inner product

$$\langle f, g \rangle_{L^2(G)} := \int_G f(x)\overline{g(x)} \, dx.$$

For every group element $y \in G$, the translation operator $\tau(y) : L^2(G) \to L^2(G)$ is defined by

$$\tau(y)f(x) := f(y^{-1}x).$$

One easily verifies that $\tau(y^{-1})$ is both the inverse and the adjoint of $\tau(y)$, and so $\tau(y)$ is a unitary operator. The map $\tau : y \mapsto \tau(y)$ is then a continuous homomorphism from G to the unitary group $U(L^2(G))$ of $L^2(G)$ (where we give the latter group the *strong operator topology*), and is known as the *regular representation* of G.

For our purposes, the regular representation is too "big" of a representation to work with because the underlying Hilbert space $L^2(G)$ is usually infinite-dimensional. However, we can find smaller representations by locating *left-invariant* closed subspaces V of $L^2(G)$, i.e., closed linear subspaces of $L^2(G)$ with the property that $\tau(y)V \subset V$ for all $y \in G$. Then the restriction of τ to V becomes a representation $\tau \restriction_V : G \to U(V)$ to the unitary group of V. In particular, if V has some finite dimension n, this gives a representation of G by a unitary group $U_n(\mathbf{C})$ after expressing V in coordinates.

We can build invariant subspaces from applying spectral theory to an invariant operator, and more specifically to a *convolution operator*. If $f, g \in L^2(G)$, we define the *convolution* $f * g : G \to \mathbf{C}$ by the formula

$$f * g(x) = \int_G f(y)g(y^{-1}x) \, d\mu(y).$$

Exercise 4.2.1. Show that if $f, g \in L^2(G)$, then $f * g$ is well-defined and lies in $C(G)$ and, in particular, also lies in $L^2(G)$.

For $g \in L^2(G)$, let $T_g : L^2(G) \to L^2(G)$ denote the right-convolution operator $T_g f := f * g$. This is easily seen to be a bounded linear operator on $L^2(G)$. Using the properties of Haar measure, we also observe that T_g will be self-adjoint if g obeys the condition

(4.7) $$g(x^{-1}) = \overline{g(x)}$$

and it also commutes with left-translations:

$$T_g \rho(y) = \rho(y) T_g.$$

In particular, for any $\lambda \in \mathbf{C}$, the *eigenspace*

$$V_\lambda := \{f \in L^2(G) : T_g f = \lambda f\}$$

will be a closed invariant subspace of $L^2(G)$. Thus we see that we can generate a large number of representations of G by using the eigenspace of a convolution operator.

Another important fact about these operators, is that the T_g are *compact*, i.e., they map bounded sets to precompact sets. This is a consequence of the following more general fact:

Exercise 4.2.2 (Compactness of integral operators). Let (X, μ) and (Y, ν) be σ-finite measure spaces, and let $K \in L^2(X \times Y, \mu \times \nu)$. Define an integral operator $T : L^2(X, \mu) \to L^2(Y, \nu)$ by the formula

$$Tf(y) := \int_X K(x, y) f(x) \, d\mu(x).$$

(i) Show that T is a bounded linear operator, with operator norm $\|T\|_{\mathrm{op}}$ bounded by $\|K\|_{L^2(X \times Y, \mu \times \nu)}$. (*Hint:* Use duality.)

(ii) Show that T is a compact linear operator. (*Hint:* Approximate K by a linear combination of functions of the form $a(x)b(y)$ for $a \in L^2(X, \mu)$ and $b \in L^2(Y, \nu)$, plus an error which is small in $L^2(X \times Y, \mu \times \nu)$ norm, so that T becomes approximated by the sum of a *finite rank operator* and an operator of small operator norm.)

Note that T_g is an integral operator with kernel $K(x, y) := g(x^{-1}y)$; from the invariance properties of Haar measure we see that $K \in L^2(G \times G)$ if $g \in L^2(G)$ (note here that we crucially use the fact that G is compact, so that $\mu(G) = 1$). Thus we conclude that the convolution operator T_g is compact when G is compact.

Exercise 4.2.3. Show that if $g \in C_c(\mathbf{R})$ is nonzero, then T_g is not compact on $L^2(\mathbf{R})$. This example demonstrates that compactness of G is needed in order to ensure compactness of T_g.

We can describe self-adjoint compact operators in terms of their eigenspaces:

Theorem 4.2.1 (Spectral theorem). *Let $T : H \to H$ be a compact self-adjoint operator on a complex Hilbert space H. Then there exists an at most countable sequence $\lambda_1, \lambda_2, \ldots$ of nonzero reals that converge to zero and an orthogonal decomposition*

$$H = V_0 \oplus \bigoplus_n V_{\lambda_n}$$

of H into the 0 eigenspace (or kernel) V_0 of T, and the λ_n-eigenspaces V_{λ_n}, which are all finite-dimensional.

Proof. From self-adjointness we see that all the eigenspaces V_λ are orthogonal to each other, and only nontrivial for λ real. If $r > 0$, then $\bigoplus_{\lambda \in \mathbf{R}:|\lambda|>r} V_\lambda$ has an orthonormal basis of eigenfunctions v, each of which is enlarged by a factor of at least r by T. In particular, this basis cannot be infinite, because otherwise the image of this basis by T would have no convergent subsequence, contradicting compactness. Thus $\bigoplus_{\lambda \in \mathbf{R}:|\lambda|>r} V_\lambda$ is finite-dimensional for any r, which implies that V_λ is finite-dimensional for every nonzero λ, and those nonzero λ with nontrivial V_λ can be enumerated to either be finite, or countable and go to zero.

Let W be the orthogonal complement of $V_0 \oplus \bigoplus_n V_{\lambda_n}$. If W is trivial, then we are done, so suppose for sake of contradiction that W is nontrivial. As all of the V_λ are invariant, and T is self-adjoint, W is also invariant, with T being self-adjoint on W. As W is orthogonal to the kernel V_0 of T, T has trivial kernel in W. More generally, T has no eigenvectors in W.

Let B be the unit ball in W. As T has trivial kernel and W is nontrivial, $\|T\|_{\mathrm{op}} > 0$. Using the identity

(4.8) $$\|T\|_{\mathrm{op}} = \sup_{W:\|x\|\leq 1} |\langle Tx, x\rangle|$$

valid for all self-adjoint operators T (see Exercise 4.2.4 below). Thus, we may find a sequence x_n of vectors of norm at most 1 such that

$$\langle Tx_n, x_n\rangle \to \lambda$$

for some $\lambda = \pm\|T\|_{\mathrm{op}}$. Since $\|Tx_n\|^2 \leq \|T\|_{\mathrm{op}}^2 \|x_n\|^2 \leq \lambda^2$, we conclude that

$$0 \leq \|Tx_n - \lambda x_n\|^2 = \|Tx_n\|^2 + \lambda^2\|x_n\|^2 - 2\langle Tx_n, x_n\rangle \leq 2\lambda^2 - 2\langle Tx_n, x_n\rangle$$

and hence

(4.9) $$Tx_n - \lambda x_n \to 0;$$

applying T we conclude that

$$T(Tx_n) - \lambda Tx_n \to 0.$$

By compactness of T, we may pass to a subsequence so that Tx_n converges to a limit y, and thus $Ty - \lambda y = 0$. As T has no eigenvectors, y must be trivial; but then $\langle Tx_n, x_n\rangle$ converges to zero, a contradiction. $\square$

Exercise 4.2.4. Establish (4.9) whenever $T : W \to W$ is a bounded self-adjoint operator on W. (*Hint:* Bound $|\langle Tx, y\rangle|$ by the right-hand side of (4.8) whenever x, y are vectors of norm at most 1, by playing with $\langle T(ax + by), (ax + by)\rangle$ for various choices of scalars a, b, in the spirit of the proof of the Cauchy-Schwarz inequality.)

This leads to the consequence that we can find nontrivial finite-dimensional representations on at least a single nonidentity element:

Theorem 4.2.2 (Baby Peter-Weyl theorem). *Let G be a compact Hausdorff group with Haar measure μ, and let $y \in G$ be a nonidentity element of G. Then there exists a finite-dimensional invariant subspace of $L^2(G)$ on which $\tau(y)$ is not the identity.*

Proof. Suppose for contradiction that $\tau(y)$ is the identity on every finite-dimensional invariant subspace of $L^2(G)$, thus $\tau(y)-1$ annihilates every such subspace. By Theorem 4.2.1, we conclude that $\tau(y) - 1$ has range in the kernel of every convolution operator T_g with $g \in L^2$, thus $T_g(\tau(y) - 1)f = 0$ for any $f, g \in L^2(G)$ with g obeying (4.7), i.e.,

$$\tau(y)(f * g) = (f * g)$$

for any such f, g. But one may easily construct f, g such that $f * g$ is nonzero at the identity and vanishing at y (e.g., one can set $f = g = 1_U$ where U is an open symmetric neighbourhood of the identity, small enough that y lies outside U^2). This gives the desired contradiction. $\square$

Remark 4.2.3. The full *Peter-Weyl theorem* describes rather precisely all the invariant subspaces of $L^2(G)$. Roughly speaking, the theorem asserts that for each irreducible finite-dimensional representation $\rho_\lambda : G \to U(V_\lambda)$ of G, $\dim(V_\lambda)$ different copies of V_λ (viewed as an invariant G-space) appear in $L^2(G)$, and that they are all orthogonal and make up all of $L^2(G)$; thus, one has an orthogonal decomposition

$$L^2(G) \equiv \bigoplus_\lambda V_\lambda^{\dim(V_\lambda)}$$

of G-spaces. Actually, this is not the sharpest form of the theorem, as it only describes the left G-action and not the right G-action; see Chapter 18 for a precise statement and proof of the Peter-Weyl theorem in its strongest form. This form is of importance in Fourier analysis and representation theory, but in this text we will only need the baby form of the theorem (Theorem 4.2.2), which is an easy consequence of the full Peter-Weyl theorem (since, if g is not the identity, then $\tau(g)$ is clearly nontrivial on $L^2(G)$ and hence on at least one of the V_λ factors).

The Peter-Weyl theorem leads to the following structural theorem for compact groups:

Theorem 4.2.4 (Gleason-Yamabe theorem for compact groups). *Let G be a compact Hausdorff group, and let U be a neighbourhood of the identity. Then there exists a compact normal subgroup H of G contained in U such* *that G/H is isomorphic to a linear group (i.e., a closed subgroup of a general linear group $\mathrm{GL}_n(\mathbf{C})$).*

Note from Cartan's theorem (Theorem 3.0.14) that every linear group is Lie; thus, compact Hausdorff groups are "almost Lie" in some sense.

Proof. Let g be an element of $G\backslash U$. By the baby Peter-Weyl theorem, we can find a finite-dimensional invariant subspace V of $L^2(G)$ on which $\tau(g)$ is nontrivial. Identifying such a subspace with $\mathbf{C}^n$ for some finite n, we thus have a continuous homomorphism $\rho : G \to \mathrm{GL}_n(\mathbf{C})$ such that $\rho(g)$ is nontrivial. By continuity, $\rho(g)$ will also be nontrivial for some open neighbourhood of g. Using the compactness of $G\backslash U$, one can then find a finite number $\rho_1, \ldots, \rho_k$ of such continuous homomorphisms $\rho_i : G \to \mathrm{GL}_{n_i}(\mathbf{C})$ such that for each $g \in G\backslash U$, at least one of $\rho_1(g), \ldots, \rho_k(g)$ is nontrivial. If we then form the direct sum

$$\rho := \bigoplus_{i=1}^k \rho_i : G \to \bigoplus_{i=1}^k \mathrm{GL}_{n_i}(\mathbf{C}) \subset \mathrm{GL}_{n_1 + \cdots + n_k}(\mathbf{C}),$$

then ρ is still a continuous homomorphism, which is now nontrivial for any $g \in G\backslash U$; thus the kernel H of ρ is a compact normal subgroup of G contained in U. There is thus a continuous bijection from the compact space G/H to the Hausdorff space $\rho(G)$, and so the two spaces are homeomorphic. As $\rho(G)$ is a compact (hence closed) subgroup of $\mathrm{GL}_{n_1 + \cdots + n_k}(\mathbf{C})$, the claim follows. $\square$

Exercise 4.2.5. Show that the hypothesis that G is Hausdorff can be omitted from Theorem 4.2.4. (*Hint:* Use Exercise 4.1.2.)

Exercise 4.2.6. Show that any compact Lie group is isomorphic to a linear group. (*Hint:* First find a neighbourhood of the identity that is so small that it does not contain any nontrivial subgroups.) The property of having *no small subgroups* will be an important one in later sections.

One can rephrase the Gleason-Yamabe theorem for compact groups in terms of the machinery of *inverse limits* (also known as *projective limits*).

Definition 4.2.5 (Inverse limits of groups). Let $(G_\alpha)_{\alpha \in A}$ be a family of groups G_α indexed by a partially ordered set $A = (A, <)$. Suppose that for each $\alpha < \beta$ in A, there is a surjective homomorphism $\pi_{\alpha \leftarrow \beta} : G_\beta \to G_\alpha$ which obeys the composition law $\pi_{\alpha \leftarrow \beta} \circ \pi_{\beta \leftarrow \gamma} = \pi_{\alpha \leftarrow \gamma}$ for all $\alpha < \beta < \gamma$. (If one wishes, one can take a *category-theoretic perspective* and view these surjections as describing a *functor* from the partially ordered set A to the category of groups.) We then define the *inverse limit* $G = \lim_{\leftarrow} G_\alpha$ to be the set of all tuples $(g_\alpha)_{\alpha \in A}$ in the product set $\prod_{\alpha \in A} G_\alpha$ such that

$\pi_{\alpha \leftarrow \beta}(g_\beta) = g_\alpha$ for all $\alpha < \beta$; one easily verifies that this is also a group. We let $\pi_\alpha : G \to G_\alpha$ denote the coordinate projection maps $\pi_\alpha : (g_\beta)_{\beta \in A} \mapsto g_\alpha$.

If the G_α are topological groups and the $\pi_{\alpha \leftarrow \beta}$ are continuous, we can give G the topology induced from $\prod_{\alpha \in A} G_\alpha$; one easily verifies that this makes G a topological group, and that the π_α are continuous homomorphisms.

Exercise 4.2.7 (Universal description of inverse limit). Let $(G_\alpha)_{\alpha \in A}$ be a family of groups G_α with the surjective homomorphisms $\pi_{\alpha \leftarrow \beta}$ as in Definition 4.2.5. Let $G = \lim_{\leftarrow} G_\alpha$ be the inverse limit, and let H be another group. Suppose that one has homomorphisms $\phi_\alpha : H \to G_\alpha$ for each $\alpha \in A$ such that $\phi_{\alpha \leftarrow \beta} \circ \phi_\alpha = \phi_\beta$ for all $\alpha < \beta$. Show that there exists a unique homomorphism $\phi : H \to G$ such that $\phi_\alpha = \pi_\alpha \circ \phi$ for all $\alpha \in A$.

Establish the same claim with "group" and "homomorphism" replaced by "topological group" and "continuous homomorphism" throughout.

Exercise 4.2.8. Let p be a prime. Show that $\mathbf{Z}_p$ is isomorphic to the inverse limit $\lim_{\leftarrow} \mathbf{Z}/p^n \mathbf{Z}$ of the cyclic groups $\mathbf{Z}/p^n \mathbf{Z}$ with $n \in \mathbf{N}$ (with the usual ordering), using the obvious projection homomorphisms from $\mathbf{Z}/p^m \mathbf{Z}$ to $\mathbf{Z}/p^n \mathbf{Z}$ for $m > n$.

Exercise 4.2.9. Show that every compact Hausdorff group is isomorphic (as a topological group) to an inverse limit of linear groups. (*Hint:* Take the index set A to be the set of all nonempty finite collections of open neighbourhoods U of the identity, indexed by inclusion.) If the compact Hausdorff group is metrisable, show that one can take the inverse limit to be indexed instead by the natural numbers with the usual ordering.

Exercise 4.2.10. Let G be an abelian group with a homomorphism $\rho : G \mapsto U(V)$ into the unitary group of a finite-dimensional space V. Show that V can be decomposed as the vector space sum of one-dimensional G-invariant spaces. (*Hint:* By the spectral theorem for unitary matrices, any unitary operator T on V decomposes V into eigenspaces, and any operator commuting with T must preserve each of these eigenspaces. Now induct on the dimension of V.)

Exercise 4.2.11 (Fourier analysis on compact abelian groups). Let G be a compact abelian Hausdorff group with Haar probability measure μ. Define a *character* to be a continuous homomorphism $\chi : G \mapsto S^1$ to the unit circle $S^1 := \{z \in \mathbf{C} : |z| = 1\}$, and let $\hat{G}$ be the collection of all such characters.

  (i) Show that for every $g \in G$ not equal to the identity, there exists a character χ such that $\chi(g) \neq 1$. (*Hint:* Combine the baby Peter-Weyl theorem with the preceding exercise.)

(ii) Show that every function in $C(G)$ is the limit in the uniform topology of finite linear combinations of characters. (*Hint:* Use the *Stone-Weierstrass theorem.*)

(iii) Show that the characters χ for $\chi \in \hat{G}$ form an orthonormal basis of $L^2(G, d\mu)$.

4.3. The structure of locally compact abelian groups

We now use the above machinery to analyse locally compact abelian groups. We follow some combinatorial arguments of Pontryagin, as presented in the text of Montgomery and Zippin [**MoZi1974**].

We first make a general observation that locally compact groups contain open subgroups that are "finitely generated modulo a compact set". Call a subgroup Γ of a topological group G *cocompact* if the quotient space is compact.

Lemma 4.3.1. *Let G be a locally compact group. Then there exists an open subgroup G' of G which has a cocompact finitely generated subgroup Γ.*

Proof. Let K be a compact neighbourhood of the identity. Then K^2 is also compact and can thus be covered by finitely many copies of K, thus

$$K^2 \subset KS$$

for some finite set S, which we may assume without loss of generality to be contained in $K^{-1}K^2$. In particular, if Γ is the group generated by S, then

$$K^2\Gamma \subset K\Gamma.$$

Multiplying this on the left by powers of K and inducting, we conclude that

$$K^n\Gamma \subset K\Gamma$$

for all $n \geq 1$. If we then let G' be the group generated by K, then Γ lies in G' and $G' \subset K\Gamma \subset G'$. Thus G'/Γ is the image of the compact set K under the quotient map, and the claim follows. $\qquad\square$

In the abelian case, we can improve this lemma by combining it with the following proposition:

Proposition 4.3.2. *Let G be a locally compact Hausdorff abelian group with a cocompact finitely generated subgroup. Then G has a cocompact discrete finitely generated subgroup.*

To prove this proposition, we need the following lemma.

Lemma 4.3.3. *Let G be a locally compact Hausdorff group, and let $g \in G$. Then the group $\langle g \rangle$ generated by g is either precompact or discrete (or both).*

Proof. By replacing G with the closed subgroup $\overline{\langle g \rangle}$ we may assume without loss of generality that $\langle g \rangle$ is dense in G.

We may assume of course that $\langle g \rangle$ is not discrete. This implies that the identity element is not an isolated point in $\langle g \rangle$, and thus for any neighbourhood of the identity U, there exist arbitrarily large n such that $g^n \in U$; since $g^{-n} = (g^n)^{-1}$ we may take these n to be large and positive rather than large and negative.

Let U be a precompact symmetric neighbourhood of the identity, then U^3 (say) is covered by a finite number $g_j U$ of left-translates of U. As $\langle g \rangle$ is dense, we conclude that U^3 is covered by a finite number of translates $g^{n_j} U^2$ of left-translates of U by powers of g. Using the fact that there are arbitrarily large n with $g^n \in U$, we may thus cover U^3 by a finite number of translates $g^{m_j} U^3$ of U^3 with $m_j > 0$. In particular, if $g^n \in U^3$, then there exists an m_j such that $g^{n-m_j} \in U^3$. Iterating this, we see that the set $\{n \in \mathbf{Z} : g^n \in U^3\}$ is *left-syndetic*, in that it has bounded gaps as one goes to $-\infty$. Similarly one can argue that this set is right-syndetic and thus syndetic. This implies that the entire group $\langle g \rangle$ is covered by a bounded number of translates of U^3 and is thus precompact as required. $\square$

Now we can prove Proposition 4.3.2.

Proof. Let us say that a locally compact Hausdorff abelian group has *rank at most r* if it has a cocompact subgroup generated by at most r generators. We will induct on the rank r. If G has rank 0, then the cocompact subgroup is trivial, and the claim is obvious; so suppose that G has some rank $r \geq 1$, and the claim has already been proven for all smaller ranks.

By hypothesis, G has a cocompact subgroup Γ generated by r generators $e_1, \ldots, e_r$. By Lemma 4.3.3, the group $\langle e_r \rangle$ is either precompact or discrete. If it is discrete, then we can quotient out by that group to obtain a locally compact Hausdorff abelian group $G/\langle e_r \rangle$ of rank at most $r - 1$; by induction hypothesis, $G/\langle e_r \rangle$ has a cocompact discrete subgroup, and so G does also. Hence we may assume that $\langle e_r \rangle$ is precompact, and more generally that $\langle e_i \rangle$ is precompact for each i. But as we are in an abelian group, Γ is the product of all the $\langle e_i \rangle$, and is thus also precompact, so $\overline{\Gamma}$ is compact. But $G/\overline{\Gamma}$ is a quotient of G/Γ and is also compact, and so G itself is compact, and the claim follows in this case. $\square$

We can then combine this with the Gleason-Yamabe theorem for compact groups to obtain

Theorem 4.3.4 (Gleason-Yamabe theorem for abelian groups). *Let G be a locally compact abelian Hausdorff group, and let U be a neighbourhood of the*

identity. Then there exists a compact normal subgroup H of G contained in U such that G/H is isomorphic to a Lie group.

Proof. By Lemma 4.3.1 and Proposition 4.3.2, we can find an open subgroup G' of G and discrete cocompact subgroup Γ of G'. By shrinking U as necessary, we may assume that U is symmetric and U^2 only intersects Γ at the identity. Let $\pi : G' \to G'/\Gamma$ be the projection to the compact abelian group G'/Γ, then $\pi(U)$ is a neighbourhood of the identity in G'/Γ. By Theorem 4.2.4, one can find a compact normal subgroup H' of G'/Γ in $\pi(U)$ such that $(G'/\Gamma)/H'$ is isomorphic to a linear group, and thus to a Lie group. If we set $H := \pi^{-1}(H') \cap U$, it is not difficult to verify that H is also a compact normal subgroup of G'. If $\phi : G' \to G'/H$ is the quotient map, then $\phi(\Gamma)$ is a discrete subgroup of G'/H and from abstract nonsense one sees that $(G'/H)/\phi(\Gamma)$ is isomorphic to the Lie group $(G/\Gamma)/H'$. Thus G'/H is locally Lie. Since G' is an open subgroup of the abelian group G, G/H is locally Lie also, and is thus G/H is isomorphic to a Lie group by Exercise 2.4.7. $\square$

Exercise 4.3.1. Show that the Hausdorff hypothesis can be dropped from the above theorem.

Exercise 4.3.2 (Characters separate points)**.** Let G be a locally compact Hausdorff abelian group, and let $g \in G$ be not equal to the identity. Show that there exists a character $\chi : G \to S^1$ (see Exercise 4.2.11) such that $\chi(g) \neq 1$. This result can be used as the foundation of the theory of *Pontryagin duality* in abstract *harmonic analysis*, but we will not pursue this here; see for instance [**Ru1962**].

Exercise 4.3.3. Show that every locally compact abelian Hausdorff group is isomorphic to the inverse limit of abelian Lie groups.

Thus, in principle at least, the study of locally compact abelian group is reduced to that of abelian Lie groups, which are more or less easy to classify:

Exercise 4.3.4.

    (i) Show that every discrete subgroup of $\mathbf{R}^d$ is isomorphic to $\mathbf{Z}^{d'}$ for some $0 \leq d' \leq d$.

    (ii) Show that every connected abelian Lie group G is isomorphic to $\mathbf{R}^d \times (\mathbf{R}/\mathbf{Z})^{d'}$ for some natural numbers d, d'. (*Hint:* First show that the kernel of the exponential map is a discrete subgroup of the Lie algebra.) Conclude in particular the *divisibility property* that if $g \in G$ and $n \geq 1$, then there exists $h \in G$ with $h^n = g$.

    (iii) Show that every compact abelian Lie group G is isomorphic to $(\mathbf{R}/\mathbf{Z})^d \times H$ for some natural number d and a H which is a finite product of finite cyclic groups. (You may need the *classification of finitely generated abelian groups*, and will also need the divisibility property to lift a certain finite group from a certain quotient space back to G.)

(iv) Show that every abelian Lie group contains an open subgroup that is isomorphic to $\mathbf{R}^d \times (\mathbf{R}/\mathbf{Z})^{d'} \times \mathbf{Z}^{d''} \times H$ for some natural numbers d, d', d'' and a finite product H of finite cyclic groups.

Remark 4.3.5. Despite the quite explicit description of (most) abelian Lie groups, some interesting behaviour can still occur in locally compact abelian groups after taking inverse limits; consider for instance the solenoid example (Exercise 1.1.3).

Chapter 5

Building metrics on groups, and the Gleason-Yamabe theorem

In this chapter we will be able to finally prove the Gleason-Yamabe theorem (Theorem 1.1.13). In the next section, we will combine the Gleason-Yamabe theorem with some topological analysis (and, in particular, using the *invariance of domain* theorem) to establish some further control on locally compact groups and, in particular, obtaining a solution to Hilbert's fifth problem.

To prove the Gleason-Yamabe theorem, we will use three major tools developed in previous chapters. The first is Theorem 3.0.21, which provided a criterion for Lie structure in terms of a special type of metric, namely a Gleason metric (Definition 3.0.20). The second tool is the existence of a left-invariant Haar measure on any locally compact group; see Theorem 4.1.6. Finally, we will also need the compact case of the Gleason-Yamabe theorem (Theorem 4.2.4), which was proven via the Peter-Weyl theorem.

To finish the proof of the Gleason-Yamabe theorem, we have to somehow use the available structures on locally compact groups (such as Haar measure) to build good metrics on those groups (or on suitable subgroups or quotient groups). The basic construction is as follows:

Definition 5.0.6 (Building metrics out of test functions). Let G be a topological group, and let $\psi : G \to \mathbf{R}^+$ be a bounded nonnegative function.

Then we define the pseudometric $d_\psi : G \times G \to \mathbf{R}^+$ by the formula

$$d_\psi(g, h) := \sup_{x \in G} |\tau(g)\psi(x) - \tau(h)\psi(x)|$$

$$= \sup_{x \in G} |\psi(g^{-1}x) - \psi(h^{-1}x)|$$

and the semi-norm $\|\|_\psi : G \to \mathbf{R}^+$ by the formula

$$\|g\|_\psi := d_\psi(g, \mathrm{id}).$$

Note that one can also write

$$\|g\|_\psi = \sup_{x \in G} |\partial_g \psi(x)|$$

where $\partial_g \psi(x) := \psi(x) - \psi(g^{-1}x)$ is the "derivative" of ψ in the direction g.

Exercise 5.0.5. Let the notation and assumptions be as in the above definition. For any $g, h, k \in G$, establish the metric-like properties

(1) (Identity) $d_\psi(g, h) \geq 0$, with equality when $g = h$.

(2) (Symmetry) $d_\psi(g, h) = d_\psi(h, g)$.

(3) (Triangle inequality) $d_\psi(g, k) \leq d_\psi(g, h) + d_\psi(h, k)$.

(4) (Continuity) If $\psi \in C_c(G)$, then the map $d_\psi : G \times G \to \mathbf{R}^+$ is continuous.

(5) (Boundedness) One has $d_\psi(g, h) \leq \sup_{x \in G} |\psi(x)|$. If $\psi \in C_c(G)$ is supported in a set K, then equality occurs unless $g^{-1}h \in KK^{-1}$.

(6) (Left-invariance) $d_\psi(g, h) = d_\psi(kg, kh)$. In particular, $d_\psi(g, h) = \|h^{-1}g\|_\psi = \|g^{-1}h\|_\psi$.

In particular, we have the norm-like properties

(1) (Identity) $\|g\|_\psi \geq 0$, with equality when $g = \mathrm{id}$.

(2) (Symmetry) $\|g\|_\psi = \|g^{-1}\|_\psi$.

(3) (Triangle inequality) $\|gh\|_\psi \leq \|g\|_\psi + \|h\|_\psi$.

(4) (Continuity) If $\psi \in C_c(G)$, then the map $\|\|_\psi : G \to \mathbf{R}^+$ is continuous.

(5) (Boundedness) One has $\|g\|_\psi \leq \sup_{x \in G} |\psi(x)|$. If $\psi \in C_c(G)$ is supported in a set K, then equality occurs unless $g \in KK^{-1}$.

We remark that the first three properties of d_ψ in the above exercise ensure that d_ψ is indeed a *pseudometric*.

To get good metrics (such as Gleason metrics) on groups G, it thus suffices to obtain test functions ψ that obey suitably good "regularity" properties. We will achieve this primarily by means of two tricks. The first trick is to obtain high-regularity test functions by convolving together two low-regularity test functions, taking advantage of the existence of a left-invariant Haar measure μ on G. The second trick is to obtain low-regularity test functions by means of a metric-like object on G. This latter trick may seem circular, as our whole objective is to get a metric on G in the first place, but the key point is that the metric one starts with does not need to have as many "good properties" as the metric one ends up with, thanks to the regularity-improving properties of convolution. As such, one can use a "bootstrap argument" (or induction argument) to create a good metric out of almost nothing. It is this bootstrap miracle which is at the heart of the proof of the Gleason-Yamabe theorem (and hence to the solution of Hilbert's fifth problem).

The arguments here are based on the nonstandard analysis arguments used to establish Hilbert's fifth problem by Hirschfeld [**Hi1990**] and by Goldbring [**Go2010**] (and also some unpublished lecture notes of Goldbring and van den Dries). However, we will not explicitly use any nonstandard analysis in this chapter.

5.1. Warmup: the Birkhoff-Kakutani theorem

To illustrate the basic idea of using test functions to build metrics, let us first establish a classical theorem on topological groups, which gives a necessary and sufficient condition for metrisability. Recall that a topological space is metrisable if there is a metric on that space that generates the topology.

Theorem 5.1.1 (Birkhoff-Kakutani theorem). *A topological group is metrisable if and only if it is Hausdorff and* first countable[1].

Remark 5.1.2. The group structure is crucial; for instance, the *long line* is Hausdorff and first countable, but not metrisable.

This theorem, together with the existence of Haar measure from the previous section (and the argument in Exercise 5.1.3 below), completes the second arrow of Figure 1.

We now prove Theorem 5.1.1 (following the arguments in [**MoZi1974**]). The "only if" direction is easy, so it suffices to establish the "if" direction. The key lemma is

Lemma 5.1.3 (Urysohn-type lemma). *Let G be a Hausdorff first countable group. Then there exists a bounded continuous function $\psi : G \to [0,1]$ with the following properties:*

(i) *(Unique maximum)* $\psi(\mathrm{id}) = 1$, *and* $\psi(x) < 1$ *for all* $x \neq \mathrm{id}$.

---

[1] A topological space is first countable if every point has a countable neighbourhood base.

(ii) *(Neighbourhood base) The sets $\{x \in G : \psi(x) > 1 - 1/n\}$ for $n = 1, 2, \ldots$ form a* neighbourhood base *at the identity.*

(iii) *(Uniform continuity) For every $\varepsilon > 0$, there exists an open neighbourhood U of the identity such that $|\psi(gx) - \psi(x)| \leq \varepsilon$ for all $g \in U$ and $x \in G$.*

Note that if G had a left-invariant metric, then the function $\psi(x) := \max(1 - d(x, \mathrm{id}), 0)$ would suffice for this lemma, which already gives some indication as to why this lemma is relevant to the Birkhoff-Kakutani theorem.

Exercise 5.1.1. Let G be a Hausdorff first countable group, and let ψ be as in Lemma 5.1.3. Show that d_ψ is a metric on G (so, in particular, $d_\psi(g, h)$ only vanishes when $g = h$) and that d_ψ generates the topology of G (thus every set which is open with respect to d_ψ is open in G, and vice versa).

In view of the above exercise, we see that to prove the Birkhoff-Kakutani theorem, it suffices to prove Lemma 5.1.3, which we now do. By first countability, we can find a countable neighbourhood base

$$V_1 \supset V_2 \supset \cdots \supset \{\mathrm{id}\}$$

of the identity. As G is Hausdorff, we must have

$$\bigcap_{n=1}^{\infty} V_n = \{\mathrm{id}\}.$$

Using the continuity of the group operations, we can recursively find a sequence of nested open neighbourhoods of the identity

$$(5.1) \qquad U_1 \supset U_{1/2} \supset U_{1/4} \supset \cdots \supset \{\mathrm{id}\}$$

such that each $U_{1/2^n}$ is symmetric (i.e., $g \in U_{1/2^n}$ if and only if $g^{-1} \in U_{1/2^n}$), is contained in V_n, and is such that $U_{1/2^{n+1}} \cdot U_{1/2^{n+1}} \subset U_{1/2^n}$ for each $n \geq 0$. In particular, the $U_{1/2^n}$ are also a neighbourhood base of the identity with

$$(5.2) \qquad \bigcap_{n=1}^{\infty} U_{1/2^n} = \{\mathrm{id}\}.$$

For every dyadic rational $a/2^n$ in $(0, 1)$, we can now define the open sets $U_{a/2^n}$ by setting

$$U_{a/2^n} := U_{1/2^{n_k}} \cdot \cdots \cdot U_{1/2^{n_1}}$$

where $a/2^n = 2^{-n_1} + \cdots + 2^{-n_k}$ is the binary expansion of $a/2^n$ with $1 \leq n_1 < \cdots < n_k$. By repeated use of the hypothesis $U_{1/2^{n+1}} \cdot U_{1/2^{n+1}} \subset U_{1/2^n}$ we see that the $U_{a/2^n}$ are increasing in $a/2^n$; indeed, we have the inclusion

$$(5.3) \qquad U_{1/2^n} \cdot U_{a/2^n} \subset U_{(a+1)/2^n}$$

for all $n \geq 1$ and $1 \leq a < 2^n$.

We now set

$$\psi(x) := \sup \left\{ 1 - \frac{a}{2^n} : n \geq 1; 1 \leq a < 2^n; x \in U_{a/2^n} \right\}$$

with the understanding that $\psi(x) = 0$ if the supremum is over the empty set. One easily verifies using (5.3) that ψ is continuous, and furthermore obeys the uniform continuity property. The neighbourhood base property follows since the $U_{1/2^n}$ are a neighbourhood base of the identity, and the unique maximum property follows from (5.2). This proves Lemma 5.1.3, and the Birkhoff-Kakutani theorem follows.

Exercise 5.1.2. Let G be a topological group. Show that G is *completely regular*, that is to say, for every closed subset F in G and every $x \in G \backslash F$, there exists a continuous function $f : G \to \mathbf{R}$ that equals 1 on F and vanishes on x.

Exercise 5.1.3 (Reduction to the metrisable case). Let G be a locally compact group, let U be an open neighbourhood of the identity, and let G' be the group generated by U.

(i) Construct a sequence of open neighbourhoods of the identity

$$U \supset U_1 \supset U_2 \supset \dots$$

with the property that $U_{n+1}^2 \subset U_n$ and $U_{n+1}^U \subset U_n$ for all $n \geq 1$, where $A^B := \{a^b : a \in A, b \in B\}$ and $a^b := b^{-1}ab$.

(ii) If we set $H := \bigcap_{n=1}^{\infty} U_n$, show that H is a closed normal subgroup G' in U, and the quotient group G'/H is Hausdorff and first countable (and thus metrisable, by the Birkhoff-Kakutani theorem).

(iii) Conclude that to prove the Gleason-Yamabe theorem (Theorem 1.1.13), it suffices to do so under the assumption that G is metrisable.

The above arguments are essentially in [**Gl1952**].

Exercise 5.1.4 (Birkhoff-Kakutani theorem for local groups). Let G be a local group which is Hausdorff and first countable. Show that there exists an open neighbourhood V_0 of the identity which is metrisable.

5.2. Obtaining the commutator estimate via convolution

We now return to the main task of constructing Gleason metrics (filling out the remaining arrows in Figure 1). The first thing we will do is dispense with the commutator property (3.1). To this end, let us temporarily define a *weak Gleason metric* on a topological group G to be a left-invariant metric

$d : G \times G \to \mathbf{R}^+$ which generates the topology on G and obeys the escape property for some constant $C > 0$, thus one has

(5.4) $\|g^n\| \geq \frac{1}{C} n \|g\|$ whenever $g \in G, n \geq 1,$ and $n\|g\| \leq \frac{1}{C}.$

In this section we will show

Theorem 5.2.1. *Every weak Gleason metric is a Gleason metric (possibly after adjusting the constant C).*

This theorem represents the leftmost arrow on the second row of Figure 1.

We now prove Theorem 5.2.1. The key idea here is to involve a bump function ϕ formed by convolving together two Lipschitz functions. The escape property (5.4) will be crucial in obtaining quantitative control of the metric geometry at very small scales, as one can study the size of a group element g very close to the origin through its powers g^n, which are further away from the origin.

Specifically, let $\varepsilon > 0$ be a small quantity to be chosen later, and let $\psi \in C_c(G)$ be a nonnegative Lipschitz function supported on the ball $B(0, \varepsilon)$ which is not identically zero. For instance, one could use the explicit function

$$\psi(x) := \left(1 - \frac{\|x\|}{\varepsilon}\right)_+$$

where $y_+ := \max(y, 0)$, although the exact form of ψ will not be important for our argument. Being Lipschitz, we see that

(5.5) $\|\partial_g \psi\|_{C_c(G)} \ll \|g\|$

for all $g \in G$ (where we allow implied constants to depend on G, ε, and ψ), where $\|\|_{C_c(G)}$ denotes the sup norm.

Let μ be a left-invariant Haar measure on G, the existence of which was established in Theorem 4.1.6. We then form the convolution $\phi := \psi * \psi$, with convolution defined using the formula

(5.6) $f * g(x) := \int_G f(y) g(y^{-1} x) \, d\mu(y).$

This is a continuous function supported in $B(0, 2\varepsilon)$, and gives a metric d_ϕ and a norm $\|\|_\phi$ as usual.

We now prove a variant of the commutator estimate (3.1), namely that

(5.7) $\|\partial_g \partial_h \phi\|_{C_c(G)} \ll \|g\| \|h\|$

whenever $g, h \in B(0, \varepsilon)$. To see this, we first use the left-invariance of Haar measure to write

(5.8) $\partial_h \phi = (\partial_h \psi) * \psi,$

thus

$$\partial_h \phi(x) = \int_G (\partial_h \psi)(y) \psi(y^{-1}x) \, d\mu(y).$$

We would like to similarly move the ∂_g operator over to the second factor, but we run into a difficulty due to the nonabelian nature of G. Nevertheless, we can still do this provided that we twist that operator by a conjugation. More precisely, we have

$$(5.9) \qquad \partial_g \partial_h \phi(x) = \int_G (\partial_h \psi)(y)(\partial_{g^y} \psi)(y^{-1}x) \, d\mu(y)$$

where $g^y := y^{-1}gy$ is g conjugated by y. If $h \in B(0, \varepsilon)$, the integrand is only nonzero when $y \in B(0, 2\varepsilon)$. Applying (5.5), we obtain the bound

$$\|\partial_g \partial_h \phi\|_{C_c(g)} \ll \|h\| \sup_{y \in B(0, 2\varepsilon)} \|g^y\|.$$

To finish the proof of (5.7), it suffices to show that

$$\|g^y\| \ll \|g\|$$

whenever $g \in B(0, \varepsilon)$ and $y \in B(0, 2\varepsilon)$.

We can achieve this by the escape property (5.4). Let n be a natural number such that $n\|g\| \le \varepsilon$, then $\|g^n\| \le \varepsilon$ and so $g^n \in B(0, \varepsilon)$. Conjugating by y, this implies that $(g^y)^n \in B(0, 5\varepsilon)$, and so by (5.4), we have $\|g^y\| \ll \frac{1}{n}$ (if ε is small enough), and the claim follows.

Next, we claim that the norm $\|\,\|_\phi$ is locally comparable to the original norm $\|\,\|$. More precisely, we claim:

(1) If $g \in G$ with $\|g\|_\phi$ sufficiently small, then $\|g\| \ll \|g\|_\phi$.

(2) If $g \in G$ with $\|g\|$ sufficiently small, then $\|g\|_\phi \ll \|g\|$.

Claim 2 follows easily from (5.8) and (5.5), so we turn to Claim 1. Let $g \in G$, and let n be a natural number such that

$$n\|g\|_\phi < \|\phi\|_{C_c(G)}.$$

Then by the triangle inequality

$$\|g^n\|_\phi < \|\phi\|_{C_c(G)}.$$

This implies that ϕ and $\tau_{g^n}\phi$ have overlapping support, and hence g^n lies in $B(0, 4\varepsilon)$. By the escape property (5.4), this implies (if ε is small enough) that $\|g\| \ll \frac{1}{n}$, and the claim follows.

Combining Claim 2 with (5.7) we see that

$$\|\partial_g \partial_h \phi\|_{C_c(G)} \ll \|g\|_\phi \|h\|_\phi$$

whenever $\|g\|_\phi, \|h\|_\phi$ are small enough. Now we use the identity

$$\|[g,h]\|_\phi = \|\tau([g,h])\phi - \phi\|_{C_c(G)}$$
$$= \|\tau(g)\tau(h)\phi - \tau(h)\tau(g)\phi\|_{C_c(G)}$$
$$= \|\partial_g\partial_h\phi - \partial_h\partial_g\phi\|_{C_c(G)}$$

and the triangle inequality to conclude that

$$\|[g,h]\|_\phi \ll \|g\|_\phi\|h\|_\phi$$

whenever $\|g\|_\phi, \|h\|_\phi$ are small enough. Theorem 5.2.1 then follows from Claim 1 and Claim 2.

5.3. Building metrics on NSS groups

We will now be able to build metrics on groups using a set of hypotheses that do not explicitly involve any metric at all. The key hypothesis will be the *no small subgroups* (NSS) property:

Definition 5.3.1 (No small subgroups). A topological group G has the *no small subgroups* (or NSS) property if there exists an open neighbourhood U of the identity which does not contain any subgroup of G other than the trivial group.

Exercise 5.3.1. Show that any Lie group is NSS.

Exercise 5.3.2. Show that any group with a weak Gleason metric is NSS.

For an example of a group which is not NSS, consider the infinite-dimensional torus $(\mathbf{R}/\mathbf{Z})^{\mathbf{N}}$. From the definition of the product topology, we see that any neighbourhood of the identity in this torus contains an infinite-dimensional subtorus, and so this group is not NSS.

Exercise 5.3.3. Show that for any prime p, the p-adic groups $\mathbf{Z}_p$ and $\mathbf{Q}_p$ are not NSS. What about the solenoid group $\mathbf{R} \times \mathbf{Z}_p/\mathbf{Z}^\Delta$?

Exercise 5.3.4. Show that an NSS group is automatically Hausdorff. (*Hint:* Use Exercise 4.1.2.)

Exercise 5.3.5. Show that an NSS locally compact group is automatically metrisable. (*Hint:* Use Exercise 5.1.3.)

Exercise 5.3.6 (NSS implies escape property). Let G be a locally compact NSS group. Show that if U is a sufficiently small neighbourhood of the identity, then for every $g \in G\backslash\{\mathrm{id}\}$, there exists a positive integer n such that $g^n \notin U$. Furthermore, for any other neighbourhood V of the identity, there exists a positive integer N such that if $g, \ldots, g^N \in U$, then $g \in V$.

We can now prove the following theorem (first proven in full generality by Yamabe [**Ya1953b**]), which is a key component in the proof of the Gleason-Yamabe theorem and in the wider theory of Hilbert's fifth problem, and a crucial arrow in Figure 1.

Theorem 5.3.2. *Every NSS locally compact group admits a weak Gleason metric. In particular, by Theorem 5.2.1 and Theorem 3.0.21, every NSS locally compact group is isomorphic to a Lie group.*

In view of this theorem and Exercise 5.3.1, we see that for locally compact groups, we obtain

Corollary 5.3.3. *A locally compact group is NSS if and only if it is isomorphic to a Lie group.*

This is a major advance towards both the Gleason-Yamabe theorem and Hilbert's fifth problem, as it has reduced the property of being a Lie group into a condition that is almost purely algebraic in nature.

We now prove Theorem 5.3.2. An important concept will be that of an *escape norm* associated to an open neighbourhood U of a group G, defined by the formula

$$(5.10) \qquad \|g\|_{e,U} := \inf\left\{ \frac{1}{n+1} : g, g^2, \ldots, g^n \in U \right\}$$

for any $g \in G$, where n ranges over the natural numbers (thus, for instance $\|g\|_{e,U} \leq 1$, with equality iff $g \notin U$). Thus, the longer it takes for the orbit $g, g^2, \ldots$ to escape U, the smaller the escape norm.

Strictly speaking, the escape norm is not necessarily a norm, as it need not obey the symmetry, nondegeneracy, or triangle inequalities; however, we shall see that in many situations, the escape norm behaves similarly to a norm, even if it does not exactly obey the norm axioms. Also, as the name suggests, the escape norm will be well suited for establishing the escape property (5.4).

It is possible for the escape norm $\|g\|_{e,U}$ of a nonidentity element $g \in G$ to be zero, if U contains the group $\langle g \rangle$ generated by U. But if the group G has the NSS property, then we see that this cannot occur for all sufficiently small U (where "sufficiently small" means "contained in a suitably chosen open neighbourhood U_0 of the identity"). In fact, more is true: if U, U' are two sufficiently small open neighbourhoods of the identity in a locally compact NSS group G, then the two escape norms are comparable, thus we have

$$(5.11) \qquad \|g\|_{e,U} \ll \|g\|_{e,U'} \ll \|g\|_{e,U}$$

for all $g \in G$ (where the implied constants can depend on U, U').

By symmetry, it suffices to prove the second inequality in (5.11). By (5.10), it suffices to find an integer m such that whenever $g \in G$ is such that $g, g^2, \ldots, g^m \in U$, then $g \in U'$. But this follows from Exercise 5.3.6. This concludes the proof of (5.11).

Exercise 5.3.7. Let G be a locally compact group. Show that if d is a left-invariant metric on G obeying the escape property (5.4) that generates the topology, then G is NSS, and $\|g\|$ is comparable to $\|g\|_{e,U}$ for all sufficiently small U and for all sufficiently small g. (In particular, any two left-invariant metrics obeying the escape property and generating the topology are locally comparable to each other.)

Henceforth, G is a locally compact NSS group. We now establish a metric-like property on the escape norm $\|\|_{e,U_0}$.

Proposition 5.3.4 (Approximate triangle inequality). *Let U_0 be a sufficiently small open neighbourhood of the identity. Then for any n and any $g_1, \ldots, g_n \in G$, one has*

$$\|g_1 \ldots g_n\|_{e,U_0} \ll \sum_{i=1}^{n} \|g_i\|_{e,U_0}$$

(where the implied constant can depend on U_0).

Of course, in view of (5.11), the exact choice of U_0 is irrelevant, so long as it is small. It is slightly convenient to take U_0 to be symmetric (thus $U_0 = U_0^{-1}$), so that $\|g\|_{e,U_0} = \|g^{-1}\|_{e,U_0}$ for all g.

Proof. We will use a bootstrap argument. To start with, assume that we somehow already have a weaker form of the conclusion, namely

$$(5.12) \qquad \|g_1 \ldots g_n\|_{e,U_0} \leq M \sum_{i=1}^{n} \|g_i\|_{e,U_0}$$

for all $n, g_1, \ldots, g_n$ and some huge constant M; we will then deduce the same estimate with a smaller value of M. Afterwards we will show how to remove the hypothesis (5.12).

Now suppose we have (5.12) for some M. Motivated by the argument in the previous section, we now try to convolve together two "Lipschitz" functions. For this, we will need some metric-like functions. Define the modified escape norm $\|g\|_{*,U_0}$ by the formula

$$\|g\|_{*,U_0} := \inf\left\{ \sum_{i=1}^{n} \|g_i\|_{e,U_0} : g = g_1 \ldots g_n \right\}$$

where the infimum is over all possible ways to split g as a finite product of group elements. From (5.12), we have

$$\frac{1}{M}\|g\|_{e,U_0} \leq \|g\|_{*,U_0} \leq \|g\|_{e,U_0} \tag{5.13}$$

and we have the triangle inequality

$$\|gh\|_{*,U_0} \leq \|g\|_{*,U_0} + \|h\|_{*,U_0}$$

for any $g, h \in G$. We also have the symmetry property $\|g\|_{*,U_0} = \|g^{-1}\|_{*,U_0}$. Thus $\|\|_{*,U_0}$ gives a left-invariant semi-metric on G by defining

$$\mathrm{dist}_{*,U_0}(g,h) := \|g^{-1}h\|_{*,U_0}.$$

We can now define a "Lipschitz" function $\psi : G \to \mathbf{R}$ by setting

$$\psi(x) := (1 - M\,\mathrm{dist}_{*,U_0}(x,U_0))_+ \,.$$

On the one hand, we see from (5.13) that this function takes values in $[0,1]$ obeys the Lipschitz bound

$$|\partial_g \psi(x)| \leq M\|g\|_{e,U_0} \tag{5.14}$$

for any $g, x \in G$. On the other hand, it is supported in the region where $\mathrm{dist}_{*,U_0}(x,U_0) \leq 1/M$, which by (5.13) (and (5.10)) is contained in U_0^2.

We could convolve ψ with itself in analogy to the preceding section, but in doing so, we will eventually end up establishing a much worse estimate than (5.12) (in which the constant M is replaced with something like M^2). Instead, we will need to convolve ψ with another function η, that we define as follows. We will need a large natural number L (independent of M) to be chosen later, then a small open neighbourhood $U_1 \subset U_0$ of the identity (depending on L, U_0) to be chosen later. We then let $\eta : G \to \mathbf{R}$ be the function

$$\eta(x) := \sup\{1 - \frac{j}{L} : x \in U_1^j U_0; j = 0, \ldots, L\} \cup \{0\}.$$

Similarly to ψ, we see that η takes values in $[0,1]$ and obeys the Lipschitz-type bound

$$|\partial_g \eta(x)| \leq \frac{1}{L} \tag{5.15}$$

for all $g \in U_1$ and $x \in G$. Also, η is supported in $U_1^L U_0$, and hence (if U_1 is sufficiently small depending on L, U_0) is supported in U_0^2, just as ψ is.

The functions ψ, η need not be continuous, but they are compactly supported, bounded, and Borel measurable, and so one can still form their convolution $\phi := \psi * \eta$, which will then be continuous and compactly supported; indeed, ϕ is supported in U_0^4.

We have a lower bound on how big ϕ is, since

$$\phi(0) \geq \mu(U_0) \gg 1$$

(where we allow implied constants to depend on μ, U_0, but remain independent of L, U_1, or M). This gives us a way to compare $\|\cdot\|_\phi$ with $\|\cdot\|_{e,U_0}$. Indeed, if $n\|g\|_\phi < \phi(0)$, then (as in the proof of Claim 1 in the previous section) we have $g^n \in U_0^8$; this implies that

$$\|g\|_{e,U_0^8} \ll \|g\|_\phi$$

for all $g \in G$, and hence by (5.11) we have

(5.16) $$\|g\|_{e,U_0} \ll \|g\|_\phi$$

also. In the converse direction, we have

(5.17) $$\begin{aligned} \|g\|_\phi &= \|\partial_g(\psi * \eta)\|_{C_c(G)} \\ &= \|(\partial_g\psi) * \eta\|_{C_c(G)} \\ &\ll M\|g\|_{e,U_0} \end{aligned}$$

thanks to (5.14). But we can do better than this, as follows. For any $g, h \in G$, we have the analogue of (5.9), namely

$$\partial_g\partial_h\phi(x) = \int_G (\partial_h\psi)(y)(\partial_{g^y}\eta)(y^{-1}x) \, d\mu(y)$$

If $h \in U_0$, then the integrand vanishes unless $y \in U_0^3$. By continuity, we can find a small open neighbourhood $U_2 \subset U_1$ of the identity such that $g^y \in U_1$ for all $g \in U_2$ and $y \in U_0^3$; we conclude from (5.14), (5.15) that

$$|\partial_g\partial_h\phi(x)| \ll \frac{M}{L}\|h\|_{e,U_0},$$

whenever $h \in U_0$ and $g \in U_2$. To use this, we observe the telescoping identity

$$\partial_{g^n} = n\partial_g + \sum_{i=0}^{n-1} \partial_g\partial_{g^i}$$

for any $g \in G$ and natural number n, and thus by the triangle inequality

(5.18) $$\|g^n\|_\phi = n\|g\|_\phi + O\left(\sum_{i=0}^{n-1} \|\partial_g\partial_{g^i}\phi\|_{C_c(G)}\right).$$

We conclude that

$$\|g^n\|_\phi = n\|g\|_\phi + O\left(n\frac{M}{L}\|g\|_{e,U_0}\right)$$

whenever $n \geq 1$ and $g, \ldots, g^n \in U_2$. Using the trivial bound $\|g^n\|_\phi = O(1)$, we then have

$$\|g\|_\phi \ll \frac{1}{n} + \frac{M}{L}\|g\|_{e,U_0};$$

optimising in n we obtain

$$\|g\|_\phi \ll \|g\|_{e,U_2} + \frac{M}{L}\|g\|_{e,U_0}$$

and hence by (5.11)

$$\|g\|_\phi \ll \left(\frac{M}{L} + O_{U_2}(1)\right)\|g\|_{e,U_0}$$

where the implied constant in $O_{U_2}(1)$ can depend on U_0, U_1, U_2, L, but is crucially independent of M. Note the essential gain of $\frac{1}{L}$ here compared with (5.17). We also have the norm inequality

$$\|g_1 \cdots g_n\|_\phi \le \sum_{i=1}^n \|g_i\|_\phi.$$

Combining these inequalities with (5.16) we see that

$$\|g_1 \cdots g_n\|_{e,U_0} \ll \left(\frac{1}{L}M + O_{U_2}(1)\right)\sum_{i=1}^n \|g_i\|_{e,U_0}.$$

Thus we have improved the constant M in the hypothesis (5.12) to $O(\frac{1}{L}M) + O_{U_2}(1)$. Choosing L large enough and iterating, we conclude that we can bootstrap any finite constant M in (5.12) to $O(1)$.

Of course, there is no reason why there has to be a finite M for which (5.12) holds in the first place. However, one can rectify this by the usual trick of creating an epsilon of room. Namely, one replaces the escape norm $\|g\|_{e,U_0}$ by, say, $\|g\|_{e,U_0} + \varepsilon$ for some small $\varepsilon > 0$ in the definition of $\|\|\|_{*,U_0}$ and in the hypothesis (5.12). Then the bound (5.12) will be automatic with a finite M (of size about $O(1/\varepsilon)$). One can then run the above argument with the requisite changes and conclude a bound of the form

$$\|g_1 \cdots g_n\|_{e,U_0} \ll \sum_{i=1}^n (\|g_i\|_{e,U_0} + \varepsilon)$$

uniformly in ε; we omit the details. Sending $\varepsilon \to 0$, we have thus shown Proposition 5.3.4. $\qquad\square$

Now we can finish the proof of Theorem 5.3.2. Let G be a locally compact NSS group, and let U_0 be a sufficiently small neighbourhood of the identity. From Proposition 5.3.4, we see that the escape norm $\|\|\|_{e,U_0}$ and the modified escape norm $\|\|\|_{*,U_0}$ are comparable. We have seen d_{*,U_0} is a left-invariant pseudometric. As G is NSS and U_0 is small, there are no non-identity elements with zero escape norm, and hence no nonidentity elements with zero modified escape norm either; thus d_{*,U_0} is a genuine metric.

We now claim that d_{*,U_0} generates the topology of G. Given the left-invariance of d_{*,U_0}, it suffices to establish two things: first, that any open neighbourhood of the identity contains a ball around the identity in the d_{*,U_0} metric; and conversely, any such ball contains an open neighbourhood around the identity.

To prove the first claim, let U be an open neighbourhood around the identity, and let $U' \subset U$ be a smaller neighbourhood of the identity. From (5.11) we see (if U' is small enough) that $\|\|\cdot\|\|_{*,U_0}$ is comparable to $\|\|\cdot\|\|_{e,U'}$, and U' contains a small ball around the origin in the d_{*,U_0} metric, giving the claim. To prove the second claim, consider a ball $B(0,r)$ in the d_{*,U_0} metric. For any positive integer m, we can find an open neighbourhood U_m of the identity such that $U_m^m \subset U_0$, and hence $\|g\|_{e,U_0} \le \frac{1}{m}$ for all $g \in U_m$. For m large enough, this implies that $U_m \subset B(0,r)$, and the claim follows.

To finish the proof of Theorem 5.3.2, we need to verify the escape property (5.4). Thus, we need to show that if $g \in G$, $n \ge 1$ are such that $n\|g\|_{*,U_0}$ is sufficiently small, then we have $\|g^n\|_{*,U_0} \gg n\|g\|_{*,U_0}$. We may of course assume that g is not the identity, as the claim is trivial otherwise. As $\|\|\cdot\|\|_{*,U_0}$ is comparable to $\|\|\cdot\|\|_{e,U_0}$, we know that there exists a natural number $m \ll 1/\|g\|_{*,U_0}$ such that $g^m \notin U_0$.

Let U_1 be a neighbourhood of the identity small enough that $U_1^2 \subset U_0$. We have $\|g^i\|_{*,U_0} \le n\|g\|_{*,U_0}$ for all $i = 1, \ldots, n$, so $g^i \in U_1$ and hence $m > n$. Let $m + i$ be the first multiple of n larger than n, then $i \le n$ and so $g^i \in U_1$. Since $g^m \notin U_0$, this implies $g^{m+i} \notin U_1$. Since $m + i$ is divisible by n, we conclude that $\|g^n\|_{e,U_1} \ge \frac{n}{m+i} \gg n\|g\|_{*,U_0}$, and the claim follows from (5.11).

5.4. NSS from subgroup trapping

In view of Theorem 5.3.2, the only remaining task in the proof of the Gleason-Yamabe theorem is to locate "big" subquotients G'/H of a locally compact group G with the NSS property. We will need some further notation. Given a neighbourhood V of the identity in a topological group G, let $Q[V]$ denote the union of all the subgroups of G that are contained in V. Thus, a group is NSS if $Q[V]$ is trivial for all sufficiently small V.

We will need a property that is weaker than NSS:

Definition 5.4.1 (Subgroup trapping). A topological group has the *subgroup trapping property* if, for every open neighbourhood U of the identity, there exists another open neighbourhood V of the identity such that $Q[V]$ generates a subgroup $\langle Q[V] \rangle$ contained in U.

Clearly, every NSS group has the subgroup trapping property. Informally, groups with the latter property do have small subgroups, but one cannot get very far away from the origin just by combining together such subgroups.

Example 5.4.2. The infinite-dimensional torus $(\mathbf{R}/\mathbf{Z})^{\mathbf{N}}$ does not have the NSS property, but it does have the subgroup trapping property.

It is difficult to produce an example of a group that does not have the subgroup trapping property; the reason for this will be made clear in the next section. For now, we establish the following key result (another arrow of Figure 1):

Proposition 5.4.3 (From subgroup trapping to NSS). *Let G be a locally compact group with the subgroup trapping property, and let U be an open neighbourhood of the identity in G. Then there exists an open subgroup G' of G, and a compact subgroup N of G' contained in U, such that G'/N is locally compact and NSS. In particular, by Theorem 5.3.2, G'/N is isomorphic to a Lie group.*

Intuitively, the idea is to use the subgroup trapping property to find a small compact normal subgroup N that contains $Q[V]$ for some small V, and then quotient this group out to get an NSS group. Unfortunately, because N is not necessarily contained in V, this quotienting operation may create some additional small subgroups. To fix this, we need to pass from the compact subgroup N to a smaller one. In order to understand the subgroups of compact groups, the main tool will be Gleason-Yamabe theorem for compact groups (Theorem 4.2.4).

For us, the main reason why we need the compact case of the Gleason-Yamabe theorem is that Lie groups automatically have the NSS property, even though G need not. Thus, one can view Theorem 4.2.4 as giving the compact case of Proposition 5.4.3.

We now prove Proposition 5.4.3, using an argument of Yamabe [**Ya1953**]. Let G be a locally compact group with the subgroup trapping property, and let U be an open neighbourhood of the identity. We may find a smaller neighbourhood U_1 of the identity with $U_1^2 \subset U$ which, in particular, implies that $\overline{U_1} \subset U$; by shrinking U_1 if necessary, we may assume that $\overline{U_1}$ is compact. By the subgroup trapping property, one can find an open neighbourhood U_2 of the identity such that $\langle Q(U_2) \rangle$ is contained in U_1, and thus $H := \overline{\langle Q(U_2) \rangle}$ is a compact subgroup of G contained in U. By shrinking U_2 if necessary we may assume $U_2 \subset U_1$.

Ideally, if H were normal and contained in U_2, then the quotient group $\langle U_2 \rangle / H$ would have the NSS property. Unfortunately, H need not be normal, and need not be contained in U_2, but we can fix this as follows. Applying Theorem 4.2.4, we can find a compact normal subgroup N of H contained in $U_2 \cap H$ such that H/N is isomorphic to a Lie group and, in particular, is NSS. In particular, we can find an open symmetric neighbourhood U_3 of the identity in G such that $U_3 N U_3 \subset U_2$ and that the quotient space

$\pi(U_3 N U_3 \cap H)$ has no nontrivial subgroups in H/N, where $\pi : H \to H/N$ is the quotient map.

We now claim that N is normalised by U_3. Indeed, if $g \in U_3$, then the conjugate $N^g := g^{-1}Ng$ of N is contained in $U_3 N U_3$ and hence in U_2. As N^g is a group, it must thus be contained in $Q(U_2)$ and hence in H. But then $\pi(N^g)$ is a subgroup of H/N that is contained in $\pi(U_3 N U_3 \cap H)$, and is hence trivial by construction. Thus $N^g \subset N$, and so N is normalised by U_3. If we then let G' be the subgroup of G generated by N and U_3, we see that G' is an open subgroup of G, with N a compact normal subgroup of G'.

To finish the job, we need to show that G'/N has the NSS property. It suffices to show that $U_3 N U_3/N$ has no nontrivial subgroups. But any subgroup in $U_3 N U_3/N$ pulls back to a subgroup in $U_3 N U_3$, hence in U_2, hence in $Q(U_2)$, hence in H; since $(U_3 N U_3 \cap H)/N$ has no nontrivial subgroups, the claim follows. This concludes the proof of Proposition 5.4.3.

5.5. The subgroup trapping property

In view of Theorem 5.3.2, Proposition 5.4.3, and Exercise 5.1.3, we see that the Gleason-Yamabe theorem (Theorem 1.1.13) now reduces to the following claim.

Proposition 5.5.1. *Every locally compact metrisable group has the subgroup trapping property.*

This proposition represents the final two arrows of Figure 1.

We now prove Proposition 5.5.1, which is the hardest step of the entire proof and uses almost all the tools already developed. In particular, it requires both Theorem 4.2.4 and Gleason's convolution trick, as well as some of the basic theory of Hausdorff distance; as such, this is perhaps the most "infinitary" of all the steps in the argument.

The Gleason-type arguments can be encapsulated in the following proposition, which is a weak version of the subgroup trapping property:

Proposition 5.5.2 (Finite trapping). *Let G be a locally compact group, let U be an open precompact neighbourhood of the identity, and let $m \geq 1$ be an integer. Then there exists an open neighbourhood V of the identity with the following property: if $Q \subset Q[V]$ is a symmetric set containing the identity, and $n \geq 1$ is such that $Q^n \subset U$, then $Q^{mn} \subset U^8$.*

Informally, Proposition 5.5.2 asserts that subsets of $Q[V]$ grow much more slowly than "large" sets such as U. We remark that if one could replace U^8 in the conclusion here by U, then a simple induction on n (after first shrinking V to lie in U) would give Proposition 5.5.1. It is the loss of 8 in the exponent that necessitates some nontrivial additional arguments.

Proof of Proposition 5.5.2. Let V be small enough to be chosen later, and let Q, n be as in the proposition. Once again we will convolve together two "Lipschitz" functions ψ, η to obtain a good bump function $\phi = \psi * \eta$ which generates a useful metric for analysing the situation. The first bump function $\psi : G \to \mathbf{R}$ will be defined by the formula

$$\psi(x) := \sup\{1 - \frac{j}{n} : x \in Q^j U; j = 0, \ldots, n\} \cup \{0\}.$$

Then ψ takes values in $[0, 1]$, equals 1 on U, is supported in U^2, and obeys the Lipschitz type property

(5.19) $$|\partial_q \psi(x)| \leq \frac{1}{n}$$

for all $q \in Q$. The second bump function $\eta : G \to \mathbf{R}$ is similarly defined by the formula

$$\eta(x) := \sup\{1 - \frac{j}{M} : x \in (V^{U^4})^j U; j = 0, \ldots, M\} \cup \{0\},$$

where $V^{U^4} := \{g^{-1}xg : x \in V, g \in U^4\}$, where M is a quantity depending on m and U to be chosen later. If V is small enough depending on U and m, then $(V^{U^4})^M \subset U$, and so η also takes values in $[0, 1]$, equals 1 on U, is supported in U^2, and obeys the Lipschitz type property

(5.20) $$|\partial_g \psi(x)| \leq \frac{1}{M}$$

for all $g \in V^{U^4}$.

Now let $\phi := \psi * \eta$. Then ϕ is supported on U^4 and $\|\phi\|_{C_c(G)} \gg 1$ (where implied constants can depend on U, μ). As before, we conclude that $g \in U^8$ whenever $\|g\|_\phi$ is sufficiently small.

Now suppose that $q \in Q[V]$; we will estimate $\|q\|_\phi$. From (5.18) one has

$$\|q\|_\phi \ll \frac{1}{n}\|q^n\|_\phi + \sup_{0 \leq i \leq n} \|\partial_{q^i}\partial_q \phi\|_{C_c(G)}$$

(note that ∂_{q^i} and ∂_q commute). For the first term, we can compute

$$\|q^n\|_\phi = \sup_x |\partial_{q^n}(\psi * \eta)(x)|$$

and

$$\partial_{q^n}(\psi * \eta)(x) = \int_G \psi(y)\partial_{(q^n)^y}(y^{-1}x)d\mu(y).$$

Since $q \in Q[V]$, $q^n \in V$, so by (5.20) we conclude that

$$\|q^n\|_\phi \ll \frac{1}{M}.$$

For the second term, we similarly expand

$$\partial_{q^i}\partial_{q^i}\phi(x) = \int_G (\partial_q\psi)(y)\partial_{(q^n)^y}(y^{-1}x)d\mu(y).$$

Using (5.20), (5.19) we conclude that

$$|\partial_{q^i}\partial_{q^i}\phi(x)| \ll \frac{1}{Mn}.$$

Putting this together we see that

$$\|q\|_\phi \ll \frac{1}{Mn}$$

for all $q \in Q[V]$ which, in particular, implies that

$$\|g\|_\phi \ll \frac{m}{M}$$

for all $g \in Q^{mn}$. For M sufficiently large, this gives $Q^{mn} \subset U^8$ as required.

$\square$

We will also need the following compactness result in the *Hausdorff distance*

$$d_H(E, F) := \max(\sup_{x \in E} \text{dist}(x, F), \sup_{y \in F} \text{dist}(E, y))$$

between two nonempty closed subsets E, F of a metric space (X, d).

Example 5.5.3. In $\mathbf{R}$ with the usual metric, the finite sets $\{\frac{i}{n} : i = 1, \ldots, n\}$ converge in Hausdorff distance to the closed interval $[0, 1]$.

Exercise 5.5.1. Show that the space $K(X)$ of nonempty closed subsets of a compact metric space X is itself a compact metric space (with the Hausdorff distance as the metric). (*Hint:* Use the *Heine-Borel theorem.*)

Now we can prove Proposition 5.5.1. Let G be a locally compact group endowed with some metric d, and let U be an open neighbourhood of the identity; by shrinking U we may assume that U is precompact. Let V_i be a sequence of balls around the identity with radius going to zero, then $Q[V_i]$ is a symmetric set in V_i that contains the identity. If, for some i, $Q[V_i]^n \subset U$ for every n, then $\langle Q[V_i]\rangle \subset U$ and we are done. Thus, we may assume for the sake of contradiction that there exists n_i such that $Q[V_i]^{n_i} \subset U$ and $Q[V_i]^{n_i+1} \not\subset U$; since the V_i go to zero, we have $n_i \to \infty$. By Proposition 5.5.2, we can also find $m_i \to \infty$ such that $Q[V_i]^{m_i n_i} \subset U^8$.

The sets $\overline{Q[V_i]}^{n_i}$ are closed subsets of $\overline{U}$; by Exercise 5.5.1, we may pass to a subsequence and assume that they converge to some closed subset E of $\overline{U}$. Since the $Q[V_i]$ are symmetric and contain the identity, E is also symmetric and contains the identity. For any fixed m, we have $Q[V_i]^{mn_i} \subset U^8$ for all sufficiently large i, which on taking Hausdorff limits implies that $E^m \subset \overline{U^8}$.

In particular, the group $H := \overline{\langle E \rangle}$ is a compact subgroup of G contained in $\overline{U^8}$.

Let U_1 be a small neighbourhood of the identity in G to be chosen later. By Theorem 4.2.4, we can find a normal subgroup N of H contained in $U_1 \cap H$ such that H/N is NSS. Let B be a neigbourhood of the identity in H/N so small that B^{10} has no small subgroups. A compactness argument then shows that there exists a natural number k such that for any $g \in H/N$ that is not in B, at least one of $g, \dots, g^k$ must lie outside of B^{10}.

Now let $\varepsilon > 0$ be a small parameter. Since $Q[V_i]^{n_i+1} \not\subset U$, we see that $Q[V_i]^{n_i+1}$ does not lie in the ε-neighbourhood $\pi^{-1}(B)_\varepsilon$ of $\pi^{-1}(B)$ if ε is small enough, where $\pi : H \to H/N$ is the projection map. Let n_i' be the first integer for which $Q[V_i]^{n_i'}$ does not lie in $\pi^{-1}(B)_\varepsilon$, then $n_i' \leq n_i + 1$ and $n_i' \to \infty$ as $i \to \infty$ (for fixed ε). On the other hand, as $Q[V_i]^{n_i'-1} \subset \pi^{-1}(B)_\varepsilon$, we see from another application of Proposition 5.5.2 that $Q[V_i]^{kn_i'} \subset (\pi^{-1}(B)_\varepsilon)^8$ if i is sufficiently large depending on ε.

On the other hand, since $Q[V_i]^{n_i}$ converges to a subset of H in the Hausdorff distance, we know that for i large enough, $Q[V_i]^{2n_i}$ and hence $Q[V_i]^{n_i'}$ is contained in the ε-neighbourhood of H. Thus we can find an element g_i of $Q[V_i]^{n_i'}$ that lies within ε of a group element h_i of H, but does not lie in $\pi^{-1}(B)_\varepsilon$; thus h_i lies inside $H \backslash \pi^{-1}(B)$. By construction of B, we can find $1 \leq j_i \leq k$ such that $h_i^{j_i}$ lies in $H \backslash \pi^{-1}(B^{10})$. But $h_i^{j_i}$ also lies within $o(1)$ of $g_i^{j_i}$, which lies in $Q[V_i]^{kn_i'}$ and hence in $(\pi^{-1}(B)_\varepsilon)^8$, where $o(1)$ denotes a quantity depending on ε that goes to zero as $\varepsilon \to 0$. We conclude that $H \backslash \pi^{-1}(B^{10})$ and $\pi^{-1}(B^8)$ are separated by $o(1)$, which leads to a contradiction if ε is sufficiently small (note that $\overline{\pi^{-1}(B^8)}$ and $H \backslash \pi^{-1}(B^{10})$ are compact and disjoint, and hence separated by a positive distance), and the claim follows.

Exercise 5.5.2. Let X be a compact metric space, $K_c(X)$ denote the space of nonempty closed and *connected* subsets of X. Show that $K_c(X)$ with the Hausdorff metric is also a compact metric space.

Exercise 5.5.3. Show that the metrisability condition in Proposition 5.5.1 can be dropped; in other words, show that every locally compact group has the subgroup trapping property.

5.6. The local group case

In [**Go2009**], [**Go2010**], [**vdDrGo2010**], the above theory was extended to the setting of local groups. In fact, there is relatively little difficulty (other than some notational difficulties) in doing so, because the analysis in the previous sections can be made to take place on a small neighbourhood of the origin. This extension to local groups is not simply a generalisation for its own sake; it will turn out that it will be natural to work with local groups when we classify approximate groups in later sections.

One technical issue that comes up in the theory of local groups is that basic cancellation laws such as $gh = gk \implies h = k$, which are easily verified for groups, are not always true for local groups. However, this is a minor issue as one can always recover the cancellation laws by passing to a slightly smaller local group, as follows.

Definition 5.6.1 (Cancellative local group). A local group G is said to be *symmetric* if the inverse operation is always well-defined. It is said to be *cancellative* if it is symmetric, and the following axioms hold:

(i) Whenever $g, h, k \in G$ are such that gh and gk are well-defined and equal to each other, then $h = k$. (Note that this implies, in particular, that $(g^{-1})^{-1} = g$.)

(ii) Whenever $g, h, k \in G$ are such that hg and kg are well-defined and equal to each other, then $h = k$.

(iii) Whenever $g, h \in G$ are such that gh and $h^{-1}g^{-1}$ are well-defined, then $(gh)^{-1} = h^{-1}g^{-1}$. (In particular, if $U \subset G$ is symmetric and U^m is well-defined in G for some $m \geq 1$, then U^m is also symmetric.)

Clearly, all global groups are cancellative, and more generally the restriction of a global group to a symmetric neighbourhood of the identity is cancellative. While not all local groups are cancellative, we have the following substitute:

Exercise 5.6.1. Let G be a local group. Show that there is a neighbourhood U of the identity which is cancellative (thus, the restriction $G \!\upharpoonright_U$ of G to U is cancellative).

Note that any symmetric neighbourhood of the identity in a cancellative local group is again a cancellative local group. Because of this, it turns out in practice that we may restrict to the cancellative setting without much loss of generality.

Next, we need to localise the notion of a quotient G/H of a global group G by a normal subgroup H. Recall that in order for a subset H of a global group G to be a normal subgroup, it has to be symmetric, contain the identity, be closed under multiplication (thus $h_1 h_2 \in H$ whenever $h_1, h_2 \in H$), and closed under conjugation (thus $h^g := g^{-1}hg \in H$ whenever $h \in H$ and $g \in G$). We now localise this concept as follows:

Definition 5.6.2 (Sub-local groups). Given two symmetric local groups G' and G, we say that G' is a *sub-local group* of G if G' is the restriction of G to a symmetric neighbourhood of the identity, and there exists an open neighbourhood V of G' with the property that whenever $g, h \in G'$ are such that gh is defined in V, then $gh \in G'$; we refer to V as an *associated neighbourhood* for G'. If G' is also a global group, we say that G' is a *subgroup* of G.

If G' is a sub-local group of G, we say that G' is *normal* if there exists an associated neighbourhood V for G' with the additional property that whenever $g' \in G', h \in V$ are such that $hg'h^{-1}$ is well-defined and lies in V, then $hg'h^{-1} \in G'$. We call V a *normalising neighbourhood* of G'.

Example 5.6.3. If G, G' are the (additive) local groups $G := \{-2, -1, 0, +1, +2\}$ and $G' := \{-1, 0, +1\}$, then G' is a sub-local group of G (with associated neighbourhood $V = G'$). Note that this is despite G' not being closed with respect to addition in G; thus we see why it is necessary to allow the associated neighbourhood V to be strictly smaller than G. In a similar vein, the open interval $(-1, 1)$ is a sub-local group of $(-2, 2)$.

The interval $(-1, 1) \times \{0\}$ is also a sub-local group of $\mathbf{R}^2$; here, one can take for instance $(-1, 1)^2$ as the associated neighbourhood. As all these examples are abelian, they are clearly normal.

Example 5.6.4. Let $T : V \to V$ be a linear transformation on a finite-dimensional vector space V, and let $G := \mathbf{Z} \ltimes_T V$ be the associated semi-direct product. Let $G' := \{0\} \times W$, where W is a subspace of V that is not preserved by T. Then G' is not a normal subgroup of G, but it is a normal sub-local group of G, where one can take $\{0\} \times V$ as a normalising neighbourhood of G'.

Example 5.6.5. In the global group $G = \mathbf{R}^2 = (\mathbf{R}^2, +)$, the open interval $H := (-1, 1) \times \{0\}$ is a normal sub-local subgroup if one takes (say) $V := (-1, 1) \times (-1, 1)$ as the normalising neighbourhood.

Example 5.6.6. Let $T : (\mathbf{R}/\mathbf{Z})^{\mathbf{Z}} \to (\mathbf{R}/\mathbf{Z})^{\mathbf{Z}}$ be the shift map $T(a_n)_{n \in \mathbf{Z}} := (a_{n-1})_{n \in \mathbf{Z}}$, and let $\mathbf{Z} \ltimes_T (\mathbf{R}/\mathbf{Z})^{\mathbf{Z}}$ be the semidirect product of $\mathbf{Z}$ and $(\mathbf{R}/\mathbf{Z})^{\mathbf{Z}}$. Then if H is any (global) subgroup of $(\mathbf{R}/\mathbf{Z})^{\mathbf{Z}}$, the set $\{0\} \times H$ is a normal sub-local subgroup of $\mathbf{Z} \ltimes_T (\mathbf{R}/\mathbf{Z})^{\mathbf{Z}}$ (with normalising neighbourhood $\{0\} \times (\mathbf{R}/\mathbf{Z})^{\mathbf{Z}}$). This is despite the fact that H will, in general, not be normal in $\mathbf{Z} \ltimes_T (\mathbf{R}/\mathbf{Z})^{\mathbf{Z}}$ in the classical (global) sense.

It is easy to see that if H is a normal sub-local group of G, then H is itself a cancellative local group, using the topology and group structure formed by restriction from G. (Note how the open neighbourhood V is needed to ensure that the domain of the multiplication map in H remains open.) One also easily verifies that if $\phi : U \to H$ is a local homomorphism from G to H for some open neighbourhood U of the identity in G, then $\ker(\phi)$ is a normal sub-local group of U, and hence of G. Note that the kernel of a local morphism is well-defined up to local identity. If H is Hausdorff, then the kernel $\ker(\phi)$ will also be closed.

As observed by Goldbring [**Go2010**], one can define the operation of quotienting a local group by a normal sub-local group, provided that one restricts to a sufficiently small neighbourhood of the origin:

Exercise 5.6.2 (Quotient spaces). Let G be a cancellative local group, and let H be a normal sub-local group with normalising neighbourhood V. Let W be a symmetric open neighbourhood of the identity such that $W^6 \subset V$. Show that there exists a cancellative local group W/H and a surjective continuous homomorphism $\phi : W \to W/H$ such that, for any $g, h \in W$, one has $\phi(g) = \phi(h)$ if and only if $gh^{-1} \in H$, and for any $E \subset W/H$, one has E open if and only if $\phi^{-1}(E)$ is open.

It is not difficult to show that the quotient W/H defined by the above exercise is unique up to local isomorphism, so we will abuse notation and talk about "the" quotient space W/H given by the above construction.

Example 5.6.7. Let G be the additive local group $G := (-2, 2)^2$, and let H be the sub-local group $H := \{0\} \times (-1, 1)$, with normalising neighbourhood $V := (-1, 1)^2$. If we then set $W := (-0.1, 0.1)^2$, then the hypotheses of Exercise 5.6.2 are obeyed, and W/H can be identified with $(-0.1, 0.1)$, with the projection map $\phi : (x, y) \mapsto x$.

Example 5.6.8. Let G be the torus $(\mathbf{R}/\mathbf{Z})^2$, and let H be the sub-local group $H = \{(x, \alpha x) \bmod \mathbf{Z}^2 : x \in (-0.1, 0.1)\}$, where $0 < \alpha < 1$ is an irrational number, with normalising neighbourhood $(-0.1, 0.1)^2 \bmod \mathbf{Z}^2$. Set $W := (-0.01, 0.01)^2 \bmod \mathbf{Z}^2$. Then the hypotheses of Exercise 5.6.2 are again obeyed, and W/H can be identified with the interval $I := (-0.01(1 + \alpha), 0.01(1 + \alpha))$, with the projection map $\phi : (x, y) \bmod \mathbf{Z}^2 \mapsto y - \alpha x$ for $(x, y) \in (-0.01, 0.01)^2$. Note, in contrast, that if one quotiented G by the *global* group $\langle H \rangle = \{(x, \alpha x) \bmod \mathbf{Z}^2 : x \in \mathbf{R}\}$ generated by H, the quotient would be a non-Hausdorff space (and would also contain a dense set of torsion points, in contrast to the interval I which is "locally torsion free"). It is because of this pathological behaviour of quotienting by global groups that we need to work with local group quotients instead.

We can now state the local version of the Gleason-Yamabe theorem, first proven in [**Go2009**] (and by a slightly different method in [**vdDrGo2010**]): different method:

Theorem 5.6.9 (Local Gleason-Yamabe theorem). *Let G be a locally compact local group. Then there exists an open symmetric neighbourhood G' of* *the identity, and a compact global group H in G' that is normalised by G', such that G'/H is well-defined and isomorphic to a local Lie group.*

The proofs of this theorem by Goldbring and Goldbring-van den Dries were phrased in the language of nonstandard analysis. However, it is possible to translate those arguments to standard analysis arguments, which closely follow the arguments given in previous sections[2]. We briefly sketch the main points here.

As in the global case, the route to obtaining (local) Lie structure is via Gleason metrics. On a local group G, we define a *local Gleason metric* to be a metric $d : U \times U \to \mathbf{R}^+$ defined on some symmetric open neighbourhood U of the identity with (say) U^{100} well-defined (to avoid technical issues), which generates the topology of U, and which obeys the following version of the left-invariance, escape and commutator properties:

(1) (Left-invariance) If $g, h, k \in U$ are such that $gh, gk \in U$, then $d(h, k) = d(gh, gk)$.

(2) (Escape property) If $g \in U$ and $n\|g\| \leq \frac{1}{C}$, then $g, \ldots, g^n$ are well-defined in U and $\|g^n\| \geq \frac{1}{C} n\|g\|$.

(3) (Commutator estimate) If $g, h \in U$ are such that $\|g\|, \|h\| \leq \frac{1}{C}$, then $[g, h]$ is well-defined in U and (3.1) holds.

One can then verify (by localisation of the arguments in Chapter 3) that any locally compact local Lie group with a local Gleason metric is locally Lie (i.e., some neighbourhood of the identity is isomorphic to a local Lie group); see Exercise 3.3.4. Next, one can define the notion of a weak local Gleason metric by dropping the commutator estimate, and one can verify an analogue of Theorem 5.2.1, namely that any weak local Gleason metric is automatically a local Gleason metric, after possibly shrinking the neighbourhood U and adjusting the constant C as necessary. The proof of this statement is essentially the same as that in Theorem 5.2.1 (which is already localised to small neighbourhoods of the identity), but uses a local Haar measure instead of a global Haar measure, and requires some preliminary shrinking of the neighbourhood U to ensure that all group-theoretic operations (and convolutions) are well-defined. We omit the (rather tedious) details.

Now we define the concept of an NSS local group as a local group which has an open neighbourhood of the identity that contains no nontrivial global subgroups. The proof of Theorem 5.3.2 is already localised to small neighbourhoods of the identity, and it is possible (after being sufficiently careful

---

[2] Actually, our arguments are not a verbatim translation of those in Goldbring and Goldbring-van den Dries, as we have made a few simplifications in which the role of Gleason metrics is much more strongly emphasised.

with the notation) to translate that argument to the local setting, and conclude that any NSS local group admits a weak Gleason metric on some open neighbourhood of the identity, and is hence locally Lie. (A typical example of being "sufficiently careful with the notation": to define the escape norm (5.10), one adopts the convention that a statement such as $g, \ldots, g^n \in U$ is automatically false if $g, \ldots, g^n$ are not all well-defined. The induction hypothesis (5.12) will play a key role in ensuring that all expressions involved are well-defined and localised to a suitably small neighbourhood of the identity.) Again, we omit the details.

The next step is to obtain a local version of Proposition 5.4.3. Here we encounter a slight difficulty because in a general local group G, we do not have a good notion of the group $\langle A \rangle$ generated by a set A of generators in G. As such, the subgroup trapping property does not automatically translate to the local group setting as defined in Definition 5.4.1. However, this difficulty can be easily avoided by rewording the definition:

Definition 5.6.10 (Subgroup trapping). A local group has the *subgroup trapping property* if, for every open neighbourhood U of the identity, there exists another open neighbourhood V of the identity such that $Q[V]$ is contained in a global subgroup H that is in turn contained in U. (Here, $Q[V]$ is, as before, the union of all the global subgroups contained in V.)

Because $Q[V]$ is now contained in a global group H, the group $\langle Q[V] \rangle$ generated by H is well-defined. As H is in the open neighbourhood U, one can then also form the closure $\overline{\langle Q[V] \rangle}$; if we choose U small enough to be precompact, then this is a compact global group (and thus describable by the Gleason-Yamabe theorem for such groups, Theorem 4.2.4). Because of this, it is possible to adapt Proposition 5.4.3 without much difficulty to the local setting to conclude that given any locally compact local group G with the subgroup trapping property, there exists an open symmetric neighbourhood G' of the identity, and a compact global group H in G' that is normalised by G', such that G'/H is well-defined and NSS (and thus locally isomorphic to a local Lie group).

Finally, to finish the proof of Theorem 5.6.9, one has to establish the analogue of Proposition 5.5.1, namely that one has to show that every locally compact metrisable local group has the subgroup trapping property. (It is not difficult to adapt Exercise 5.1.3 to the local group setting to reduce to the metrisable case.) The first step is to prove the local group analogue of Proposition 5.5.2 (again adopting the obvious convention that a statement such as $Q^n \subset U$ is only considered true if Q^n is well-defined, and adding the additional hypothesis that U is sufficiently small in order to ensure that all manipulations are justified). This can be done by a routine modification of the proof. But then one can modify the rest of the argument in

Proposition 5.5.1 to hold in the local setting as well (note, as in the proof of Proposition 5.4.3, that the compact set H generated in the course of this argument remains a *global* group rather than a local one, and so one can again use Theorem 4.2.4 without difficulty). Again, we omit the details.

Chapter 6

The structure of locally compact groups

In the previous chapters, we established the *Gleason-Yamabe theorem*, Theorem 1.1.13. Roughly speaking, this theorem asserts the "mesoscopic" structure of a locally compact group (after restricting to an open subgroup G' to remove the macroscopic structure, and quotienting out by K to remove the microscopic structure) is always of Lie type.

In this chapter, we combine the Gleason-Yamabe theorem with some additional tools from point-set topology to improve the description of locally compact groups in various situations.

We first record some easy special cases of this. If the locally compact group G has the no small subgroups property, then one can take K to be trivial; thus G' is Lie, which implies that G is locally Lie and thus Lie as well. Thus the assertion that all locally compact NSS groups are Lie (Theorem 5.3.2) is a special case of the Gleason-Yamabe theorem.

In a similar spirit, if the locally compact group G is connected, then the only open subgroup G' of G is the full group G; in particular, by arguing as in the treatment of the compact case (Exercise 4.2.9), we conclude that any connected locally compact Hausdorff group is the inverse limit of Lie groups.

Now we return to the general case, in which G need not be connected or NSS. One slight defect of Theorem 1.1.13 is that the group G' can depend on the open neighbourhood U. However, by using a basic result from the theory of totally disconnected groups known as *van Dantzig's theorem*, one can make G' independent of U:

Theorem 6.0.11 (Gleason-Yamabe theorem, stronger version). *Let G be a locally compact group. Then there exists an open subgroup G' of G such that, for any open neighbourhood U of the identity in G', there exists a compact normal subgroup K of G' in U such that G'/K is isomorphic to a Lie group.*

We will prove this theorem later in this section. As in previous sections, if G is Hausdorff, the group G' is thus an inverse limit of Lie groups (and if G (and hence G') is first countable, it is the inverse limit of a *sequence* of Lie groups).

Exercise 6.0.3. By working with the locally compact group

$$
G = \begin{pmatrix} 1 & \mathbf{Q}_p & \mathbf{Q}_p & \mathbf{Q}_p \\ 0 & 1 & \mathbf{Q}_p & \mathbf{Q}_p \\ 0 & 0 & 1 & \mathbf{Q}_p \\ 0 & 0 & 0 & 1 \end{pmatrix},
$$

where $\mathbf{Q}_p$ is a p-adic group, show that one cannot demand in Theorem 6.0.11 that the open subgroup G' be normal.

It remains to analyse inverse limits of Lie groups. To do this, it helps to have some control on the dimensions of the Lie groups involved. A basic tool for this purpose is the *invariance of domain theorem*:

Theorem 6.0.12 (Brouwer invariance of domain theorem). *Let U be an open subset of $\mathbf{R}^n$, and let $f : U \to \mathbf{R}^n$ be a continuous injective map. Then $f(U)$ is also open.*

We prove this theorem later in this chapter also. It has an important corollary:

Corollary 6.0.13 (Topological invariance of dimension). *If $n > m$, and U is a nonempty open subset of $\mathbf{R}^n$, then there is no continuous injective mapping from U to $\mathbf{R}^m$. In particular, $\mathbf{R}^n$ and $\mathbf{R}^m$ are not homeomorphic.*

Exercise 6.0.4 (Uniqueness of dimension). Let X be a nonempty topological space. If X is a manifold of dimension d_1, and also a manifold of dimension d_2, show that $d_1 = d_2$. Thus, we may define the dimension $\dim(X)$ of a nonempty manifold in a well-defined manner.

If X, Y are nonempty manifolds, and there is a continuous injection from X to Y, show that $\dim(X) \leq \dim(Y)$.

Remark 6.0.14. Note that the analogue of the above exercise for surjections is false: the existence of a continuous surjection from one nonempty manifold X to another Y does *not* imply that $\dim(X) \geq \dim(Y)$, thanks to the existence of *space-filling* curves. Thus we see that invariance of domain, while intuitively plausible, is not an entirely trivial observation.

As we shall see, we can use Corollary 6.0.13 to bound the dimension of the Lie groups L_n in an inverse limit $G = \lim_{n \to \infty} L_n$ by the "dimension" of the inverse limit G. Among other things, this can be used to obtain a positive resolution to Hilbert's fifth problem, Theorem 1.1.9. Again, this theorem will be proven later in this section.

Another application of this machinery is the following variant of Hilbert's fifth problem, which was used in Gromov's original proof of Gromov's theorem on groups of polynomial growth, although we will not actually need it here:

Proposition 6.0.15. *Let G be a locally compact σ-compact group that acts transitively, faithfully, and continuously on a connected topological manifold X. Then G is isomorphic to a Lie group.*

Recall that a *continuous action* of a topological group G on a topological space X is a continuous map $\cdot : G \times X \to X$ which obeys the associativity law $(gh)x = g(hx)$ for $g, h \in G$ and $x \in X$, and the identity law $1x = x$ for all $x \in X$. The action is *transitive* if, for every $x, y \in X$, there is a $g \in G$ with $gx = y$, and *faithful* if, whenever $g, h \in G$ are distinct, one has $gx \neq hx$ for at least one x.

The σ-compact hypothesis is a technical one, and can likely be dropped, but we retain it for this discussion (as in most applications we can reduce to this case).

Exercise 6.0.5. Show that Proposition 6.0.15 implies Theorem 1.1.9.

Remark 6.0.16. It is conjectured that the transitivity hypothesis in Proposition 6.0.15 can be dropped; this is known as the *Hilbert-Smith conjecture*. It remains open, with the key remaining difficulty being the need to figure out a way to eliminate the possibility that G is a p-adic group $\mathbf{Z}_p$. See Chapter 17 for more discussion.

6.1. Van Dantzig's theorem

Recall that a (nonempty) topological space X is *connected* if the only *clopen* (i.e., closed and open) subsets of X are the whole space X and the empty set $\emptyset$; a nonempty topological space is *disconnected* if it is not connected. (By convention, the empty set is considered to be neither connected nor disconnected, somewhat analogously to how the natural number 1 is neither considered prime nor composite.)

At the opposite extreme to connectedness is the property of being a *totally disconnected* space. This is a space whose only connected subsets are the singleton sets. Typical examples of totally disconnected spaces include discrete spaces (e.g., the integers $\mathbf{Z}$ with the discrete topology) and Cantor spaces (such as the standard Cantor set).

Most topological spaces are neither connected nor totally disconnected, but some intermediate combination of both. In the case of topological *groups* G, this rather vague assertion can be formalised as follows.

Exercise 6.1.1.

(i) Define a *connected component* of a topological space X to be a maxmial connected set. Show that the connected components of X form a partition of X, thus every point in X belongs to exactly one connected component.

(ii) Let G be a topological group, and let G° be the connected component of the identity. Show that G° is a closed normal subgroup of G, and that G/G° is a totally disconnected subgroup of G. Thus, one has a *short exact sequence*

$$0 \to G^\circ \to G \to G/G^\circ \to 0$$

of topological groups that describes G as an extension of a totally disconnected group by a connected group.

(iii) Conversely, if one has a short exact sequence

$$0 \to H \to G \to K \to 0$$

of topological groups, with H connected and K totally disconnected, show that H is isomorphic to G°, and K is isomorphic to G/G°.

(iv) If G is locally compact, show that G° and G/G° are also locally compact.

In principle at least, the study of locally compact groups thus splits into the study of connected locally compact groups, and the study of totally disconnected locally compact groups. Note, however, that even if one has a complete understanding of the factors H, K of a short exact sequence $0 \to H \to G \to K \to 0$, it may still be a nontrivial issue to fully understand the combined group G, due to the possible presence of nontrivial *group cohomology*. See for instance [**Ta2011c**, §2.4] for more discussion.

For totally disconnected locally compact groups, one has the following fundamental theorem of van Dantzig [**vDa1936**]:

Theorem 6.1.1 (Van Dantzig's theorem). *Every totally disconnected locally compact group G contains a compact open subgroup H (which will of course still be totally disconnected).*

Example 6.1.2. Let p be a prime. Then the *p-adic field* $\mathbf{Q}_p$ (with the usual p-adic valuation) is totally disconnected locally compact, and the p-adic integers $\mathbf{Z}_p$ are a compact open subgroup.

Of course, this situation is the polar opposite of what occurs in the connected case, in which the only open subgroup is the whole group.

To prove van Dantzig's theorem, we first need a lemma from point set topology, which shows that totally disconnected spaces contain enough *clopen sets* to separate points:

Lemma 6.1.3. *Let X be a totally disconnected compact Hausdorff space, and let x, y be distinct points in X. Then there exists a clopen set that contains x but not y.*

Proof. Let K be the intersection of all the clopen sets that contain x (note that X is obviously clopen). Clearly K is closed and contains x. Our objective is to show that K consists solely of $\{x\}$. As X is totally disconnected, it will suffice to show that K is connected.

Suppose this is not the case, then we can split $K = K_1 \cup K_2$ where K_1, K_2 are disjoint nonempty closed sets; without loss of generality, we may assume that x lies in K_1. As all compact Hausdorff spaces are *normal*, we can thus enclose K_1, K_2 in disjoint open subsets U_1, U_2 of X. In particular, the topological boundary ∂U_2 is compact and lies outside of K. By definition of K, we thus see that for every $y \in \partial U_2$, we can find a clopen neighbourhood of x that avoids y; by compactness of ∂U_2 (and the fact that finite intersections of clopen sets are clopen), we can thus find a clopen neighbourhood L of x that is disjoint from ∂U_2. One then verifies that $L \backslash U_2 = L \backslash \overline{U_2}$ is a clopen neighbourhood of x that is disjoint from K_2, contradicting the definition of K, and the claim follows. $\square$

Now we can prove van Dantzig's theorem. We will use an argument from [**HeRo1979**]. Let G be totally disconnected locally compact (and thus Hausdorff). Then we can find a compact neighbourhood K of the identity. By Lemma 6.1.3, for every $y \in \partial K$, we can find a clopen neighbourhood of the identity that avoids y; by compactness of ∂K, we may thus find a clopen neighbourhood of the identity that avoids ∂K. By intersecting this neighbourhood with K, we may thus find a compact clopen neighbourhood F of the identity. As F is both compact and open, we may then use the continuity of the group operations to find a symmetric neighbourhood U of the identity such that $UF \subset F$. In particular, if we let G' be the group generated by U, then G' is an open subgroup of G contained in F and is thus compact as required.

Remark 6.1.4. The same argument shows that a totally disconnected locally compact group contains arbitrarily small compact open subgroups, or in other words, the compact open subgroups form a neighbourhood base for the identity.

In view of van Dantzig's theorem, we see that the "local" behaviour of totally disconnected locally compact groups can be modeled by the compact totally disconnected groups, which are better understood. Thanks to the Gleason-Yamabe theorem for compact groups, such groups are the inverse limits of compact totally disconnected Lie groups. But it is easy to see that a compact totally disconnected Lie group must be finite, and so compact totally disconnected groups are necessarily *profinite* (i.e., the inverse limit of finite groups). The global behaviour however remains more complicated, in part because the compact open subgroup given by van Dantzig's theorem need not be normal, and so does not necessarily induce a splitting of G into compact and discrete factors.

Example 6.1.5. Let p be a prime, and let G be the semi-direct product $\mathbf{Z} \ltimes \mathbf{Q}_p$, where the integers $\mathbf{Z}$ act on $\mathbf{Q}_p$ by the map $m : x \mapsto p^m x$, and we give G the product of the discrete topology of $\mathbf{Z}$ and the p-adic topology on $\mathbf{Q}_p$. One easily verifies that G is a totally disconnected locally compact group. It certainly has compact open subgroups, such as $\{0\} \times \mathbf{Z}_p$. However, it is easy to show that G has no nontrivial compact normal subgroups (the problem is that the conjugation action of $\mathbf{Z}$ on $\mathbf{Q}_p$ has all nontrivial orbits unbounded).

We can pull van Dantzig's theorem back to more general locally compact groups:

Exercise 6.1.2. Let G be a locally compact group.

(i) Show that G contains an open subgroup G' which is "compact-by-connected" in the sense that $G'/(G')^\circ$ is compact. (*Hint:* Apply van Dantzig's theorem to G/G°.)

(ii) If G is compact-by-connected, and U is an open neighbourhood of the identity, show that there exists a compact subgroup K of G in U such that G/K is isomorphic to a Lie group. (*Hint:* Use Theorem 1.1.13, and observe that any open subgroup of the compact-by-connected group G has finite index and thus has only finitely many conjugates.) Conclude Theorem 6.0.11.

(iii) Show that any locally compact Hausdorff group G contains an open subgroup G' that is isomorphic to an inverse limit of Lie groups $(L_\alpha)_{\alpha \in A}$, in which each Lie group L_α has at most finitely many connected components. Furthermore, each L_α is isomorphic to G'/K_α

for some compact normal subgroup K_α of G', with $K_\beta \leq K_\alpha$ for $\alpha < \beta$. If G is first countable, show that this inverse limit can be taken to be a sequence (so that the index set A is simply the natural numbers $\mathbf{N}$ with the usual ordering), and the K_n then shrink to zero in the sense that they lie inside any given open neighbourhood of the identity for n large enough.

Exercise 6.1.3. Let G be a totally disconnected locally compact group. Show that every compact subgroup K of G is contained in a compact open subgroup. (*Hint:* van Dantzig's theorem provides a compact open subgroup, but it need not contain K. But is there a way to modify it so that it is normalised by K? Why would being normalised by K be useful?)

6.2. The invariance of domain theorem

In this section we give a proof of the invariance of domain theorem. The main topological tool for this is Brouwer's famous fixed point theorem:

Theorem 6.2.1 (Brouwer fixed point theorem). *Let $f : B^n \to B^n$ be a continuous function on the unit ball $B^n := \{x \in \mathbf{R}^n : \|x\| \leq 1\}$ in a Euclidean space $\mathbf{R}^n$. Then f has at least one fixed point, thus there exists $x \in B^n$ with $f(x) = x$.*

This theorem has many proofs. We quickly sketch one of these proofs (based on one from [**Ku1998**]) as follows:

Exercise 6.2.1. For this exercise, suppose for the sake of contradiction that Theorem 6.2.1 is false, thus there is a continuous map from B^n to B^n with no fixed point.

(i) Show that there exists a *smooth* map from B^n to B^n with no fixed point.

(ii) Show that there exists a smooth map from B^n to the unit sphere S^{n-1}, which equals the identity function on S^{n-1}.

(iii) Show that there exists a smooth map ϕ from B^n to the unit sphere S^{n-1}, which equals the map $x \mapsto \frac{x}{\|x\|}$ on a neighbourhood of S^{n-1}.

(iv) By computing the integral $\int_{B^n} \det(\partial_1 \phi, \dots, \partial_n \phi)$ in two different ways (one by using Stokes' theorem, and the other by using the $n-1$-dimensional nature of the sphere S^{n-1}), establish a contradiction.

Now we prove Theorem 6.0.12. By rescaling and translation invariance, it will suffice to show the following claim:

Theorem 6.2.2 (Invariance of domain, again). *Let $f : B^n \to \mathbf{R}^n$ be an continuous injective map. Then $f(0)$ lies in the interior of $f(B^n)$.*

Let f be as in Theorem 6.2.2. The map $f : B^n \to f(B^n)$ is a continuous bijection between compact Hausdorff spaces and is thus a homeomorphism. In particular, the inverse map $f^{-1} : f(B^n) \to B^n$ is continuous. Using the *Tietze extension theorem*, we can find a continuous function $G : \mathbf{R}^n \to \mathbf{R}^n$ that extends f^{-1}.

The function G has a zero on $f(B^n)$, namely at $f(0)$. We can use the Brouwer fixed point theorem to show that this zero is stable:

Lemma 6.2.3 (Stability of zero). *Let $\tilde{G} : f(B^n) \to \mathbf{R}^n$ be a continuous function such that $\|G(y) - \tilde{G}(y)\| \le 1$ for all $y \in f(B^n)$. Then $\tilde{G}$ has at least one zero (i.e., there is a $y \in f(B^n)$ such that $\tilde{G}(y) = 0$).*

Proof. Apply Theorem 6.2.1 to the function

$$x \mapsto x - \tilde{G}(f(x)) = G(f(x)) - \tilde{G}(f(x)). \qquad \square$$

Now suppose that Theorem 6.2.2 failed, so that $f(0)$ is not an interior point of $f(B^n)$. We will use this to locate a small perturbation of G that no longer has a zero on $f(B^n)$, contradicting Lemma 6.2.3.

We turn to the details. Let $\varepsilon > 0$ be a small number. By continuity of G, we see (if ε is chosen small enough) that we have $\|G(y)\| \le 0.1$ whenever $y \in \mathbf{R}^n$ and $\|y - f(0)\| \le 2\varepsilon$.

On the other hand, since $f(0)$ is not an interior point of $f(B^n)$, there exists a point $c \in \mathbf{R}^n$ with $\|c - f(0)\| < \varepsilon$ that lies outside $f(B^n)$. By translating f if necessary, we may take $c = 0$; thus $f(B^n)$ avoids zero, $\|f(0)\| < \varepsilon$, and we have

$$(6.1) \qquad \|G(y)\| \le 0.1 \text{ whenever } \|y\| \le \varepsilon.$$

Let Σ denote the set $\Sigma := \Sigma_1 \cup \Sigma_2$, where

$$\Sigma_1 := \{y \in f(B^n) : \|y\| \ge \varepsilon\}$$

and

$$\Sigma_2 := \{y \in \mathbf{R}^n : \|y\| = \varepsilon\}.$$

By construction, Σ is compact but does not contain $f(0)$. Crucially, there is a continuous map $\Phi : f(B^n) \to \Sigma$ defined by setting

$$(6.2) \qquad \Phi(y) := \max(\frac{\varepsilon}{\|y\|}, 1)y.$$

Note that Φ is continuous and well-defined since $f(B^n)$ avoids zero. Informally, Σ is a perturbation of $f(B^n)$ caused by pushing $f(B^n)$ out a small distance away from the origin 0 (and hence also away from $f(0)$), with Φ being the "pushing" map.

By construction, G is nonzero on Σ_1; since Σ_1 is compact, G is bounded from below on Σ_1 by some $\delta > 0$. By shrinking δ if necessary we may assume that $\delta < 0.1$.

By the *Weierstrass approximation theorem*, we can find a polynomial $P : \mathbf{R}^n \to \mathbf{R}^n$ such that

$$(6.3) \qquad\qquad \|P(y) - G(y)\| < \delta$$

for all $y \in \Sigma$; in particular, P does not vanish on Σ_1. At present, it is possible that P vanishes on Σ_2. But as P is smooth and Σ_2 has measure zero, $P(\Sigma_2)$ also has measure zero; so[1] by shifting P by a small generic constant we may assume without loss of generality that P also does not vanish on Σ_2.

Now consider the function $\tilde{G} : f(B^n) \to \mathbf{R}^n$ defined by

$$\tilde{G}(y) := P(\Phi(y)).$$

This is a continuous function that is never zero. From (6.3), (6.2) we have

$$\|G(y) - \tilde{G}(y)\| < \delta$$

whenever $y \in f(B^n)$ is such that $\|y\| > \varepsilon$. On the other hand, if $\|y\| \le \varepsilon$, then from (6.2), (6.1) we have

$$\|G(y)\|, \|G(\Phi(y))\| \le 0.1$$

and hence by (6.3) and the triangle inequality

$$\|G(y) - \tilde{G}(y)\| \le 0.2 + \delta.$$

Thus in all cases we have

$$\|G(y) - \tilde{G}(y)\| \le 0.2 + \delta \le 0.3$$

for all $y \in f(B^n)$. But this, combined with the nonvanishing nature of $\tilde{G}$, contradicts Lemma 6.2.3.

6.3. Hilbert's fifth problem

We now establish Theorem 1.1.9. Let G be a locally Euclidean group. By Exercise 1.1.2, G is Hausdorff; it is also locally compact and first countable. Thus, by Exercise 6.1.2, such a group contains an open subgroup G' which is isomorphic to the inverse limit $\lim_{n \to \infty} L_n$ of Lie groups L_n, each of which has only finitely many components. Clearly, G' is also locally Euclidean. If it is Lie, then G is locally Lie and thus Lie, by Exercise 2.4.7. Thus, by replacing G with G' if necessary, we may assume without loss of generality that G is the inverse limit $G = \lim_{n \to \infty} L_n$, each of which has only finitely many components.

---

[1] If one wishes, one can use an algebraic geometry argument here instead of a measure-theoretic one, noting that $P(\Sigma_2)$ lies in an algebraic hypersurface and can thus be generically avoided by perturbation. A purely topological way to avoid zeroes in Σ_2 is also given in [**Ku1998**].

By Exercise 6.1.2, each L_n is isomorphic to the quotient of G by some compact normal subgroup K_n with $K_{n+1} \subset K_n$. In particular, L_n is isomorphic to the quotient of L_{n+1} by a compact normal subgroup $H_n \equiv K_n/K_{n+1}$. By Cartan's theorem (Theorem 3.0.14), H_n is also a Lie group. Among other things, this implies that the quotient homomorphism from the Lie algebra $\mathfrak{l}_{n+1}$ of L_{n+1} to the Lie algebra $\mathfrak{l}_n$ of L_n is surjective; indeed, it is the quotient map by the Lie algebra $\mathfrak{h}_n$ of H_n. This implies that there is a continuous map from $\mathfrak{l}_n$ to $\mathfrak{l}_{n+1}$ that inverts the quotient map; in other words, we have a continuous map $\eta_{n \leftarrow n+1} : L(L_n) \to L(L_{n+1})$ from the one-parameter subgroups $\phi_n : \mathbf{R} \to L_n$ of L_n to the one-parameter subgroups $\phi_{n+1} : \mathbf{R} \to L_{n+1}$ of L_{n+1}, such that $\eta_{n \leftarrow n+1}(\phi_n) \bmod H_n = \phi_n$ for all $\phi_n \in L(L_n)$.

Exercise 6.3.1. By iterating these maps and passing to the inverse limit, conclude that for each $n \in \mathbf{N}$, there is a continuous map $\eta_n : L(L_n) \to L(G)$ such that $\eta_n(\phi_n) \bmod K_n = \phi_n$ for all $\phi_n \in L(L_n)$.

Because L_n is a Lie group, the exponential map $\phi_n \mapsto \phi_n(1)$ is a homeomorphism from a neighbourhood of the origin in $L(L_n)$ to a neighbourhood of the identity in L_n. We can thus obtain a continuous map $\phi_n(1) \mapsto \eta_n(\phi_n)(1)$ from a neighbourhood of the identity in L_n to G. Since $\eta_n(\phi_n)(1) \bmod K_n = \phi_n(1)$, this map is injective.

Now we use the hypothesis that G is locally Euclidean (and in particular, has a well-defined dimension $\dim(G)$). By Exercise 6.0.4, we have

$$\dim(L_n) \leq \dim(G)$$

for all n. On the other hand, since each L_n is a quotient of the next Lie group L_{n+1}, one has

$$\dim(L_n) \leq \dim(L_{n+1}).$$

Since there are only finitely many possible values for the (necessarily integral) dimension $\dim(L_n)$ between 0 and $\dim(G)$, we conclude that the dimension must eventually stabilise, i.e., one has

$$\dim(L_n) = \dim(L_{n+1})$$

for all sufficiently large n. By discarding the first few terms in the sequence and relabeling, we may thus assume that the dimension is constant for *all* n. Since $L_n \equiv L_{n+1}/H_n$, this implies that the Lie groups H_n have dimension zero for all n. As the H_n are also compact, they are thus finite. Thus each K_{n+1} is a finite extension of K_n. As K_n is the inverse limit of the K_n/K_m as $m \to \infty$, we conclude that K_n is a *profinite group*, i.e., the inverse limit of finite groups. In particular, K_n is totally disconnected.

We now study the short exact squence

$$0 \to K_n \to G \to G_n \to 0,$$

playing off the locally connected nature of the Lie group G_n against the totally disconnected nature of K_n.

As discussed earlier, we have a continuous injective map ψ_n from a neighbourhood U_n of the identity in G_n to G that partially inverts the quotient map. By translation, we may normalise $\psi_n(1) = 1$. As G_n is locally connected, we can find a connected neighborhood V_n of the identity in G_n such that $(V_n \cup V_n^{-1})^2 \subset U_n$.

Now consider the set $\{\psi_n(g)\psi_n(g^{-1}) : g \in V_n\}$. On the one hand, this set is contained in K_n and contains 1; on the other hand, it is connected. As K_n is totally disconnected, this set must equal $\{1\}$, thus $\psi_n(g^{-1}) = \psi_n(g)^{-1}$ for all $g \in V_n$. A similar argument based on consideration of the set $\{\psi_n(g)\psi_n(h)\psi_n(gh)^{-1} : g, h \in V_n\}$ shows that $\psi_n(gh) = \psi_n(g)\psi_n(h)$ for all $g \in V_n$. Thus ψ_n is a homomorphism from the local group V_n to G.

Finally, for any $k \in K_n$, a consideration of the set $\{\psi_n(g)k\psi_n(g)^{-1} : g \in V_n\}$ reveals that $\psi_n(g)$ commutes with K_n. As a consequence, we see that the preimage $\pi_n^{-1}(V_n)$ of V_n under the quotient map $\pi_n : G \to G_n$ is isomorphic as a local group to $V_n \times K_n$, after identifying $\psi_n(g)k$ with (g, k) for any $g \in V_n$ and $k \in K_n$.

On the other hand, G is locally Euclidean, and hence $V_n \times K_n$ is locally Euclidean also and, in particular, locally connected. This implies that K_n is locally connected; but as K_n is also totally disconnected, it must be discrete. This G is now locally isomorphic to V_n and hence to G_n, and is thus locally Lie and hence Lie as required. (Here, we say that two groups are *locally isomorphic* if they have neighbourhoods of the identity which are isomorphic to each other as local groups.)

Exercise 6.3.2. Let G be a locally compact Hausdorff first-countable group which is "finite-dimensional" in the sense that it does not contain continuous injective images of nontrivial open sets of Euclidean spaces $\mathbf{R}^n$ of arbitrarily large dimension. Show that G is locally isomorphic to the direct product $L \times K$ of a Lie group L and a totally disconnected compact group K. (Note that this local isomorphism does not necessarily extend to a global isomorphism, as the example of the solenoid group $\mathbf{R} \times \mathbf{Z}_p / \mathbf{Z}^{\Delta}$ from Example 1.1.3 shows.)

Remark 6.3.1. Of course, it is possible for locally compact groups to be infinite-dimensional; a simple example is the infinite-dimensional torus $(\mathbf{R}/\mathbf{Z})^{\mathbf{N}}$, which is compact, abelian, metrisable, and locally connected, but infinite-dimensional. (It will still be an inverse limit of Lie groups, though.)

Exercise 6.3.3. Show that a topological group is Lie if and only if it is locally compact, Hausdorff, first-countable, locally connected, and finite-dimensional.

Remark 6.3.2. It is interesting to note that this characterisation barely uses the real numbers $\mathbf{R}$, which are of course fundamental in defining the smooth structure of a Lie group; the only remaining reference to $\mathbf{R}$ comes through the notion of finite dimensionality. It is also possible, using *dimension theory*, to obtain alternate characterisations of finite dimensionality (e.g., finite *Lebesgue covering dimension*) that avoid explicit mention of the real line, thus capturing the concept of a Lie group using only the concepts of point-set topology (and the concept of a group, of course).

6.4. Transitive actions

We now prove Proposition 6.0.15. As this is a stronger statement than Theorem 1.1.9, it will not be surprising that we will be using a very similar argument to prove the result.

Let G be a locally compact σ-compact group that acts transitively, faithfully, and continuously on a connected manifold X. The advantage of transitivity is that one can now view X as a homogeneous space $X = G/H$ of G, where $H = \mathrm{Stab}(x_0)$ is the stabiliser of a point x_0 (and is thus a closed subgroup of G). Note that *a priori*, we only know that X and G/H are identifiable as *sets*, with the identification map $\iota : G/H \to X$ defined by setting $\iota(gH) := gx_0$ being continuous; but thanks to the σ-compact hypothesis, we can upgrade ι to a homeomorphism. Indeed, as G is σ-compact, G/H is also; and so given any compact neighbourhood of the identity K in G, G/H can be covered by countably many translates of KH/H. By the *Baire category theorem*, one of these translates gK has an image $\iota(gKH/H) = gKx_0$ in X with nonempty interior, which implies that $K^{-1}Kx_0$ has x_0 as an interior point. From this it is not hard to see that the map ι is open; as it is also a continuous bijection, it is therefore a homeomorphism. (This argument dates back at least to [**Fr1936**].)

By the Gleason-Yamabe theorem (Theorem 6.0.11), G has an open subgroup G' that is the inverse limit of Lie groups. (Note that G is Hausdorff because it acts faithfully on the Hausdorff space X.) G' acts transitively on $G'H/H$, which is an open subset of G/H and thus also a manifold. Thus, we may assume without loss of generality that G is itself the inverse limit of Lie groups.

As G is σ-compact, the manifold X is also. As G acts faithfully on X, this makes G first-countable; and so (by Exercise 6.1.2) G is the inverse limit of a *sequence* of Lie groups $G_n = G/K_n$, with each G_{n+1} projecting surjectively onto G_n, and with the K_n shrinking to the identity.

Let H_n be the projection of H onto G_n; this is a closed subgroup of the Lie group G_n, and each H_{n+1} projects surjectively onto H_n. Then G_n/H_n are manifolds, and G/H is the inverse limit of the $G_n/H_n = G/HK_n$.

Exercise 6.4.1. Show that the dimensions of the G_n/H_n are nondecreasing, and bounded above by the dimension of G/H. (*Hint:* Repeat the arguments of the previous section. The H_n need no longer be compact, but they are still closed, and this still suffices to make the preceding arguments go through.)

Thus, for n large enough, the dimensions of G_n/H_n must be constant; by renumbering, we may assume that *all* the G_n/H_n have the same dimension. As each G_{n+1}/H_{n+1} is a cover of G_n/H_n with structure group K_n/K_{n+1}, we conclude that the K_n/K_{n+1} are zero-dimensional and compact, and thus finite. On the other hand, G/H is locally connected, which implies that the K_n/K_{n+1} are eventually trivial. Indeed, if we pick a simply connected neighbbourhood U_1 of the identity in G_1/H_1, then by local connectedness of G/H, there exists a connected neighbourhood U of the identity in G/H whose projection to G_1/H_1 is contained in U_1. Being open, U must contain one of the K_n. If K_n/K_m is nontrivial for any $m > n$, then the projection of U to G_m/H_m will then be disconnected (as this projection will be contained in a neighbourhood with the topological structure of $U \times K_1/K_m$, and its intersection with the latter fibre is at least as large as K_n/K_m. We conclude that K_n is trivial for n large enough, and so $G = G_n$ is a Lie group as required.

Note that while the manifold X in Proposition 6.0.15 is initially only required *a priori* to be a topological manifold, it automatically acquires a smooth structure also:

Exercise 6.4.2. Let G and X be as in Proposition 6.0.15. Show that one can endow X with the structure of a smooth manifold, such that the action of G is also smooth. (*Hint:* Apply Cartan's theorem to the stabiliser of a point in X.)

Chapter 7

Ultraproducts as a bridge between hard analysis and soft analysis

Roughly speaking, mathematical analysis can be divided into two major styles, namely *hard analysis* and *soft analysis*. The precise distinction between the two types of analysis is imprecise (and in some cases one may use a blend of the two styles), but some key differences can be listed as follows.

(1) Hard analysis tends to be concerned with *quantitative* or *effective* properties such as estimates, upper and lower bounds, convergence rates, and growth rates or decay rates. In contrast, soft analysis tends to be concerned with *qualitative* or *ineffective* properties such as existence and uniqueness, finiteness, measurability, continuity, differentiability, connectedness, or compactness.

(2) Hard analysis tends to be focused on *finitary, finite-dimensional* or *discrete* objects, such as finite sets, finitely generated groups, finite Boolean combination of boxes or balls, or "finite-complexity" functions, such as polynomials or functions on a finite set. In contrast, soft analysis tends to be focused on *infinitary, infinite-dimensional,* or *continuous* objects, such as arbitrary measurable sets or measurable functions, or abstract locally compact groups.

(3) Hard analysis tends to involve explicit use of many parameters such as ε, δ, N, etc. In contrast, soft analysis tends to rely instead on

properties such as continuity, differentiability, compactness, etc.,
which *implicitly* are defined using a similar set of parameters, but
whose parameters often do not make an *explicit* appearance in arguments.

(4) In hard analysis, it is often the case that a key lemma in the literature is not quite optimised for the application at hand, and one
has to reprove a slight variant of that lemma (using a variant of
the proof of the original lemma) in order for it to be suitable for
applications. In contrast, in soft analysis, key results can often
be used as "black boxes", without need of further modification or
inspection of the proof.

(5) The properties in soft analysis tend to enjoy precise closure properties; for instance, the composition or linear combination of continuous functions is again continuous, and similarly for measurability,
differentiability, etc. In contrast, the closure properties in hard
analysis tend to be fuzzier, in that the parameters in the conclusion are often different from the parameters in the hypotheses. For
instance, the composition of two Lipschitz functions with Lipschitz
constant K is still Lipschitz, but now with Lipschitz constant K^2
instead of K. These changes in parameters mean that hard analysis arguments often require more "bookkeeping" than their soft
analysis counterparts, and are less able to utilise algebraic constructions (e.g., quotient space constructions) that rely heavily on
precise closure properties.

In the text so far, focusing on the theory surrounding Hilbert's fifth
problem, the results and techniques have fallen well inside the category of
soft analysis. However, we will now turn to the theory of approximate
groups, which is a topic which is traditionally studied using the methods
of hard analysis. (Later we will also study groups of polynomial growth,
which lies on an intermediate position in the spectrum between hard and
soft analysis, and which can be profitably analysed using both styles of
analysis.)

Despite the superficial differences between hard and soft analysis, though,
there are a number of important *correspondences* between results in hard
analysis and results in soft analysis. For instance, if one has some sort of
uniform quantitative bound on some expression relating to finitary objects,
one can often use limiting arguments to then conclude a qualitative bound
on analogous expressions on infinitary objects, by viewing the latter objects as some sort of "limit" of the former objects. Conversely, if one has a
qualitative bound on infinitary objects, one can often use *compactness and*

contradiction arguments to recover uniform quantitative bounds on finitary objects as a corollary.

Remark 7.0.1. Another type of correspondence between hard analysis and soft analysis, which is "syntactical" rather than "semantical" in nature, arises by taking the *proofs* of a soft analysis result, and translating such a qualitative proof somehow (e.g., by carefully manipulating quantifiers) into a quantitative proof of an analogous hard analysis result. This type of technique is sometimes referred to as *proof mining* in the proof theory literature, and is discussed in [**Ta2008**, §1.3]. We will, however, not employ systematic proof mining techniques here, although in later sections we will informally borrow arguments from infinitary settings (such as the methods used to construct Gleason metrics) and adapt them to finitary ones.

Let us illustrate the correspondence between hard and soft analysis results with a simple example.

Proposition 7.0.2. *Let X be a sequentially compact topological space, let S be a dense subset of X, and let $f : X \to [0, +\infty]$ be a continuous function (giving the extended half-line $[0, +\infty]$ the usual order of topology). Then the following statements are equivalent:*

(i) *(Qualitative bound on infinitary objects) For all $x \in X$, one has $f(x) < +\infty$.*

(ii) *(Quantitative bound on finitary objects) There exists $M < +\infty$ such that $f(x) \leq M$ for all $x \in S$.*

In applications, S is typically a (noncompact) set of "finitary" (or "finite complexity") objects of a certain class, and X is some sort of "completion" or "compactification" of S which admits additional "infinitary" objects that may be viewed as limits of finitary objects.

Proof. To see that (ii) implies (i), observe from density that every point x in X is adherent to S, and so given any neighbourhood U of x, there exists $y \in S \cap U$. Since $f(y) \leq M$, we conclude from the continuity of f that $f(x) \leq M$ also, and the claim follows.

Conversely, to show that (i) implies (ii), we use the "compactness and contradiction" argument. Suppose for the sake of contradiction that (ii) failed. Then for any natural number n, there exists[1] $x_n \in S$ such that $f(x_n) \geq n$. Using sequential compactness, and passing to a subsequence if necessary, we may assume that the x_n converge to a limit $x \in X$. By continuity of f, this implies that $f(x) = +\infty$, contradicting (i). □

---

[1] Here we have used the axiom of choice, which we will assume throughout the text.

Remark 7.0.3. Note that the above deduction of (ii) from (i) is *ineffective* in that it gives no explicit bound on the uniform bound M in (ii). Without any further information on *how* the qualitative bound (i) is proven, this is the best one can do in general (and this is one of the most significant weaknesses of infinitary methods when used to solve finitary problems); but if one has access to the proof of (i), one can often *finitise* or *proof mine* that argument to extract an effective bound for M, although often the bound one obtains in the process is quite poor (particularly if the proof of (i) relied extensively on infinitary tools, such as limits). See [**Ta2008**, §1.2] for some related discussion.

The above simple example illustrates that in order to get from an "infinitary" statement such as (i) to a "finitary" statement such as (ii), a key step is to be able to take a sequence $(x_n)_{n \in \mathbf{N}}$ (or in some cases, a more general *net* $(x_\alpha)_{\alpha \in A}$) of finitary objects and extract a suitable infinitary limit object x. In the literature, there are three main ways in which one can extract such a limit:

(1) (Topological limit) If the x_n are all elements of some topological space S (e.g., an incomplete function space) which has a suitable "compactification" or "completion" X (e.g., a Banach space), then (after passing to a subsequence if necessary) one can often ensure the x_n converge in a topological sense (or in a metrical sense) to a limit x. The use of this type of limit to pass between quantitative/finitary and qualitative/infinitary results is particularly common in the more analytical areas of mathematics (such as ergodic theory, asymptotic combinatorics, or PDE), due to the abundance of useful compactness results in analysis such as the (sequential) *Banach-Alaoglu theorem, Prokhorov's theorem,* the *Helly selection theorem,* the *Arzelá-Ascoli theorem,* or even the humble *Bolzano-Weierstrass theorem.* However, one often has to take care with the nature of convergence, as many compactness theorems only guarantee convergence in a weak sense rather than in a strong one.

(2) (Categorical limit) If the x_n are all objects in some category (e.g., metric spaces, groups, fields, etc.) with a number of morphisms between the x_n (e.g., morphisms from x_{n+1} to x_n, or vice versa), then one can often form a *direct limit* $\lim_{\to} x_n$ or *inverse limit* $\lim_{\leftarrow} x_n$ of these objects to form a limiting object x. The use of these types of limits to connect quantitative and qualitative results is common in subjects such as algebraic geometry that are particularly amenable to categorical ways of thinking. (We already have seen inverse limits appear in the discussion of Hilbert's fifth problem, although in

that context they were not really used to connect quantitative and qualitative results together.)

(3) (Logical limit) If the x_n are all distinct spaces (or elements or subsets of distinct spaces), with few morphisms connecting them together, then topological and categorical limits are often unavailable or unhelpful. In such cases, however, one can still tie together such objects using an *ultraproduct* construction (or similar device) to create a limiting object $\lim_{n \to \alpha} x_n$ or limiting space $\prod_{n \to \alpha} x_n$ that is a *logical limit* of the x_n, in the sense that various properties of the x_n (particularly those that can be phrased using the language of first-order logic) are preserved in the limit. As such, logical limits are often very well suited for the task of connecting finitary and infinitary mathematics together. Ultralimit type constructions are of course used extensively in logic (particularly in model theory), but are also popular in metric geometry. They can also be used in many of the previously mentioned areas of mathematics, such as algebraic geometry (as discussed in [**Ta2011b**, §2.1]).

The three types of limits are analogous in many ways, with a number of connections between them. For instance, in the study of groups of polynomial growth, both topological limits (using the metric notion of *Gromov-Hausdorff convergence*) and logical limits (using the *ultralimit* construction) are commonly used, and to some extent the two constructions are at least partially interchangeable in this setting. (See also [**Ta2011c**, §4.4-4.5] for an example of the use of ultralimits as a substitute for topological limits.) In the theory of approximate groups, though, it was observed by Hrushovski [**Hr2012**] that logical limits (and, in particular, *ultraproducts*) are the most useful type of limit to connect finitary approximate groups to their infinitary counterparts. One reason for this is that one is often interested in obtaining results on approximate groups A that are uniform in the choice of ambient group G. As such, one often seeks to take a limit of approximate groups A_n that lie in completely unrelated ambient groups G_n, with no obvious morphisms or metrics tying the G_n to each other. As such, the topological and categorical limits are not easily usable, whereas the logical limits can still be employed without much difficulty.

Logical limits are closely tied with *nonstandard analysis*. Indeed, by applying an ultraproduct construction to standard number systems such as the natural numbers $\mathbf{N}$ or the reals $\mathbf{R}$, one can obtain nonstandard number systems such as the nonstandard natural numbers $^*\mathbf{N}$ or the nonstandard real numbers (or *hyperreals*) $^*\mathbf{R}$. These nonstandard number systems behave very similarly to their standard counterparts, but also enjoy the advantage of containing the standard number systems as proper subsystems (e.g., $\mathbf{R}$ is a subring of ${}^*\mathbf{R}$), which allows for some convenient algebraic manipulations (such as the quotient space construction to create spaces such as ${}^*\mathbf{R}/\mathbf{R}$) which are not easily accessible in the purely standard universe. Nonstandard spaces also enjoy a useful completeness property, known as *countable saturation* (or more precisely, ω_1-saturation), which is analogous to metric completeness (as discussed in [**Ta2011c**, §4.4]) and which will be particularly useful for us in tying together the theory of approximate groups with the theory of Hilbert's fifth problem. See [**Ta2008**, §1.5] for more discussion on ultrafilters and nonstandard analysis.

In this section, we lay out the basic theory of ultraproducts and ultralimits (in particular, proving *Los's theorem*, which roughly speaking asserts that ultralimits are limits in a logical sense, as well as the countable saturation (or more precisely, ω_1-saturation) property alluded to earlier). We also lay out some of the basic foundations of nonstandard analysis, although we will not rely too heavily on nonstandard tools in this text. Finally, we apply this general theory to approximate groups, to connect finite approximate groups to an infinitary type of approximate group which we will call an *ultra approximate group*. We will then study these ultra approximate groups (and models of such groups) in more detail in the next section.

Remark 7.0.4. Throughout this text, we will assume the *axiom of choice*, in order to easily use ultrafilter-based tools. If one really wanted to expend the effort, though, one could eliminate the axiom of choice from the proofs of the final "finitary" results that one is ultimately interested in proving, at the cost of making the proofs significantly lengthier. Indeed, there is a general result of Gödel [**Go1938**] that any result which can be stated in the language of Peano arithmetic (which, roughly speaking, means that the result is "finitary" in nature), and can be proven in set theory using the axiom of choice (or more precisely, in the *ZFC axiom system*), can also be proven in set theory without the axiom of choice (i.e., in the ZF system). As this text is not focused on foundations, we shall simply assume the axiom of choice henceforth to avoid further distraction by such issues.

7.1. Ultrafilters

The concept of an *ultrafilter* will be fundamental. We will only need to consider ultrafilters on the natural numbers $\mathbf{N}$, although one can certainly consider ultrafilters on more general sets.

Definition 7.1.1 (Ultrafilter). A *filter* α on the natural numbers is a collection of sets of natural numbers obeying the following axioms:

    (1) (Monotonicity) If $E \subset F \subset \mathbf{N}$, and $E \in \alpha$, then $F \in \alpha$.

    (2) (Intersection) If $E, F \in \alpha$, then $E \cap F \in \alpha$.

(3) (Properness) $\emptyset \notin \alpha$.

An *ultrafilter* α on the natural numbers is a filter which obeys an additional axiom:

(4) (Maximality) If $E \subset \mathbf{N}$, then exactly one of E and $\mathbf{N} \backslash E$ lies in α.

A *nonprincipal ultrafilter* (also known as a *free ultrafilter*) is an ultrafilter α that obeys one final axiom:

(5) (Nonprincipality) No finite set belongs to α. (Equivalently: any *cofinite* set will belong to α.)

Given a nonprincipal ultrafilter α, we say that a set $E \subset \mathbf{N}$ is α-*large* if it is contained in α, and α-*small* otherwise. A property $P(n)$ of the natural numbers $n \in \mathbf{N}$ is said *to hold for all n sufficiently close to* α if it holds in an α-large set.

The most basic fact about nonprincipal ultrafilters is that they exist — a result due to Tarski, and known as the *ultrafilter lemma*:

Exercise 7.1.1 (Ultrafilter lemma). Show that there exists at least one nonprincipal ultrafilter. (*Hint:* First construct a nonprincipal filter, and then use *Zorn's lemma* to place it inside a maximal nonprincipal filter.)

Exercise 7.1.2. Call an ultrafilter *principal* if it is not nonprincipal. Show that for any natural number n_0, the set $\{E \subset \mathbf{N} : n_0 \in E\}$ is a principal ultrafilter, and conversely every principal ultrafilter is of this form. Thus, the space of principal ultrafilters can be identified with $\mathbf{N}$.

The space of all ultrafilters is denoted $\beta\mathbf{N}$, and so by the preceding exercise one can view $\mathbf{N}$ as a subset of $\beta\mathbf{N}$, with $\beta\mathbf{N} \backslash \mathbf{N}$ denoting the nonprincipal ultrafilters.

Ultrafilters can be interpreted in a number of different ways, as indicated by the exercises below.

Exercise 7.1.3 (Ultrafilters as finitely additive probability measures). If α is an ultrafilter, show that the map $\mu : 2^{\mathbf{N}} \to \{0,1\}$ that maps α-large subsets of $\mathbf{N}$ to 1 and α-small subsets of $\mathbf{N}$ to 0, is a finitely additive probability measure (thus $\mu(E) + \mu(F) = \mu(E \cup F)$ whenever $E, F \subset \mathbf{N}$ are disjoint). Conversely, show that every finitely additive probability measure $\mu : 2^{\mathbf{N}} \to \{0,1\}$ that takes values in $\{0,1\}$ arises in this manner.

In view of the above exercise, one may view "α-large" as analogous to "full measure", and "holding for all n sufficiently close to α" as analogous to "holding for almost all n", except that the measure is only finitely additive instead of countably additive, and so concepts such as α-largeness are only stable under finitely many intersections, rather than countably many intersections.

Exercise 7.1.4 (Ultrafilters as Stone-Cech compactification of $\mathbf{N}$). We view $\beta\mathbf{N}$ as a subset of the power set $2^{2^{\mathbf{N}}}$, and give it the induced topology from the product topology on $2^{2^{\mathbf{N}}}$.

(i) If α is a nonprincipal ultrafilter, show that every neighbourhood of α in $\beta\mathbf{N}$ intersects $\mathbf{N}$ in an α-large set, and that every α-large set arises in this manner. (This explains the notation "n sufficiently close to α".)

(ii) Show that this makes $\beta\mathbf{N}$ a *compactification* of $\mathbf{N}$ in the sense that it is compact Hausdorff and contains $\mathbf{N}$ as a dense subset.

(iii) Show that $\beta\mathbf{N}$ is a *universal compactification* (or *Stone-Cech compactification*), thus if X is any other compactification of $\mathbf{N}$, then there is a unique continuous map $f : \beta\mathbf{N} \to X$ that is the identity on $\mathbf{N}$.

(iv) If Y is any compact Hausdorff space, show that every function $f : \mathbf{N} \to Y$ has a unique continuous extension $\beta f : \beta\mathbf{N} \to Y$ to the space $\beta\mathbf{N}$.

Remark 7.1.2. More discussion on the Stone-Cech compactification can be found in [**Ta2010**, §2.5]. The compact space $\beta\mathbf{N}$ can also be endowed with an interesting semigroup structure, which is of importance in Ramsey theory and ergodic theory; see [**Ta2009**, §2.3] for more details.

Exercise 7.1.5 (Ultrafilters as Banach limits). Recall that a functional $\lambda : \ell^{\infty}(\mathbf{N}) \to \mathbf{C}$ from the **-algebra* $\ell^{\infty}(\mathbf{N})$ of bounded complex sequences $(a_n)_{n \in \mathbf{N}}$ to the complex numbers is a **-homomorphism* if it is complex-linear and obeys the conjugation law

$$\lambda\left((\overline{a_n})_{n \in \mathbf{N}}\right) = \overline{\lambda\left((a_n)_{n \in \mathbf{N}}\right)}$$

and the homomorphism law

$$\lambda\left((a_n b_n)_{n \in \mathbf{N}}\right) = \lambda\left((a_n)_{n \in \mathbf{N}}\right)\lambda\left((b_n)_{n \in \mathbf{N}}\right)$$

for all bounded sequences a_n, b_n, while mapping the constant sequence $(1)_{n \in \mathbf{N}}$ to 1.

(i) Show that if α is an ultrafilter, show that there exists a unique **-homomorphism $\alpha-\lim : \ell^{\infty}(\mathbf{N}) \to \mathbf{C}$ with the property that for every $E \subset \mathbf{N}$, one has $\alpha-\lim 1_E(n) = 1$ if and only if E is α-large.

(ii) Conversely, show that all **-homomorphisms λ from $\ell^{\infty}(\mathbf{N})$ to $\mathbf{C}$ arise in this fashion, and that a **-homomorphism arises from a nonprincipal ultrafilter if and only if it extends the limit functional (thus $\lambda((a_n)_{n \in \mathbf{N}}) = \lim_{n \to \infty} a_n$ whenever a_n is convergent).

Remark 7.1.3. From the above exercise we see that $\beta\mathbf{N}$ can be viewed as the *Gelfand spectrum* of $\ell^\infty(\mathbf{N})$.

Exercise 7.1.6 (Ultrafilters and Arrow's theorem). Let S be a finite set consisting of at least three elements (representing "candidates"). Let $O(S)$ denote the space of all total orderings $<$ on S. Define a *voting system* to be a function $V : O(S)^{\mathbf{N}} \to O(S)$ from a sequence $(<_n)_{n\in\mathbf{N}}$ of orderings on S (representing the preferences of an infinite number of "voters") to an ordering $V((<_n)_{n\in\mathbf{N}})$ on S. If α is a nonprincipal ultrafilter, show that there is a unique voting system $V = V_\alpha : O(S)^{\mathbf{N}} \to O(S)$ with the property that for any distinct $A, B \in S$, one has $AV((<_n)_{n\in\mathbf{N}})B$ if and only if $A <_n B$ for all n sufficiently close to α. Show furthermore that this voting system obeys the following axioms for all distinct candidates A, B and all orderings $<_n \in O(S)$:

- (Consensus) If $A <_n B$ for all n, then $AV((<_n)_{n\in\mathbf{N}})B$.

- (Nondictatorship) There is no n_0 with the property that $AV((<_n)_{n\in\mathbf{N}})B$ holds if and only if $A <_{n_0} B$ holds for all choices of $A, B, <_n$.

- (Independence of irrelevant alternatives) The truth value of $AV((<_n)_{n\in\mathbf{N}})B$ depends only on the truth values of $A <_n B$ for all n. Thus, if $(<'_n)_{n\in\mathbf{N}}$ is another sequence of orderings such that $A <_n B$ if and only if $A <'_n B$, then $AV((<_n)_{n\in\mathbf{N}})B$ holds if and only if $AV((<'_n)_{n\in\mathbf{N}})B$ holds.

Conversely, show that any voting system obeying the above axioms arises from an nonprincipal ultrafilter in this fashion. Thus we see that *Arrow's theorem* [**Ar1950**] does not hold for infinite sets due to the existence of nonprincipal ultrafilters in this setting; or to put it another way, Arrow's theorem is equivalent to the assertion that finite sets have no nonprincipal ultrafilters. (This connection between ultrafilters and Arrow's theorem was first observed in [**KiSo1972**]. See [**Ta2008**, §1.5] for more discussion of the voting interpretation of an ultrafilter.)

As the above examples indicate, there are many different ways to view ultrafilters. My personal preference (at least from the perspective of forming ultraproducts and doing nonstandard analysis) is to view a nonprincipal ultrafilter as a consistent way to take limits of sequences that do not necessarily converge in the classical sense, as per Exercise 7.1.5; but other readers may prefer to use one of the other interpretations of an ultrafilter instead.

7.2. Ultrapowers and ultralimits

We now turn to the formal definition of an ultrapower and ultralimit. In order to define these two terms, we must first fix two objects:

- A *nonprincipal ultrafilter* $\alpha \in \beta\mathbf{N}\backslash\mathbf{N}$; and

- A *standard universe* $\mathfrak{U}$, which is some set of objects, the elements of which we refer to as *standard objects*. We refer to subsets of the standard universe (i.e., sets of standard objects) as *standard spaces* (or *standard sets*).

Informally, the standard universe is going to contain all the objects that we are initially interested in studying. For instance, if we are doing arithmetic, the standard universe might be the natural numbers $\mathbf{N}$. If we are doing analysis, the standard universe might include the real numbers $\mathbf{R}$, as well as the set of subsets of $\mathbf{R}$, the set of functions from $\mathbf{R}$ to $\mathbf{R}$, and so forth. If we are studying approximate groups in an ambient group G, the standard universe might include the elements of G, subsets of G, and similar such objects. We will place some more constraints on the standard universe later, but for now, the only requirement we have is that this universe forms a set. (In particular, the standard universe cannot be so huge to be a *proper class*, such as the class of all sets, or the class of all groups, etc.; although in most applications outside of set theory, this is hardly a restriction.)

In practice, the standard universe will contain many familiar mathematical spaces, such as the natural numbers, the real numbers, and so forth. To emphasise this, we will sometimes refer to the natural numbers $\mathbf{N}$ as the *standard* natural numbers, the real numbers $\mathbf{R}$ as the *standard* real numbers $\mathbf{R}$, and so forth. This is in order to distinguish such spaces from the *nonstandard* natural numbers $^*\mathbf{N}$, the *nonstandard* real numbers $^*\mathbf{R}$, etc., which we now define.

Remark 7.2.1. For the applications in this text, the precise choice of nonprincipal ultrafilter α will be completely irrelevant; one such ultrafilter will be just as good for our purposes as another. Basically, we only use the ultrafilter α in order to fix a consistent way to make choices amongst any sequence of options indexed by the natural numbers; but the precise way in which these choices are made will not be important, so long as it is consistent (in the sense that the ultrafilter axioms are obeyed). There are other applications of ultrafilters, however, in which some nonprincipal ultrafilters are decidedly superior to others; see [**Ta2009**, §2.3] for an example of this in dynamical systems.

Definition 7.2.2 (Ultralimits and ultraproducts). Two sequences $(x_n)_{n\in\mathbf{N}}$, $(y_n)_{n\in\mathbf{N}}$ of standard objects $x_n, y_n \in \mathfrak{U}$ are said to be *equivalent* if one has

$x_n = y_n$ for all n sufficiently close to α. This is easily seen to be an equivalence relation. We define the *ultralimit* $\lim_{n \to \alpha} x_n$ to be the equivalence class of a sequence $(x_n)_{n \in \mathbf{N}}$; thus $\lim_{n \to \alpha} x_n = \lim_{n \to \alpha} y_n$ if and only if $x_n = y_n$ for all n sufficiently close to α. Such ultralimits will be called *nonstandard objects*.

Given a sequence $(X_n)_{n \in \mathbf{N}}$ of standard spaces $X_n \subset \mathfrak{U}$, we define the *ultraproduct* $\prod_{n \to \alpha} X_n$ to be the space of all ultralimits $\lim_{n \to \alpha} x_n$, where $x_n \in X_n$ for each n. Such spaces will be called *nonstandard spaces* (also known as *nonstandard sets* or *internal sets*).

If X is a standard space, we define the *ultrapower* $^*X$ of X to be the ultraproduct $^*X := \prod_{n \to \alpha} X$ of the constant sequence X. We embed X inside $^*X$ by identifying each element x of X with the ultralimit $\lim_{n \to \alpha} x$. If X is the standard elements of some type, we refer to elements of $^*X$ (i.e., ultralimits of sequences in X) as nonstandard elements of the same type. Thus, for instance, elements of $^*\mathbf{N}$ are nonstandard natural numbers, elements of $^*\mathbf{R}$ are nonstandard real numbers (also known as *hyperreals*), and so forth. The ultrapower $^*\mathfrak{U}$ of the standard universe will be called the *nonstandard universe*.

Note that one can define an ultralimit $\lim_{n \to \alpha} x_n$ for sequences x_n of standard objects that are only defined for an α-large set of n, and similarly one can define ultraproducts $\prod_{n \to \alpha} X_n$ on spaces X_n that are only defined for an α-large set of n. From the nonprincipal nature of α we observe that we can modify x_n or X_n for finitely many n without affecting the ultralimit $\lim_{n \to \alpha} x_n$ or ultraproduct $\prod_{n \to \alpha} X_n$.

Remark 7.2.3. The relation between a standard space X and its ultrapower $^*X$ is analogous in some ways to the relationship between the rationals $\mathbf{Q}$ and its *metric completion*, namely the real line $\mathbf{R}$. Indeed, using the usual metric completion construction, one can interpret a real number as an equivalence class of Cauchy sequences of rationals (or as a formal limit $\lim_{n \to \infty} q_n$ of one of the Cauchy sequences in this class), and then one identifies each rational q with its formal limit $\lim_{n \to \infty} q$ in order to keep the rationals embedded inside the reals. Note, however, that whilst in the metric completion, one is only allowed to take (formal) limits of Cauchy sequences, in the ultrapower construction one may take limits of *arbitrary* sequences. This makes the ultrapower construction more "powerful" than the metric completion construction in that no *a priori* convergence (or even boundedness) hypotheses are needed in order to take limits. See [**Ta2011c**, §4.4] for more discussion of ultrapowers as a completion.

Remark 7.2.4. With our conventions, we have the slightly unfortunate consequence that every standard object is also a nonstandard object; for instance, every standard natural number is also a nonstandard natural number, every standard real is also a nonstandard real, and so forth. We will occasionally use the terminology *strictly nonstandard* to denote a nonstandard object that is not standard. For instance, an element of $^*\mathbf{N}\backslash\mathbf{N}$ would be a strictly nonstandard natural number.

Exercise 7.2.1. If X is a standard space, show that $X = {}^*X$ if and only if X is finite. In particular, if $x_n \in X$ is a sequence with finite (standard) range X, then $\lim_{n\to\alpha} x_n$ also lives in the same range X.

The above exercise shows that for finite spaces, there is no distinction between a standard space X and its nonstandard counterpart $^*X$. However, for infinite spaces, the nonstandard version of the space is usually much larger:

Exercise 7.2.2. Show that $^*\mathbf{N}$ is uncountable. (*Hint*: Adapt the *Cantor diagonal argument*.) The precise cardinality of an ultrapower is actually quite difficult to determine in general, as it can depend on the choice of ultrafilter and even on the choice of additional set theory axioms such as the *continuum hypothesis*.

Exercise 7.2.3 (Ultrapowers preserve Boolean operations). Let X, Y be standard spaces. Show that $(X \cup Y)^* = X^* \cup Y^*$, $(X \cap Y)^* = X^* \cap Y^*$, and $(X\backslash Y)^* = X^*\backslash Y^*$. Also show that $X \subset Y$ if and only if $X^* \subset Y^*$.

Strictly speaking, ultrapowers do not preserve Cartesian products: from the definitions, $(X \times Y)^*$ and $X^* \times Y^*$ are slightly different spaces. However, there is an obvious bijection between the two sets, and we shall abuse notation and implicitly make the identification $(X \times Y)^* = X^* \times Y^*$ throughout the rest of this section. More generally, given sequences X_n, Y_n of standard spaces, we make the identification $\prod_{n\to\alpha}(X_n \times Y_n) = (\prod_{n\to\alpha} X_n) \times (\prod_{n\to\alpha} Y_n)$.

Remark 7.2.5. Just as standard spaces X are subsets of the standard universe $\mathfrak{U}$, nonstandard spaces $\prod_{n\to\alpha} X_n$ are subsets of the nonstandard universe $^*\mathfrak{U}$. However, the converse is not true in general; not every subset of the nonstandard universe is necessarily a nonstandard space. For instance, we will show later that $\mathbf{N}$ is not a nonstandard subset of $^*\mathbf{N}$ (and as such, is sometimes referred to as an *external subset* of $^*\mathbf{N}$). Very roughly speaking, internal spaces in the nonstandard universe play a role similar to that of elementary sets (i.e., finite Boolean combinations of intervals) in the real line[2]; they are closed under various operations, but fall well short of exhausting all possible subsets of the ambient space. This analogy is pursued further in Chapter 20.

---

[2]Actually, an even better analogy would be the *clopen* subsets of a *Cantor space*.

Now we record an important property of nonstandard spaces, analogous to the *finite intersection property* formulation of (countable) compactness.

Lemma 7.2.6 (Countable saturation). *Let $F_1, F_2, \ldots$ be a sequence of nonstandard spaces such that any finite collection of these spaces has nonempty intersection (thus $\bigcap_{m=1}^{M} F_m \neq \emptyset$ for all $M \geq 1$). Then the entire sequence has nonempty intersection (thus $\bigcap_{m=1}^{\infty} F_m \neq \emptyset$).*

Proof. Write $F_m = \prod_{n \to \alpha} F_{n,m}$, then for each m, $\bigcap_{m=1}^{M} F_m$ is the ultraproduct of $\bigcap_{m=1}^{M} F_{m,n}$ (cf. Exercise 7.2.3). In particular, $\bigcap_{m=1}^{M} F_{m,n}$ is nonempty for all n in an α-large subset E_M of the natural numbers. By shrinking the E_M as necessary we may assume that they are monotone, thus $E_1 \supset E_2 \supset \ldots$, and such that E_M does not contain any natural number less than M; in particular, $\bigcap_{M=1}^{\infty} E_M = \emptyset$. For any $n \in E_1$, let M_n denote the largest natural number M such that $n \in E_M$, and then let x_n denote an element of $\bigcap_{m=1}^{M_n} F_{m,n}$. Then by construction, for each $m \geq 1$, we have $x_n \in F_{m,n}$ for all $n \in E_m$, and thus the nonstandard object $x := \lim_{n \to \alpha} x_n$ lies in F_m for all m, and thus $\bigcap_{m=1}^{\infty} F_m$ is nonempty as required. $\square$

Remark 7.2.7. Strictly speaking, this property is slightly stronger than countable saturation, and should technically be referred to as ω_1-*saturation*.

Exercise 7.2.4 (Countable compactness). Show that any countable cover of a nonstandard space by other nonstandard spaces has a finite subcover.

Exercise 7.2.5 (Bolzano-Weierstrass type theorem). Let $X_1, \ldots, X_k$ be a finite collection of nonstandard spaces, and let $P_i : X_1 \times \cdots \times X_k \to \{\text{true}, \text{false}\}$ be an at most countable collection of nonstandard k-ary predicates on these spaces. Let $x_{1,m}, \ldots, x_{k,m}$ for $m \in \mathbf{N}$ be sequences of nonstandard objects in $X_1, \ldots, X_k$ respectively (indexed by the standard natural numbers $\mathbf{N}$). Show that there are subsequences $x_{1,m_j}, \ldots, x_{k,m_j}$ which *converge elementarily* to some elements $x_1, \ldots, x_k$ of $X_1, \ldots, X_k$ in the sense that for each predicate P_i, one has $P_i(x_{1,m_j}, \ldots, x_{k,m_j}) = P_i(x_1, \ldots, x_k)$ for all sufficiently large j. (See [**Ta2011c**, §4.4] for more discussion on these sorts of completeness properties on nonstandard spaces.)

Remark 7.2.8. By replacing the ultraproduct construction with more sophisticated set-theoretic constructions, one can strengthen the countable saturation property by replacing "countable" by "at most cardinality κ" for any given cardinal κ; however, as one increases κ, the size of the ultraproduct-type object must also necessarily increase (although one can make the size of the model only slightly larger than κ, if one is willing to use *inaccessible cardinals*). Such (huge) saturated models are of importance in model theory, but for our applications we will only need to work with the countably saturated models provided by the ultraproduct construction.

We have taken ultralimits and ultraproducts of standard objects and standard spaces; now we perform a similar construction to standard functions and relations.

Definition 7.2.9 (Ultralimits of functions and relations). A *standard function* is a function $f : X \to Y$ between two standard spaces X, Y. Given a sequence $f_n : X_n \to Y_n$ of standard functions, we define the *ultralimit* $\lim_{n \to \alpha} f_n : \prod_{n \to \alpha} X_n \to \prod_{n \to \alpha} Y_n$ to be the function defined by the formula

$$(\lim_{n \to \alpha} f_n)(\lim_{n \to \alpha} x_n) := \lim_{n \to \alpha} f_n(x_n)$$

whenever $x_n \in X_n$. One can easily verify that this does indeed define a function, which we refer to as a *nonstandard function* (also called an *internal function* in the literature). The ultralimit $^*f := \lim_{n \to \alpha} f$ of a single standard function is thus a nonstandard function from $^*X$ to $^*Y$ that extends f; by abuse of notation we shall also refer to $^*f$ as f (analogously to how we identify $\lim_{n \to \alpha} x$ with x rather than giving it a different name such as $^*x$).

Similarly, for a (standard) natural number $k \in \mathbf{N}$, define a *standard k-ary relation* (or *standard k-ary predicate*) to be a standard function $R : X_1 \times \cdots \times X_k \to \{\text{true}, \text{false}\}$ from the product of k standard spaces to the Boolean space $\{\text{true}, \text{false}\}$ (which we will treat as part of the standard universe $\mathfrak{U}$). By the preceding construction (and Exercise 7.2.1), the ultralimit $\lim_{n \to \alpha} R_n$ of a sequence of standard k-ary relations $R_n : X_{1,n} \times \cdots \times X_{k,n} \to \{\text{true}, \text{false}\}$ (with k independent of n) is another k-ary relation

$$\lim_{n \to \alpha} R_n : \lim_{n \to \alpha} X_{1,n} \times \lim_{n \to \alpha} X_{k,n} \to \{\text{true}, \text{false}\},$$

which we will call a *nonstandard k-ary relation* (also known as an *internal k-ary relation*). Thus $\lim_{n \to \alpha} R_n(\lim_{n \to \alpha} x_{1,n}, \ldots, \lim_{n \to \alpha} x_{k,n})$ is true if and only if $R_n(x_{1,n}, \ldots, x_{k,n})$ is true for all n sufficiently close to α. Again, we identify each relation R with its nonstandard extension $^*R := \lim_{n \to \alpha} R$.

In a very similar spirit, given a sequence $O_n : X_{1,n} \times \cdots \times X_{k,n} \to Y_n$ of standard k-ary operators, one can define the ultralimit

$$\lim_{n \to \alpha} O_n : \lim_{n \to \alpha} X_{1,n} \times \lim_{n \to \alpha} X_{k,n} \to \lim_{n \to \alpha} Y_n$$

which we will call a *nonstandard k-ary operator* (also known as an *internal k-ary operator*). Again, we identify each standard operator O with its nonstandard extension $^*O := \lim_{n \to \alpha} O$.

Example 7.2.10. Let $a = \lim_{n \to \alpha} a_n$ and $b = \lim_{n \to \alpha} b_n$ be two nonstandard natural numbers (thus a_n, b_n are sequences of standard natural numbers). Then, by definition,

(1) $a + b$ is the nonstandard natural number $a + b := \lim_{n \to \alpha} a_n + b_n$.

(2) ab is the nonstandard natural number $ab := \lim_{n \to \alpha} a_n b_n$.

(3) One has $a < b$ if $a_n < b_n$ for all n sufficiently close to α.

(4) a is even if a_n is even for all n sufficiently close to α.

(5) a is prime if a_n is prime for all n sufficiently close to α.

A basic fact about the ultraproduct construction is that any elementary (or more precisely, *first-order*) statement that is true for a standard space (or a sequence of such standard spaces) is also true for the associated non-standard space (either an ultrapower or an ultraproduct). We will formalise this assertion shortly, but let us first illustrate it with some examples.

Exercise 7.2.6 (Examples of Łos's theorem).

(i) Show that addition on the nonstandard natural numbers is both commutative and associative.

(ii) Show that a nonstandard natural number a is even if and only if it is of the form $a = 2b$ for some nonstandard natural number b.

(iii) Show that a nonstandard natural number a is prime if and only if it is greater than 1, but cannot be factored as $a = bc$ for some nonstandard natural numbers $b, c > 1$.

(iv) Show that the standard *Goldbach conjecture* (every even standard natural number larger than 2 is the sum of two standard primes) is logically equivalent to the nonstandard Goldbach conjecture (every even nonstandard natural number larger than 2 is the sum of two nonstandard primes).

(v) Show that the standard *twin prime conjecture* (for any standard natural number a, there exists a standard prime $p > a$ such that $p + 2$ is also prime) is logically equivalent to the nonstandard twin prime conjecture (for any nonstandard natural number a, there exists a nonstandard prime $p > a$ such that $p+2$ is also a nonstandard prime).

Now we generalise the above examples.

Exercise 7.2.7. Let $k \geq 0$, and let $P_n, Q_n : X_{1,n} \times \cdots \times X_{k,n} \to \{\text{true}, \text{false}\}$ be two sequences of k-ary predicates. For each $1 \leq i \leq k$, let $X_i := \prod_{n \to \alpha} X_{i,n}$, and write $P := \lim_{n \to \alpha} P_n$ and $Q := \lim_{n \to \alpha} Q_n$.

(i) (Continuity of Boolean operations) Show that $\lim_{n \to \alpha}(P_n \wedge Q_n) = P \wedge Q$, $\lim_{n \to \alpha}(P_n \vee Q_n) = P \vee Q$, and $\lim_{n \to \alpha}(\neg P_n) = \neg P$.

(ii) (Continuity of existential quantifier) If $1 \leq i \leq k$, $\exists_i P_n$ denotes the $k - 1$-ary predicate $\exists x_i \in X_{i,n} : P_n(x_{1,n}, \ldots, x_{k,n})$ in the remaining $k - 1$ variables $x_{j,n} \in X_{j,n}$ for $j \neq i$, and $\exists_i P$ denotes the $k - 1$-ary predicate $\exists x_i \in X_i : P(x_1, \ldots, x_k)$ in the remaining $k - 1$ variables $x_j \in X_j$ for $j \neq i$, show that $\lim_{n \to \alpha} \exists_i P_n = \exists_i P$.

(iii) (Continuity of universal quantifier) If $1 \leq i \leq k$, $\forall_i P_n$ denotes the $k - 1$-ary predicate $\forall x_i \in X_{i,n} : P_n(x_{1,n}, \ldots, x_{k,n})$ in the remaining $k - 1$ variables $x_{j,n} \in X_{j,n}$ for $j \neq i$, and $\exists_i P$ denotes the $k - 1$-ary predicate $\forall x_i \in X_i : P(x_1, \ldots, x_k)$ in the remaining $k - 1$ variables $x_j \in X_j$ for $j \neq i$, show that $\lim_{n \to \alpha} \forall_i P_n = \forall_i P$.

By iterating the above exercise, we can obtain *Łos's theorem* [**Lo1955**], which we first state informally as follows.

Theorem 7.2.11 (Łos's theorem, informal version). *Let k, m be natural numbers. For each natural number n, suppose one has a collection $a_{1,n}, \ldots, a_{m,n}$ of standard objects, spaces, operators, and relations (of various arities), and let $a_1, \ldots, a_m$ be the corresponding nonstandard objects, spaces, operators, and relations formed via the ultralimit or ultraproduct construction. Let P be a formal k-ary predicate expressible in first-order logic using m objects, spaces, operators, and relations. Then P, when quantified over $a_1, \ldots, a_m$, is the ultralimit of P quantified over $a_{1,n}, \ldots, a_{m,n}$. (In particular, P quantified over $a_1, \ldots, a_m$ is a nonstandard predicate.)*

Specialising to the case $k = 0$, (so P is a first-order sentence involving m objects, spaces, operators, and relations), we see that P is true when quantified over $a_1, \ldots, a_m$ if and only if P is true when quantified over $a_{1,n}, \ldots, a_{m,n}$ for all n sufficiently close to α.

Corollary 7.2.12 (Special case of Łos's theorem, informal version). *Any first-order sentence that holds for a finite collection of standard objects, spaces, operators, and relations, also holds for their corresponding nonstandard counterparts (using the ultralimit or ultrapower construction).*

Before we state the theorem more formally, let is illustrate it with some more examples.

(1) **R** is an ordered field when given the usual constants $0, 1$, field operations $+, -, \cdot, ()^{-1}$, and order relation $<$. (Technically, $()^{-1}$ is not an operation on **R** because it is undefined at 0, but this can easily be fixed by artificial means, e.g., by arbitrarily defining $()^{-1}$ at 0.). The assertion that **R** is an ordered field can be expressed as a (rather lengthy) sentence in first-order logic using the indicated spaces, objects, operations, and relations. As a consequence, the ultrapower *$\mathbf{R}$ (with the same constants $0, 1$ and the nonstandard extensions of the field operations and order relation) is also an ordered field.

(2) Let $G_n = (G_n, \cdot_n)$ be a sequence of groups; then the ultraproduct $G := \prod_{n \to \alpha} G_n$ (with the ultralimit group operation $\cdot := \lim_{n \to \alpha} \cdot_n$, and similarly for the group identity and inverse operations) is also a group, because the group axioms can be phrased in first-order logic using the indicated structures. If, for each n, g_n is an element of G_n, then the ultralimit $\lim_{n \to \alpha} g_n$ is a central element of G if and only if g_n is a central element of G_n for all n sufficiently close to α. Similarly, if for each n, H_n is a subset of G_n, then the ultraproduct $\prod_{n \to \alpha} H_n$ is a normal subgroup of G if and only if H_n is a normal subgroup of G_n for all n sufficiently close to α.

(3) The standard natural numbers $\mathbf{N}$ obey the axiom of induction: if $P(n)$ is a predicate definable in the language of Peano arithmetic, and $P(0)$ is true, and $P(n)$ implies $P(n+1)$ for all $n \in \mathbf{N}$, then $P(n)$ is true for all $n \in \mathbf{N}$. As a consequence, the same axiom of induction also holds for the nonstandard natural numbers $^*\mathbf{N}$. Note, however, it is important for the nonstandard axiom of induction that the predicate be definable in the language of Peano arithmetic. For instance, the following argument is fallacious: "$0 \in \mathbf{N}$, and for any nonstandard natural number n, $n \in \mathbf{N}$ implies $n + 1 \in \mathbf{N}$. Hence, by induction, $n \in \mathbf{N}$ for all nonstandard natural numbers n". The reason is that the predicate "$n \in \mathbf{N}$" is not formalisable in the language of Peano arithmetic.

(4) Exercise 7.2.3 can be interpreted as a special case of Corollary 7.2.12.

Exercise 7.2.8.

(i) Show that $\mathbf{N}$ (viewed as a subset of $^*\mathbf{N}$) is *not* a nonstandard space. (*Hint:* If it were, the fallacious induction mentioned earlier could now be made valid.)

(ii) Establish the *overspill principle*: if a nonstandard predicate $P(n)$ on the nonstandard natural numbers is true for all standard natural numbers, then it is true for at least one strictly nonstandard natural number as well.

(iii) Show that if a nonstandard subset of $^*\mathbf{N}$ consists only of standard natural numbers, then it is finite.

(iv) Show that the nonstandard natural numbers $^*\mathbf{N}$ are not well-ordered. Why does this not contradict Los's theorem?

Exercise 7.2.9. For each n, let G_n be a group, and let S_n be a subset of G_n. Write $G := \prod_{n \to \alpha} G_n$ and $S := \prod_{n \to \alpha} S_n$.

(i) Give an example in which each S_n generates G_n as a group, but S does not generate G as a group.

(ii) Show that S generates G as a group if and only if there exists a (standard) natural number M such that $(S_n \cup \{1\} \cup S_n^{-1})^M = G_n$ for all n sufficiently close to α. (*Hint:* Apply countable saturation (Exercise 7.2.4) to the sets $(S \cup \{1\} \cup S^{-1})^M$.)

Now we make Łos's theorem more precise, by defining a *formal language* for first-order logic. (This section is optional and may be safely omitted by the reader.)

Suppose we are given a collection $\mathcal{L}$ of formal objects, spaces, operators, and relations, with each object (formally) belonging to one of the spaces, and with each of the operators and relations formally defined on some collection of these spaces (and in the case of operators, taking values in one of the spaces) as well. We will refer to elements of $\mathcal{L}$ as *primitive terms*. Initially, we do not actually identify these formal terms with concrete counterparts, whether they be standard or nonstandard. A *well-formed formula* involving $\mathcal{L}$ and one or more formal free variables $x_1, \ldots, x_n$, each of which belonging to one of the formal spaces, is formed by a finite number of applications of the following rules:

(1) Each free variable $x_1, \ldots, x_n$ is a well-formed formula (belonging to the associated formal space).

(2) Each object in $\mathcal{C}$ is a well-formed formula (belonging to the associated formal space).

(3) If $a_1, \ldots, a_k$ are well-formed formulae, and O is a k-ary operator, then $O(a_1, \ldots, a_k)$ will be a well-formed formula if the $a_1, \ldots, a_k$ belong to the formal spaces indicated by the domain of O. Of course, $O(a_1, \ldots, a_k)$ then belongs to the formal space indicated by the range of O.

A *formal predicate* $P = P(x_1, \ldots, x_k)$ involving $\mathcal{L}$ and one or more free variables $x_1, \ldots, x_k$ is then formed by a finite number of applications of the following rules:

- If X, Y are well-formed formulae belonging to the same formal space, then $X = Y$ is a predicate.

- If $X_1, \ldots, X_k$ are well-formed formulae, and R is a k-ary relation in $\mathcal{L}$, then $R(X_1, \ldots, X_k)$ is a predicate if the $X_1, \ldots, X_k$ belong to the formal spaces indicated by the domain of R.

- If P and Q are predicates, then so are $(P \vee Q)$, $(P \wedge Q)$, $(\neg P)$, $(P \implies Q)$, and $(P \iff Q)$.

- If $P(x_1, \ldots, x_k, x)$ is a predicate in $x_1, \ldots, x_k$ and an additional free variable x in some formal space X, then $(\exists x \in X : P(x_1, \ldots, x_k, x))$

> and $(\forall x \in X : P(x_1, \ldots, x_k, x))$ are predicates in $x_1, \ldots, x_k$. (Similarly for any permutation in the ordering of $x_1, \ldots, x_k, x$.)

In practice, of course, many of the parentheses in the above constructions can be removed without causing any ambiguity, and one usually does so in order to improve readability, for instance abbreviating $((P \vee Q) \vee R)$ as $P \vee Q \vee R$. However, for the purposes of studying formal logic, it can be convenient to insist on these parentheses.

Given a formal predicate involving $\mathcal{L}$ and one or more free variables $x_1, \ldots, x_n$, we may *quantify* (or *interpret*) the predicate by associating each formal object in $\mathcal{L}$ to an actual object, which may be a standard object, a nonstandard object, or something else entirely, and similarly associating an actual space, relation, or operator to each formal space, relation, or operator, making sure that all the relevant domains and ranges are respected. When one does so, the formal predicate P becomes a concrete predicate $P : X_1 \times \cdots \times X_k \to \{\text{true}, \text{false}\}$ on the appropriate quantified spaces by interpreting all the symbols in P in the obvious fashion. For instance, if the primitive terms consist of a formal space X and a formal binary operation $\cdot : X \times X \to X$, then $P(x) := (\forall y \in X : xy = yx)$ is a formal unary predicate, and if one quantifies this predicate over a concrete group $G = (G, \cdot)$ (which may be standard, nonstandard, or neither), then one obtains a concrete unary predicate $P : G \to \{\text{true}, \text{false}\}$, with $P(g)$ true precisely when g is a central element of G.

Exercise 7.2.10. With the above definitions, prove Theorem 7.2.11.

Exercise 7.2.11 (Compactness theorem). Let $\mathcal{L}$ be an at most countable collection of formal objects, spaces, operators and relations. Let $S_1, S_2, S_3, \ldots$ be a sequence of sentences involving the symbols in $\mathcal{L}$. Suppose that any finite collection $S_1, \ldots, S_n$ of these sentences is satisfiable in the standard universe, thus there exists an assignment of standard objects, spaces, operators, or relations to each element of $\mathcal{L}$ for which $S_1, \ldots, S_n$ are all true when quantified over these assignments. Show that the entire collection $S_1, S_2, \ldots$ are then satisfiable in the nonstandard universe, thus there exists an assignment of nonstandard objects. (This is the countable case of the *compactness theorem* in logic. One can also use ultraproducts over larger sets than $\mathbf{N}$ to prove the general case of the compactness theorem, but we will not do so here.)

7.3. Nonstandard finite sets and nonstandard finite sums

It will be convenient to extend the standard machinery of finite sets and finite sums to the nonstandard setting. Define a *nonstandard finite set* to be an ultraproduct $A = \prod_{n \to \alpha} A_n$ of finite sets, and a *nonstandard finite sequence of reals* $(x_m)_{m \in A}$ to be a nonstandard function $m \mapsto x_m$ from a nonstandard finite set A to the nonstandard reals, or equivalently an ultralimit of standard finite sequences $(x_{m,n})_{m \in A_n}$ of reals. We can define the (nonstandard) cardinality $|A|$ of a nonstandard finite set in the usual manner:

$$| \prod_{n \to \alpha} A_n | := \lim_{n \to \alpha} |A_n|.$$

Thus, for instance, if N is a nonstandard natural number, then the nonstandard finite set $\{n \in {}^*\mathbf{N} : 1 \le n \le N\}$ has nonstandard cardinality N.

Similarly, if $(x_m)_{m \in A} = \lim_{n \to \alpha} (x_{m,n})_{m \in A_n}$ is a nonstandard finite sequence of reals, we define the nonstandard sum $\sum_{m \in A} x_m$ as

$$\sum_{m \in A} x_m := \lim_{n \to \alpha} \sum_{m \in A_n} x_{m,n}.$$

Thus, for instance, $\sum_{m \in A} 1 = |A|$. By using Łos's theorem, all the basic laws of algebra for manipulating standard finite sums carry over to nonstandard finite sums. For instance, we may interchange summations

$$\sum_{m_1 \in A_1} \sum_{m_2 \in A_2} x_{m_1,m_2} = \sum_{m_2 \in A_2} \sum_{m_1 \in A_1} x_{m_1,m_2} = \sum_{(m_1,m_2) \in A_1 \times A_2} x_{m_1,m_2}$$

for any nonstandard finite sequence $(x_{m_1,m_2})_{(m_1,m_2) \in A_1 \times A_2}$ of reals indexed by a product set, simply because the same assertion is obviously true for standard finite sequences.

7.4. Asymptotic notation

The ultrapower and ultralimit constructions are extremely general, being basically applicable to any collection of standard objects, spaces, objects, and relations. However, when one applies these constructions to standard *number systems*, such as the natural numbers $\mathbf{N}$, the integers $\mathbf{Z}$, the real numbers $\mathbf{R}$, or the complex numbers $\mathbf{C}$ to obtain nonstandard number systems such as ${}^*\mathbf{N}$, ${}^*\mathbf{Z}$, ${}^*\mathbf{R}$, ${}^*\mathbf{C}$, then one can gain an additional tool of use in analysis, namely a clean *asymptotic notation* which is closely related to standard asymptotic notation, but has a slightly different arrangement of quantifiers that make the nonstandard asymptotic notation better suited to algebraic constructions (such as the quotient space construction) than standard asymptotic notation.

Definition 7.4.1 (Asymptotic notation)**.** Let R be one of the standard number systems ($\mathbf{N}$, $\mathbf{Z}$, $\mathbf{Q}$, $\mathbf{R}$, $\mathbf{C}$). A nonstandard number x in ${}^*R$ is said to be *bounded* if one has $|x| \le C$ for some standard real number C (or equivalently, $|x| \le n$ some standard natural number n), and *unbounded* otherwise. If y is a nonnegative nonstandard real, we write $x = O(y)$ if

$|x| \leq Cy$ for some standard real number C, thus a nonstandard number is bounded if and only if it is $O(1)$. We say that x is *infinitesimal* if $|x| \leq \varepsilon$ for every standard real number $\varepsilon > 0$ (or equivalently, if $|x| \leq \frac{1}{n}$ for all standard positive natural numbers n), and write $x = o(y)$ if $|x| \leq \varepsilon y$ for all standard real numbers $\varepsilon > 0$, thus x is infinitesimal if and only if it is $o(1)$.

Example 7.4.2. All standard numbers are bounded, and all nonzero standard numbers are noninfinitesimal. The nonstandard natural number $N :=$ $\lim_{n \to \alpha} n$ is unbounded, and the nonstandard real number $\frac{1}{N} = \lim_{n \to \alpha} \frac{1}{n}$ is nonzero but infinitesimal.

Remark 7.4.3. One can also develop analogous nonstandard asymptotic notation on other spaces, such as normed vector spaces or locally compact groups; see for instance [**Ta2011c**, §4.5] for an example of the former, and Hirschfeld's nonstandard proof [**Hi1990**] of Hilbert's fifth problem for an example of the latter. We will however not need to deploy nonstandard asymptotic notation in such generality here.

Note that we can apply asymptotic notation to individual nonstandard numbers, in contrast to standard asymptotic notation in which one needs to have the numbers involved to depend on some additional parameter if one wishes to prevent the notation from degenerating into triviality. In particular, we can form well-defined sets such as the *bounded nonstandard reals*

$$O(\mathbf{R}) := \{x \in {}^*\mathbf{R} : x = O(1)\}$$

and the *infinitesimal nonstandard reals*

$$o(\mathbf{R}) := \{x \in {}^*\mathbf{R} : x = o(1)\}.$$

Exercise 7.4.1 (Standard part). Show that $o(\mathbf{R})$ is an ideal of the commutative ring $O(\mathbf{R})$, and one has the decomposition $O(\mathbf{R}) = \mathbf{R} \oplus o(\mathbf{R})$, thus every bounded real $x = O(\mathbf{R})$ has a unique decomposition $x = \mathrm{st}(x) + (x - \mathrm{st}(x))$ into a *standard part* $\mathrm{st}(x) \in \mathbf{R}$ and an *infinitesimal part* $x - \mathrm{st}(x) \in o(\mathbf{R})$. Show that the map $x \mapsto \mathrm{st}(x)$ is a ring homomorphism from $O(\mathbf{R})$ to $\mathbf{R}$.

Exercise 7.4.2. Let N be an unbounded nonstandard natural number. Write $O_{\mathbf{Z}}(N) := \{n \in {}^*\mathbf{Z} : n = O(N)\}$ and $o_{\mathbf{Z}}(N) := \{n \in {}^*\mathbf{Z} : n = o(N)\}$. Show that $o_{\mathbf{Z}}(N)$ is a subgroup of the additive group $O_{\mathbf{Z}}(N)$, and that the quotient group $O_{\mathbf{Z}}(N)/o_{\mathbf{Z}}(N)$ is isomorphic (as an additive group) to the standard real numbers $\mathbf{R}$. (One could in fact use this construction as a *definition* of the real number system, although this would be a rather idiosyncratic choice of construction.)

Remark 7.4.4. In some texts, the standard part $x = \operatorname{st} \lim_{n \to \alpha} x_n$ of an ultralimit of a bounded sequence of numbers is denoted by $\lim_{n \to \alpha} x_n$ (with the ultralimit itself not having a specific notation assigned to it). The assertion $x = \operatorname{st} \lim_{n \to \alpha} x_n$ is also sometimes referred to as "x_n converges to x along α", or equivalently that every neighbourhood of x contains x_n for α-almost every n.

The following exercise should be compared with Proposition 7.0.2. Note the absence of a continuity hypothesis on the function f.

Exercise 7.4.3. Let $f : X \to \mathbf{R}$ be a standard function, which we extend to a nonstandard function $f : {}^*X \to {}^*\mathbf{R}$ in the usual manner. Show that the following statements are equivalent:

(i) (Qualitative bound on nonstandard objects) For all $x \in {}^*X$, one has $f(x) = O(1)$.

(ii) (Quantitative bound on standard objects) There exists a standard real $M < +\infty$ such that $|f(x)| \leq M$ for all $x \in X$.

(*Hint:* Use some version of the countable saturation property.)

One of the first motivations of nonstandard analysis was to obtain a rigorous theory of infinitesimals that could be used as an alternate (though logically equivalent) foundation for real analysis. We will not need to use this foundation here, but the following exercises are intended to give some indication of one might start setting up such foundations.

Exercise 7.4.4. Let E be a standard subset of the reals $\mathbf{R}$.

(i) Show that E is open if and only if $E + o(\mathbf{R}) \subset {}^*E$, or equivalently if $\operatorname{st}^{-1}(E) \subset {}^*E$. (Here, $A + B := \{a + b : a \in A, b \in B\}$ denotes the sumset of A and B.)

(ii) Show that E is closed if and only if ${}^*E \cap O(\mathbf{R}) \subset E + o(\mathbf{R})$, or equivalently if $\operatorname{st}({}^*E \cap O(\mathbf{R})) \subset E$.

(iii) Show that E is bounded if and only if ${}^*E \subset O(\mathbf{R})$.

(iv) Show that E is compact if and only if ${}^*E \subset E + o(\mathbf{R})$.

Exercise 7.4.5. Let $(x_n)_{n \in \mathbf{N}}$ be a standard sequence of reals, and let $(x_n)_{n \in {}^*\mathbf{N}}$ be its nonstandard extension. Let L be a real number.

(i) Show that x_n converges to L as $n \to \infty$ if and only if $x_N = L + o(1)$ for all unbounded natural numbers N.

(ii) Show that $\sum_{n=1}^{\infty} x_n$ is conditionally convergent to L if and only if $\sum_{n=1}^{N} x_N = L + o(1)$ for all unbounded natural numbers N.

Exercise 7.4.6. Let $f : \mathbf{R} \to \mathbf{R}$ be a standard function, and let $f : {}^*\mathbf{R} \to {}^*\mathbf{R}$ be its nonstandard extension. Show that the following are equivalent:

    (i) f is a continuous function.

    (ii) One has $f(y) = f(x) + o(1)$ whenever $x \in \mathbf{R}$, $y \in {}^*\mathbf{R}$, and $y = x + o(1)$.

    (iii) One has $f(\text{st}(x)) = \text{st}(f(x))$ for all $x \in O(\mathbf{R})$.

Exercise 7.4.7. Let $f : \mathbf{R} \to \mathbf{R}$ be a standard function, and let $f : {}^*\mathbf{R} \to {}^*\mathbf{R}$ be its nonstandard extension. Show that the following are equivalent:

    (i) f is a uniformly continuous function.

    (ii) One has $f(y) = f(x) + o(1)$ whenever $x, y \in {}^*\mathbf{R}$ and $y = x + o(1)$.

Exercise 7.4.8. Let $f : \mathbf{R} \to \mathbf{R}$ be a standard function, and let $f : {}^*\mathbf{R} \to {}^*\mathbf{R}$ be its nonstandard extension. Show that the following are equivalent:

    (i) f is a differentiable function.

    (ii) There exists a standard function $f' : \mathbf{R} \to \mathbf{R}$ such that $f(x + h) = f(x) + f'(x)h + o(|h|)$ for all $x \in \mathbf{R}$ and $h = o(1)$, or equivalently if $f'(x) = \text{st}\, \frac{f(x+h) - f(x)}{h}$ for all $x \in \mathbf{R}$ and nonzero $h = o(1)$.

Exercise 7.4.9. Let $f : \mathbf{R} \to \mathbf{R}$ be a standard continuous function, and let $f : {}^*\mathbf{R} \to {}^*\mathbf{R}$ be its nonstandard extension. Show that for any standard reals $a < b$ and any unbounded natural number N, one has

$$\int_a^b f(x)\, dx = \text{st}\, \frac{1}{N} \sum_{n=1}^N f\left(a + \frac{b-a}{N}n\right)$$

(note that $\sum_{n=1}^N f(a + \frac{b-a}{N}n)$ is a nonstandard finite sum). More generally, show that

$$\int_a^b f(x)\, dx = \text{st}\, \frac{1}{N} \sum_{n=1}^N f(x_n^*)(x_n - x_{n-1})$$

whenever $(x_n)_{0 \le n \le N}$ and $(x_n^*)_{1 \le n \le N}$ are nonstandard finite sequences with $a \le x_{n-1} \le x_n^* \le x_n \le b$ and $x_n = x_{n-1} + o(1)$ for all $1 \le n \le N$.

7.5. Ultra approximate groups

In this section we specialise the above ultraproduct machinery to the concept of a finite approximate group, to link such objects with a type of infinite approximate group which we call an *ultra approximate group*. These objects will be studied much more intensively in the next two lectures, but for now we focus on the correspondence between ultra approximate groups and their finitary counterparts.

We first recall the notion of an approximate group. For simplicity, we work for now in the global group setting, although later on we will also need to generalise to local groups.

Definition 7.5.1 (Approximate group). Let $K \geq 1$ be a standard real number. A (global) *approximate group* is a subset A of a global group $G = (G, \cdot)$ which is symmetric, contains the identity, and is such that $A \cdot A$ can be covered by at most K left-translates of A.

Definition 7.5.2 (Ultra approximate group). An *ultra approximate group* is an ultraproduct $A = \prod_{n \to \alpha} A_n$, where each $A_n \subset G_n$ is a standard finite K-approximate group for some K independent of n.

Example 7.5.3. If $N = \prod_{n \to \alpha} N_n$ is a nonstandard natural number, then the nonstandard arithmetic progression $\{m \in {}^*\mathbf{N} : -N \leq m \leq N\}$ is an ultra approximate group, because it is an ultraproduct of standard finite 2-approximate groups $\{m \in \mathbf{N} : -N_n \leq m \leq N_n\}$.

Example 7.5.4. If $N = \prod_{n \to \alpha} N_n$ is a nonstandard natural number, then the nonstandard cyclic group $\mathbf{Z}/N\mathbf{Z} = \prod_{n \to \alpha} \mathbf{Z}/N_n\mathbf{Z}$ is an ultra approximate group, because it is an ultraproduct of standard cyclic groups. More generally, any nonstandard finite group (i.e., an ultraproduct of standard finite groups) is an ultra approximate group.

For any fixed K, the statement of a set A in a group G being a K-approximate group can be phrased in first-order logic quantified over G using the group operations and the predicate $x \in A$ of membership in A. From this and Łos's theorem, we see that any ultraproduct of K-approximate groups is again a K-approximate group. In particular, an ultra approximate group is a K-approximate group for some (standard) finite K.

We will be able to make a correspondence between various assertions about K-approximate groups and about ultra approximate groups, so that problems about the former can be translated to problems about the latter. By itself, this correspondence is not particularly deep or substantial; but the point will be that by moving from the finitary category of finite K-approximate groups to the infinitary category of ultra approximate groups, we will be able to deploy tools from infinitary mathematics and, in particular, the topological group theory results (such as the Gleason-Yamabe theorem) from previous sections, to bear on the problem. We will execute this strategy in the next few sections, but for now, let us just see how the correspondence works. We begin with a simple example, in the bounded *torsion* case. For a standard natural number r, we say that a group G is *r-torsion* if every element has order at most r, thus for each $g \in G$ one has $g^n = 1$ for some $1 \leq n \leq r$.

Proposition 7.5.5 (Correspondence in the torsion case). *Let $r \geq 1$. The following two statements are equivalent:*

(i) *(Finitary statement)* For all $K \geq 1$, there exists $C_{K,r} \geq 1$ such that, given a finite K-approximate group A in an r-torsion group G, one can find a finite (genuine) subgroup H of G such that A^4 contains H, and A can be covered by at most $C_{K,r}$ left-translates of H.

(ii) *(Infinitary statement)* For any ultra-approximate group A in an r-torsion nonstandard group G, one can find a nonstandard finite subgroup H of G such that A^4 contains H, and A can be covered by finitely many left-translates of H.

Proof. Let us first assume (i) and establish (ii). Let $A = \prod_{n \to \alpha} A_n$ be an ultra-approximate group in an r-torsion nonstandard group $G = \prod_{n \to \alpha} G_n$. Since G is r-torsion, and the property of being r-torsion is a first-order statement in the language of groups, we see from Łos's theorem that G_n is a r-torsion group for all n sufficiently close to α. Since the A_n are all finite K-approximate groups for some K independent of n, we conclude from (i) that for all n sufficiently close to α, we can find a finite subgroup H_n of G_n such that A_n^4 contains H_n, and A_n can be covered by at most $C_{K,r}$ left-translates of H_n. Taking $H := \prod_{n \to \alpha} H_n$ and applying Łos's theorem again, we cnoclude that H is a nonstandard finite subgroup of G, that A^4 contains H, and A can be covered by at most $C_{K,r}$ left-translates of H, giving (ii) as desired.

Now we assume (ii) and establish (i). Suppose for contradiction that (i) failed. Carefully negating the quantifiers, we conclude that there exists $K \geq$ such that for each natural number n, one can find a finite K-approximate group A_n in an r-torsion group G_n for which there does *not* exist any finite subgroup H_n of G_n such that A_n^4 contains H_n *and* A_n can be covered by at most n left-translates of H_n. We then form the ultraproducts $A := \prod_{n \to \alpha} A_n$ and $G := \prod_{n \to \alpha} G_n$. By Łos's theorem, G is an r-torsion nonstandard group, and A is an ultra approximate group in G. Thus, by (ii), there is a nonstandard finite subgroup $H = \prod_{n \to \alpha} H_n$ of G such that A^4 contains H and A is covered by M left-translates of H for some (standard) natural number M. By Łos's theorem, we conclude that for all n sufficiently close to α, H_n is a finite subgroup of G_n, A_n^4 contains H_n, and A_n is covered by M left-translates of H_n. In particular, as α is nonprincipal, this assertion holds for at least one $n > M$. But this contradicts the construction of A_n. $\square$

Note how the infinitary statement in the above proposition is a qualitative version of the more quantitative finitary statement. In particular, parameters such as K and $C_{K,r}$ have been efficiently concealed in the infinitary formulation, whereas they must be made explicit in the finitary formulation. As such, the infinitary approach leads to a reduction in the amount of "epsilon management" that one often has to perform in finitary settings.

In the next section we will use the Gleason-Yamabe theorem to establish (ii) and hence conclude (i) as a corollary (this argument is essentially due to Hrushovski [**Hr2012**]). Interestingly, no purely finitary proof of (i) in full generality is currently known (other than by finitising the infinitary proof, which is in principle possible, but would create an extremely long and messy proof as a consequence).

We move from the bounded torsion setting to another special case, namely the abelian setting. Here, the finite subgroups need to be replaced by the more general concept of a *coset progression*.

Definition 7.5.6. A (standard symmetric) *generalised arithmetic progression* P of rank r in a (standard) additive group $G = (G, +)$ is a set of the form

$$P = P(v_1, \ldots, v_r; N_1, \ldots, N_r) := \{a_1 v_1 + \cdots + a_r v_r : |a_1| \leq N_1, \ldots, |a_r| \leq N_r\}$$

where $v_1, \ldots, v_r \in G$, $N_1, \ldots, N_r > 0$, and the $a_1, \ldots, a_r$ are constrained to be integers. An *ultra generalised arithmetic progression* is an ultraproduct $P := \prod_{n \to \alpha} P_n$ of standard generalised arithmetic progressions P_n of fixed rank r.

A (standard) *coset progression* Q of rank r in a (standard) additive group $G = (G, +)$ is a set of the form $Q = H + P$, where H is a finite subgroup of G and P is a generalised arithmetic progression of rank r. An *ultra coset progression* is an ultraproduct $Q := \prod_{n \to \alpha} Q_n$ of standard coset progressions Q_n of fixed rank r; equivalently, it is a set of the form $H + P$, where H is a nonstandard finite group in a nonstandard group G, and P is an ultra generalised arithmetic progression in G.

Exercise 7.5.1. Show that the following two statements are equivalent:

(i) (Finitary abelian Freiman theorem) For all $K \geq 1$, there exists $C_K, r_K \geq 1$ such that, given a finite K-approximate group A in an abelian group $G = (G, +)$, one can find a coset progression Q in G of rank at most r_K such that A^4 contains Q, and A can be covered by at most C_K left-translates of Q.

(ii) (Infinitary abelian Freiman theorem) For any ultra approximate group A in a abelian nonstandard group $G = (G, +)$, one can find an ultra coset progression Q of G such that A^4 contains Q, and A can be covered by finitely many left-translates of Q.

The statement (i) was established by Green and Ruzsa [**GrRu2007**] by finitary means (such as the finite Fourier transform). In later sections we will give an alternate proof of (i) that proceeds via (ii) and the structure theory of locally compact abelian groups.

Finally, we can give the correspondence in the full nonabelian setting.

Definition 7.5.7. Given a (standard) finite number of generators $v_1, \ldots, v_r$ in a (standard) group $G = (G, \cdot)$, and (standard) real numbers $N_1, \ldots, N_r > 0$, define the *noncommutative progression* $P(v_1, \ldots, v_r; N_1, \ldots, N_r)$ to be the set of all words in $v_1^{\pm 1}, \ldots, v_r^{\pm 1}$ involving at most N_i copies of $v_i^{\pm 1}$ for each $1 \le i \le r$. We refer to r as the *rank* of the noncommutative progression. A *nilprogression* of rank r and step at most s is a noncommutative progression that lies in a *nilpotent group* of step at most s. A *coset nilprogression* in G of rank r and step at most s is a set of the form $\pi^{-1}(P)$, where H is a finite group in G, $N(H) := \{g \in H : gH = Hg\}$ is the *normaliser* of H, $\pi : N(H) \to N(H)/H$ is the quotient map, and P is a nilprogression of rank r in $N(H)/H$.

An *ultra coset nilprogression* is an ultraproduct $Q = \prod_{n \to \alpha} Q_n$ of coset nilprogressions of rank r and step at most s, for some standard natural numbers r, s independent of n.

Exercise 7.5.2. Show that the following two statements are equivalent:

(i) (Finitary nonabelian Freiman theorem) For all $K \ge 1$, there exists $C_K, s_K, r_K \ge 1$ such that, given a finite K-approximate group A in a group $G = (G, \cdot)$, one can find a coset nilprogression Q in G of rank at most r_K and step at most s_K such that A^4 contains Q, and A can be covered by at most C_K left-translates of Q.

(ii) (Infinitary nonabelian Freiman theorem) For any ultra approximate group A in a nonstandard group $G = (G, \cdot)$, one can find an ultra coset nilprogression Q in G such that A^4 contains Q, and A can be covered by finitely many left-translates of Q.

In later sections we will establish (ii), and hence (i), thus giving a (qualitatively) satisfactory description of finite approximate groups in arbitrary groups.

Exercise 7.5.3. Let n be a standard natural number. Show that every nonstandard finite subgroup of $^* \mathrm{GL}_n(\mathbf{C})$ is virtually abelian.

Chapter 8

Models of ultra approximate groups

In Chapter 7, we introduced the notion of an *ultra approximate group* — an ultraproduct $A = \prod_{n \to \alpha} A_n$ of finite K-approximate groups A_n for some K independent of n, where each K-approximate group A_n may lie in a distinct ambient group G_n. Although these objects arise initially from the "finitary" objects A_n, it turns out that ultra approximate groups A can be profitably analysed by means of *infinitary* groups L (and, in particular, locally compact groups or Lie groups L), by means of certain *models*[1] $\rho : \langle A \rangle \to L$ of A (or of the group $\langle A \rangle$ generated by A). We will define precisely what we mean by a model later, but as a first approximation one can view a model as a representation of the ultra approximate group A (or of $\langle A \rangle$) that is "macroscopically faithful" in that it accurately describes the "large scale" behaviour of A (or equivalently, that the kernel of the representation is "microscopic" in some sense). In the next section we will see how one can use "Gleason lemma" technology to convert this macroscopic control of an ultra approximate group into microscopic control, which will be the key to classifying approximate groups.

Models of ultra approximate groups can be viewed as the multiplicative combinatorics analogue of the more well-known concept of an *ultralimit* of metric spaces, which we briefly review below the fold as motivation.

The crucial observation is that ultra approximate groups enjoy a *local compactness* property which allows them to be usefully modeled by locally compact groups (and hence, through Theorem 1.1.13, by Lie groups also).

---

[1] The terminology here comes from additive combinatorics, and is a little different from the notion of a model from model theory.

As per the *Heine-Borel theorem*, the local compactness will come from a combination of a *completeness* property and a local *total boundedness* property. The completeness property turns out to be a direct consequence of the *countable saturation* property of ultraproducts, thus illustrating one of the key advantages of the ultraproduct setting. The local total boundedness property is more interesting. Roughly speaking, it asserts that "large bounded sets" (such as A or A^{100}) can be covered by finitely many translates of "small bounded sets" S, where "small" is a topological group sense, implying in particular that large powers S^m of S lie inside a set such as A or A^4. The easiest way to obtain such a property comes from the following lemma of Sanders [**Sa2009**]:

Lemma 8.0.8 (Sanders lemma). *Let A be a finite K-approximate group in a (global) group G, and let $m \geq 1$. Then there exists a symmetric subset S of A^4 with $|S| \gg_{K,m} |A|$ containing the identity such that $S^m \subset A^4$.*

This lemma has an elementary combinatorial proof, and is the key to endowing an ultra approximate group with locally compact structure. There is also a closely related lemma of Croot and Sisask [**CrSi2010**] which can achieve similar results, and which will also be discussed below. (The locally compact structure can also be established more abstractly using the much more general methods of definability theory, as was first done by Hrushovski [**Hr2012**], but we will not discuss this approach here.)

By combining the locally compact structure of ultra approximate groups A with the Gleason-Yamabe theorem, one ends up being able to model a large "ultra approximate subgroup" A' of A by a Lie group L. Such Lie models serve a number of important purposes in the structure theory of approximate groups. First, as all Lie groups have a dimension which is a natural number, they allow one to assign a natural number "dimension" to ultra approximate groups, which opens up the ability to perform "induction on dimension" arguments. Second, Lie groups have an *escape property* (which is in fact equivalent to *no small subgroups* property): if a group element g lies outside of a very small ball B_ε, then some power g^n of it will escape a somewhat larger ball B_1. Or equivalently: if a long orbit $g, g^2, \ldots, g^n$ lies inside the larger ball B_1, one can deduce that the original element g lies inside the small ball B_ε. Because all Lie groups have this property, we will be able to show that all ultra approximate groups A "essentially" have a similar property, in that they are "controlled" by a nearby ultra approximate group which obeys a number of escape-type properties analogous to those enjoyed by small balls in a Lie group, and which we will call a *strong ultra approximate group*. This will be discussed in the next section, where we will also see how these escape-type properties can be exploited to create a metric structure on strong approximate groups analogous to the Gleason metrics studied in Chapter 3, which can in turn be exploited (together with an induction on dimension argument) to fully classify such approximate groups (in the finite case, at least).

There are some cases where the analysis is particularly simple. For instance, in the bounded torsion case, one can show that the associated Lie model L is necessarily zero-dimensional, which allows for an easy classification of approximate groups of bounded torsion.

The material in this chapter is loosely based on the papers [**Hr2012**], [**BrGrTa2011**].

8.1. Ultralimits of metric spaces (Optional)

Suppose one has a sequence (X_n, d_n) of metric spaces. Intuitively, there should be a sense in which such a sequence can (in certain circumstances) "converge" to a limit (X, d) that is another metric space. Some informal examples of this intuition:

(i) The sets $\{-n, \ldots, n\}$ (with the usual metric) should "converge" as $n \to \infty$ to the integers $\mathbf{Z}$ (with the usual metric).

(ii) The cyclic groups $\mathbf{Z}/n\mathbf{Z}$ (with the "discrete" metric $d(i, j) :=$ dist$(i + nZ, j + nZ)$) should also "converge" as $n \to \infty$ to the integers $\mathbf{Z}$ (with the usual metric).

(iii) The sets $\{0, \frac{1}{n}, \ldots, \frac{n-1}{n}\}$ (with the usual metric) should "converge" as $n \to \infty$ to the interval $[0, 1]$ (with the usual metric).

(iv) The cyclic groups $\mathbf{Z}/n\mathbf{Z}$ (with the "bounded" metric $d(i, j) :=$ $\frac{1}{n}$dist$(i + nZ, j + nZ)$) should "converge" as $n \to \infty$ to the unit circle $\mathbf{R}/\mathbf{Z}$ (with the usual metric).

(v) A Euclidean circle of radius n (such as $\{(x, y) \in \mathbf{R}^2 : x^2 + (y-n)^2 = n^2\}$) should "converge" as $n \to \infty$ to a Euclidean line (such as $\{(x, 0) \in \mathbf{R}^2 : x \in \mathbf{R}\}$).

Let us now try to formalise this intuition, proceeding in stages. The first attempt to formalise the above concepts is via the *Hausdoff distance*, which already made an appearance in Chapter 5:

Definition 8.1.1 (Hausdorff distance). Let $X = (X, d)$ be a metric space. The *Hausdorff distance* $d_H(E, F)$ between two nonempty subsets E, F of X is defined by the formula

$$d_H(E, F) := \max(\sup_{x \in E} d(x, F), \sup_{y \in F} d(E, y)).$$

Thus, if $d_H(E, F) < r$, then every point in E lies within a distance less than r of a point in F, and every point in F lies within a distance less r of a point in E (thus $E \subset N_r(F)$ and $F \subset N_r(E)$, where $N_r(F)$ is the r-neighbourhood of E); and conversely, if the latter claim holds, then $d_H(E, F) \leq r$.

This distance is always symmetric, nonnegative, and obeys the triangle inequality. If one restricts attention to nonempty compact sets E, F, then one easily verifies that $d_H(E, F) = 0$ if and only if $E = F$, so that the Hausdorff distance becomes a metric. In particular, it becomes meaningful to discuss the concept of a sequence E_n of nonempty compact subsets of X converging in the Hausdorff distance to a (unique) limit nonempty compact set E. Note that this concept captures the intuitive example (iii) given above, but not any of the others. Nevertheless, we will discuss it first as it is slightly simpler than the more general notions we will be using.

Exercise 8.1.1 (Hausdorff convergence and connectedness). Let E_n be a sequence of nonempty compact subsets of a metric space X converging to another nonempty compact set E.

    (i) If the E_n are all connected, show that E is connected also.

    (ii) If the E_n are all path-connected, does this imply that E is path-connected also? Support your answer with a proof or counter-example.

    (iii) If the E_n are all disconnected, does this imply that E is disconnected also? Support your answer with a proof or counterexample.

One of the key properties of Hausdorff distance in a compact set is that it is itself compact:

Lemma 8.1.2 (Compactness of the Hausdorff metric). *Let E_n be a sequence of compact subsets of a compact metric space X. Then there is a subsequence of the E_n that is convergent in the Hausdorff metric.*

The proof of this lemma was set as Exercise 5.5.1. It can be proven by "conventional" means (relying in particular on the Heine-Borel theorem), but we will now sketch how one can establish this result using ultralimits instead. As in Chapter 7, we fix a standard universe $\mathfrak{U}$ and a nonprincipal ultrafilter α in order to define ultrapowers.

If $X = (X, d)$ is a (standard) metric space with metric $d : X \times X \to \mathbf{R}^+$, then the ultrapower ${}^*X$ comes with a "nonstandard metric" ${}^*d : {}^*X \times {}^*X \to {}^*\mathbf{R}^+$ that extends d. (In the previous sections, we referred to this metric as d rather than ${}^*d$, but here it will be convenient to use a different symbol for this extension of d to reduce confusion.) Let us say that two elements x, y of ${}^*X$ are *infinitesimally close* if ${}^*d(x, y) = o(1)$. (The set of points infinitesimally close to a standard point x is sometimes known as the *monad* of x, although we will not use this terminology.)

Exercise 8.1.2. Let (X, d) be a (standard) compact metric space. Show that for any nonstandard point $x \in {}^*X$ there exists a unique standard point $\mathrm{st}(x) \in {}^*X$ which is infinitesimally close to x, with

$$d(\mathrm{st}(x), \mathrm{st}(y)) = \mathrm{st}({}^*d(x, y)).$$

Exercise 8.1.3 (Automatic completeness). Let (X, d) be a (standard) compact metric space, and let $E = \prod_{n \to \alpha} E_n$ be a nonstandard subset of ${}^*X$ (i.e., an ultraproduct of standard subsets E_n of X), and let $\mathrm{st} : {}^*X \to X$ be the standard part function as defined in the previous exercise. Show that $\mathrm{st}(E)$ is complete (and thus compact, by the Heine-Borel theorem). (*Hint:* Take advantage of the countable saturation property from Chapter 7.)

Exercise 8.1.4. Let (X, d) be a compact metric space, and let E_n for $n \in \mathbf{N}$ be a sequence of nonempty compact subsets of X. Write $E := \prod_{n \to \alpha} E_n$ for the ultraproduct of the E_n, and let $\mathrm{st} : {}^*X \to X$ be the standard part function as defined in the previous exercises. Show that $\mathrm{st}(E)$ is a nonempty compact subset of X, and that $\lim_{n \to \alpha} d_H(E_n, \mathrm{st}(E)) = o(1)$. Use this to give an alternate proof of Lemma 8.1.2.

One can extend some of the above theory from compact metric spaces to locally compact metric spaces.

Exercise 8.1.5. Let (X, d) be a (standard) locally compact metric space. Let $X \subset O(X) \subset {}^*X$ denote the set of ultralimits $\lim_{n \to \alpha} x_n$ of *precompact* sequences x_n in X.

 (i) Show that $O(X) = {}^*X$ if and only if X is compact.
 (ii) Show that there is a unique function $\mathrm{st} : O(X) \to X$ such that x is infinitesimally close to $\mathrm{st}(x)$ for all $x \in O(X)$.
 (iii) If E is a nonstandard subset of ${}^*X$, show that $\mathrm{st}(E \cap O(X))$ is complete.
 (iv) Show that if K is a compact set and $\varepsilon > 0$ is a (standard) real number, then $K \cap E \subset N_\varepsilon(E_n)$ and $K \cap E_n \subset N_\varepsilon(E)$ for all n sufficiently close to α.

Now we consider limits of metric spaces (X_n, d_n) that are not necessarily all embedded in a single metric space. We begin by considering the case of bounded metric spaces.

Definition 8.1.3 (Gromov-Hausdorff distance). The *Gromov-Hausdorff distance* $d_{\mathrm{GH}}(X, Y)$ between two bounded metric spaces (X, d_X), (Y, d_Y) is the infimum of the Hausdorff distance $d_H(\iota_{X \to Z}(X), \iota_{Y \to Z}(Y))$, ranging over all isometric embeddings $\iota_{X \to Z} : X \to Z$, $\iota_{Y \to Z} : Y \to Z$ from X, Y respectively to Z.

Exercise 8.1.6.

(i) Show that Gromov-Hausdorff distance is a pseudometric (i.e., symmetric, nonnegative, and obeys the triangle inequality). (*Hint:* To show that $d_{\mathrm{GH}}(X, Z) \leq d_{\mathrm{GH}}(X, Y) + d_{\mathrm{GH}}(Y, Z)$, build a metric on the disjoint union $X \uplus Z$ which intuitively captures the idea of the shortest path from X to Z via Y.)

(ii) If E is a dense subset of a bounded metric space X, show that $d_{\mathrm{GH}}(E, X) = 0$. In particular, any bounded metric space E is at a zero Gromov-Hausdorff distance from its metric completion $\overline{E}$.

(iii) Show that if X, Y are compact metric spaces, then $d_{\mathrm{GH}}(X, Y) = 0$ if and only if X and Y are isometric. (*Hint:* If $d_{\mathrm{GH}}(X, Y) = 0$, find a sequence of "approximate isometries" from X to Y and from Y to X which almost invert each other. Then adapt the Arzelá-Ascoli theorem to take a limit (or one can use ultralimits and standard parts).)

We say that a sequence $X_n = (X_n, d_n)$ of bounded metric spaces *converges in the Gromov-Hausdorff sense* to another bounded metric space X_∞ if one has $d_{\mathrm{GH}}(X_n, X_\infty) \to 0$ as $n \to \infty$. This is a more general notion of convergence than Hausdorff convergence, and encompasses examples (iii) and (iv) at the beginning of this section.

Exercise 8.1.7. Show that every compact metric space is the limit (in the Gromov-Hausdorff sense) of finite metric spaces.

Now we turn to an important compactness result on Gromov-Hausdorff convergence. We say that a sequence $X_n = (X_n, d_n)$ of metric spaces is *uniformly totally bounded* if the diameters of the X_n are bounded, and, for every $\varepsilon > 0$, there exists C_ε such that each X_n can be covered by at most C_ε balls of radius ε in the d_n metric. (Of course, this implies that each X_n is individually *totally bounded*.)

Proposition 8.1.4 (Uniformly total bounded spaces are Gromov-Hausdorff precompact). *Let X_n be a sequence of uniformly totally bounded metric spaces. Then there exists a subsequence X_{n_j} which converges in the Gromov-Hausdorff space to a compact limit X_∞.*

We will prove this proposition using ultrafilters. Let $X_n = (X_n, d_n)$ be a sequence of uniformly totally bounded metric spaces, which we will assume to be standard (by defining the standard universe in a suitable fashion). Then we can form the ultraproduct $X := \prod_{n \to \alpha} X_n$ and the ultralimit $d := \lim_{n \to \alpha} d_n$, thus $d : X \times X \to {}^*\mathbf{R}^+$ is a nonstandard metric (obeying the nonstandard symmetry, triangle inequality, and positivity properties). As the X_n are uniformly bounded, d has bounded range. If we then take the standard part $\operatorname{st}(d) : X \times X \to \mathbf{R}^+$ of d, then $\operatorname{st}(d)$ is a pseudometric on X (i.e., it obeys all the axioms of a metric except possibly for positivity.) We can then construct a quotient metric space $X_\infty := X/\sim$ in the usual manner, by declaring two points x, y in X to be equivalent, $x \sim y$, if $\operatorname{st}(d)(x, y) = 0$ (or equivalently if $d(x, y) = o(1)$). The pseudometric $\operatorname{st}(d)$ then descends to a genuine metric $d_\infty : X_\infty \times X_\infty \to \mathbf{R}^+$ on X_∞. The space (X_∞, d_∞) is known as a *metric ultralimit* (or *ultralimit* for short) of the (X_n, d_n) (note that this is a slightly different usage of the term "ultralimit" from what we have been using previously).

One can easily establish compactness:

Exercise 8.1.8 (Compactness).

(i) Show that (X_∞, d_∞) is totally bounded. (*Hint:* Use the uniform total boundedness of the X_n together with Łos's theorem.)

(ii) Show that (X_∞, d_∞) is complete. (*Hint:* Use countable saturation).

In particular, by the Heine-Borel theorem, X_∞ is compact.

Now we establish Gromov-Hausdorff convergence, in the sense that for every standard $\varepsilon > 0$, one has $d_{\mathrm{GH}}(X_n, X_\infty) < \varepsilon$ for all n sufficiently close to α. From this it is easy to extract a subsequence X_{n_j} that converges in the Gromov-Hausdorff sense to X_∞ as required.

Fix a standard $\varepsilon > 0$. As X_∞ is totally bounded, we can cover it by at most M balls $B_\infty(x_{1,\infty}, \varepsilon/10), \dots, B_\infty(x_{M,\infty}, \varepsilon/10)$ of radius $\varepsilon/10$ (say) in the metric d_∞ for some standard natural number M. Lifting back to the ultraproduct X, we conclude that we may cover X by M balls $B(x_1, 2\varepsilon/10), \dots, B(x_M, 2\varepsilon/10)$ in the nonstandard metric d.

Note that $d_\infty(x_{i,\infty}, x_{j,\infty})$ is the standard part of $d(x_i, x_j)$ and, in particular,

$$d_\infty(x_{i,\infty}, x_{j,\infty}) - \varepsilon/10 < d(x_i, x_j) < d_\infty(x_{i,\infty}, x_{j,\infty}) + \varepsilon/10.$$

Each ball centre x_i is an ultralimit $x_i = \lim_{n \to \alpha} x_{i,n}$ of points $x_{i,n}$ in X_n. By Łos's theorem, we conclude that for n sufficiently close to α, X_n is covered by the M balls $B_n(x_{1,n}, 2\varepsilon/10), \dots, B_n(x_{M,n}, 2\varepsilon/10)$ in the metric d_n, and

$$d_\infty(x_{i,\infty}, x_{j,\infty}) - \varepsilon/10 < d_n(x_{i,n}, x_{j,n}) < d_\infty(x_{i,\infty}, x_{j,\infty}) + \varepsilon/10$$

for all $1 \le i, j \le M$.

Because of this, we can embed both X_n and X_∞ in the disjoint union $X_n \uplus X_\infty$, with a metric $d^{(n)}$ extending the metrics on X_n and X_∞, with the cross distances $d^{(n)}(x, y)$ between points $x \in X_n$ and $y \in X_\infty$ defined by the formula

$$d^{(n)}(x, y) = d^{(n)}(y, x) := \inf\{d_\infty(x_{i,\infty}, y) + d_\infty(x_{i,n}, x) + \varepsilon/2 : 1 \le i \le M\};$$

informally, this connects each ball center $x_{i,\infty}$ to its counterpart $x_{i,n}$ by a path of length $\varepsilon/2$. It is a routine matter to verify that $d^{(n)}$ is indeed a metric, and that X_n and X_∞ are separated by a Hausdorff distance less than ϵ in $X_n \uplus X_\infty$, and the claim follows.

Exercise 8.1.9. Prove Proposition 8.1.4 without using ultrafilters.

Exercise 8.1.10 (Completeness). Let X_n be a sequence of bounded metric spaces which is Cauchy in the Gromov-Hausdorff sense (i.e., $d_{\mathrm{GH}}(X_n, X_m) \to 0$ as $n, m \to \infty$). Show that X_n converges in the Gromov-Hausdorff sense to some limit X.

Now we generalise Gromov-Hausdorff convergence to the setting of unbounded metric spaces. To motivate the definition, let us first give an equivalent form of Gromov-Hausdorff convergence:

Exercise 8.1.11. Let (X_n, d_n) be a sequence of bounded metric spaces, and let (X_∞, d_∞) be another bounded metric space. Show that the following are equivalent:

(i) (X_n, d_n) converges in the Gromov-Hausdorff sense to (X_∞, d_∞).

(ii) There exist maps $\phi_n : X_\infty \to X_n$ which are asymptotically isometric isomorphisms in the sense that $\sup_{x,y \in X} |d_n(\phi_n(x), \phi_n(y)) - d(x, y)| \to 0$ and $\sup_{x_n \in X_n} \mathrm{dist}_n(x_n, \phi_n(X_\infty)) \to 0$ as $n \to \infty$.

Define a *pointed metric space* to be a triplet (X, d, p), where (X, d) is a metric space and p is a point in X.

Definition 8.1.5 (Pointed Gromov-Hausdorff convergence). A sequence of pointed metric spaces (X_n, d_n, p_n) is said to *converge in the pointed Gromov-Hausdorff sense* to another pointed metric space $(X_\infty, d_\infty, p_\infty)$ if there exists a sequence of maps $\phi_n : X_\infty \to X_n$ such that

(1) $d_n(\phi_n(p_\infty), p_n) \to 0$ as $n \to \infty$;

(2) (Asymptotic isometry) For each $R > 0$, one has

$$\sup_{x,y \in B_\infty(p_\infty, R)} |d_n(\phi_n(x), \phi_n(y)) - d_\infty(x, y)| \to 0$$

as $n \to \infty$;

(3) (Asymptotic surjectivity) For each $R' > R > 0$, one has

$$\sup_{x \in B_n(p_n, R)} \mathrm{dist}_n(x, \phi_n(B_\infty(p_\infty, R'))) \to 0$$

as $n \to \infty$.

Exercise 8.1.12. Verify that all the examples (i)–(v) given at the start of the section are examples of pointed Gromov-Hausdorff convergence, once one selects a suitable point in each space.

The above definition is by no means the only definition of pointed Gromov-Hausdorff convergence. Here is another (which is basically the original definition of Gromov [**Gr1981**]):

Exercise 8.1.13. Let (X_n, d_n, p_n) be a sequence of pointed metric spaces, and let $(X_\infty, d_\infty, p_\infty)$ be another pointed metric space. Show that the following are equivalent:

(1) (X_n, d_n, p_n) converges in the pointed Gromov-Hausdorff sense to $(X_\infty, d_\infty, p_\infty)$.

(2) There exist metrics $\tilde{d}_n$ on the disjoint union $X_n \cup X_\infty$ extending the metrics d_n, d_∞ such that for some sequence $\varepsilon_n \to 0$, one has $\tilde{d}_n(p_n, p_\infty) \leq \varepsilon_n$, $B_n(p_n, 1/\varepsilon_n) \subset N_{\varepsilon_n}(X_\infty)$, and $B_\infty(p_\infty, 1/\varepsilon_n) \subset N_{\varepsilon_n}(X_n)$.

Using this equivalence, construct a pseudometric between pointed metric spaces that describes pointwise Gromov-Hausdorff convergence.

Exercise 8.1.14. Let (X_n, d_n, p_n) be a sequence of pointed metric spaces which converge in the pointed Gromov-Hausdorff sense to a limit $(X_\infty, d_\infty, p_\infty)$, and also to another limit $(X'_\infty, d'_\infty, p'_\infty)$. Suppose that these two limit spaces are *proper*, which means that the closed balls $\overline{B_\infty(p_\infty, R)}$ and $\overline{B'_\infty(p'_\infty, R)}$ are compact. Show that the two limit spaces are pointedly isometric, thus there is an *isometric isomorphism* $\phi : X_\infty \to X'_\infty$ that maps p_∞ to p'_∞.

In the compact case, Gromov-Hausdorff convergence and pointwise Gromov-Hausdorff convergence are almost equivalent:

Exercise 8.1.15. Let (X_n, d_n, p_n) be a sequence of bounded pointed metric spaces, and let $(X_\infty, d_\infty, p_\infty)$ be a compact pointed metric space.

(i) Show that if (X_n, d_n, p_n) converges in the pointed Gromov-Hausdorff sense to $(X_\infty, d_\infty, p_\infty)$, then (X_n, d_n) converges in the Gromov-Hausdorff sense to (X_∞, d_∞).

(ii) Conversely, if (X_n, d_n) converges in the Gromov-Hausdorff sense to (X_∞, d_∞), show that some subsequence $(X_{n_j}, d_{n_j}, p_{n_j})$ converges in the pointed Gromov-Hausdorff sense to (X_∞, d_∞, q) for some $q \in X_\infty$.

There is a compactness theorem for pointed Gromov-Hausdorff convergence analogous to that for ordinary Gromov-Hausdorff convergence (Proposition 8.1.4):

Proposition 8.1.6 (Locally uniformly total bounded spaces are pointed Gromov-Hausdorff precompact). *Let $X_n = (X_n, d_n, p_n)$ be a sequence of* *pointed metric spaces such that the balls $B_n(p_n, R)$ are uniformly totally bounded in n for each fixed $R > 0$. Then there exists a subsequence X_{n_j} which converges in the pointed Gromov-Hausdorff space to a limit $X_\infty = (X_\infty, d_\infty, p_\infty)$ which is proper (i.e., every closed ball is compact).*

We sketch a proof of this proposition in the exercise below.

Exercise 8.1.16. Let X_n be as in the above proposition; we assume the X_n to be standard. Let $X := \prod_{n \to \alpha} X_n$ be the ultraproduct of the X_n, and define the ultralimits $d := \prod_{n \to \alpha} d_n$ and $p := \prod_{n \to \alpha}$. Let $O(X) := \{x \in X : d(x, p) = O(1)\}$ be the points in X that are a bounded distance from p.

(i) Show that $\mathrm{st}(d) : O(X) \times O(X) \to \mathbf{R}^+$ is a pseudometric on $O(X)$.

(ii) Show that the associated quotient space $X_\infty := O(X)/\sim$ with the quotient metric d_∞ is a proper metric space.

(iii) Show that some subsequence of the (X_n, p_n) converge in the pointed Gromov-Hausdorff sense to (X_∞, p_∞), where p_∞ is the image of p under the quotient map, thus establishing Proposition 8.1.6.

Exercise 8.1.17. Establish Proposition 8.1.6 without using ultrafilters.

Exercise 8.1.18. Let G be a locally compact group with a Gleason metric d. Show that the pointed metric spaces $(G, nd, 1)$ converge in the pointed Gromov-Hausdorff sense as $n \to \infty$ to the vector space $L(G)$ with the metric $d_{L(G)}(x, y) := \|x - y\|$ and distinguished point 0, where the norm $\|\|$ on $L(G)$ was defined in Exercise 3.3.1.

Remark 8.1.7. The above exercise suggests that one could attack Hilbert's fifth problem by somehow "blowing up" the locally compact group G around the origin and extracting a Gromov-Hausdorff limit using ultraproducts. This approach can be formalised using the language of nonstandard analysis, and in particular can be used to describe Hirschfeld's nonstandard solution [**Hi1990**] to Hilbert's fifth problem, as well as the later work of Goldbring [**Go2010**]. However, we will not detail this approach extensively here (though, on some level, it contains much the same ingredients as the known "standard" solutions to that problem, such as the one given in previous sections).

8.2. Sanders-Croot-Sisask theory

We now prove Sanders' lemma (Lemma 8.0.8), which roughly speaking will be needed to establish an important "total boundedness" property of ultra approximate groups, which in turn is necessary to ensure local compactness for models of such groups.

Let A be a K-approximate group for some $K > 1$, and let m be a (large) natural number. Our task is to locate a large set S such that the iterated power S^m is contained inside A^4. Sanders' strategy for doing this is to pick a set S that nearly stabilises a set B that is comparable in some sense to A. More precisely, suppose we have nonempty finite sets S and B with the property that

$$|B \backslash sB| < \frac{1}{m}|B|$$

for all $s \in S$. Then an easy induction shows that

$$|B \backslash gB| < \frac{k}{m}|B|$$

for all $g \in S^k$ and all $k \geq 1$. In particular, B and gB are not disjoint whenever $g \in S^m$, which means that $S^m \subset BB^{-1}$. Thus, to prove Sanders' lemma, it suffices to find a nonempty set $B \subset A^2$ with the property that the set

$$(8.1) \qquad S := \left\{ s \in G : |B \backslash sB| < \frac{1}{m}|B| \right\}$$

has cardinality $\gg_{K,m} |A|$. (Note that this set S will automatically be symmetric and contain the origin.)

Remark 8.2.1. The set (8.1) is known as a *symmetry set* of B and is sometimes denoted $\mathrm{Sym}_{1/m}(B)$. One can also interpret this set as a ball, using the mapping $g \mapsto \frac{1}{|B|} 1_{gB}$ of G into $\ell^1(G)$ to pull back the $\ell^1(G)$ metric to G, in which case S is the ball of radius $\frac{1}{m}$ centred at the origin.

It remains to find a set B for which the set (8.1) is large. To motivate how we would do this, let us naively try setting $B := A^2$. If the set (8.1) associated to this set is large, we will be done, so let us informally consider the opposite case in which S is extremely small; in particular, we have

$$|A^2 \backslash gA^2| \geq \frac{1}{m}|A^2|$$

for "most" choices of $g \in G$. In particular,

$$|A^2 \cap gA^2| \leq (1 - \frac{1}{m})|A^2|.$$

Now observe that $(A \cap gA)A$ is contained in $A^2 \cap gA^2$, and so

$$|(A \cap gA)A| \leq (1 - \frac{1}{m})|A^2|.$$

We have thus achieved a dichotomy: either the choice $B := A^2$ "works", or else the set $B' := A'A$ is significantly smaller than $B = A^2$ for some $A' = A \cap gA$. We can then try to repeat this dichotomy, to show that the choice $B' = A'A$ either "works", or else the set $B'' := A''A$ is significantly smaller than $B' = A'A$ for some $A'' = A' \cap g'A'$. We can keep iterating this dichotomy, creating ever smaller sets of the form $B_n = A_nA$; but on the other hand, these sets should be at least as large as $|A|$ (provided that we choose g, g', etc., to prevent the sets A', A'', etc., from becoming completely empty). So at some point this iteration has to terminate, at which point we should get a set that "works".

The original paper of Sanders [**Sa2009**] contains a formalisation of the above argument. We present a slightly different arrangement of the argument below, which is focused on making sure that sets such as A' do not get too small.

For any $0 < t \leq 1$, define the quantity

$$f(t) := \inf \left\{ \frac{|A'A|}{|A|} : A' \subset A; |A'| \geq t|A| \right\}.$$

Then f is a nondecreasing function that takes values between 1 and K. By the pigeonhole principle, we can find $t \gg_{K,m} 1$ such that

$$(8.2) \qquad\qquad f\left(\frac{t^2}{2K} \right) > (1 - \frac{1}{m})f(t).$$

(The reasons for these particular choices of parameters will become clearer shortly.) Fix this t, and let A' attain the infimum for $f(t)$, thus $A' \subset A$, $|A'| \geq t|A|$, and

$$|A'A| = f(t)|A|.$$

(Such an A' exists since the infimum is only being taken over a finite set.) Set $B := A'A$, and let S be the set in (8.1). Observe that if $g \notin S$, then

$$|A'A \backslash gA'A| \geq \frac{1}{m}|A'A|,$$

and so by arguing as before we see that

$$|(A' \cap gA')A| \leq |A'A| - \frac{1}{m}|A'A|.$$

Since $|A^2| \geq |A|$ and $|A'A| = f(t)|A|$, we thus have

$$|(A' \cap gA')A| \leq (1 - \frac{1}{m})f(t)|A|.$$

In particular, from (8.2) and the definition of f, we conclude that

$$|A' \cap gA'| < \frac{t^2}{2K}|A|$$

for all $g \notin S$. In other words,

$$(8.3) \qquad\qquad S \supset \left\{ g \in G : |A' \cap gA'| \geq \frac{t^2}{2K}|A| \right\}.$$

So to finish the proof of Sanders' lemma, it will suffice to obtain a lower bound on the right-hand side of (8.3). But this can be done by a standard Cauchy-Schwarz argument: Starting with the identity

$$\sum_{x \in A^2} \sum_{a \in A} 1_{aA'}(x) = |A||A'| \geq t|A|^2$$

and using the Cauchy-Schwarz inequality and the bound $|A^2| \leq K|A|$, we conclude that

$$\sum_{x \in A^2} \sum_{a,b \in A} 1_{aA'}(x) 1_{bA'}(x) \geq \frac{t^2}{K} |A|^3$$

and thus

$$\sum_{a,b \in A} |aA' \cap bA'| \geq \frac{t^2}{K} |A|^3.$$

By the pigeonhole principle, we may thus find $a \in A$ such that

$$\sum_{b \in A} |aA' \cap bA'| \geq \frac{t^2}{K} |A|^2,$$

and thus (setting $g := a^{-1}b$)

$$\sum_{g \in a^{-1}A} |A' \cap gA'| \geq \frac{t^2}{K} |A|^2.$$

Since $a^{-1}A$ has cardinality $|A|$, this forces $|A' \cap gA'|$ to exceed $\frac{t^2}{2K}|A|$ for at least $\frac{t^2}{2K}|A|$ values of g, and the claim follows.

Remark 8.2.2. The lower bound on S obtained by this argument is of the shape $|S| \geq \exp(-K^{O(m)})|A|$. This is not optimal; see Remark 8.2.4 below.

We now present an alternate approach to Sanders' lemma, using the Croot-Sisask theory [**CrSi2010**] of *almost periods*. Whereas in the Sanders argument, one selected the set S to be the set of group elements that almost stabilised a set B, we now select S to be the set of group elements that almost stabilise (or are *almost periods* of) the convolution

$$1_A * 1_A(x) := \sum_{y \in G} 1_A(y) 1_A(y^{-1}x).$$

It is convenient to work in the $\ell^2(G)$ metric. Since the function $1_A * 1_A$ has an ℓ^1 norm of $|A|^2$ and is supported in A^2, which has cardinality at most $K|A|$, we see from Cauchy-Schwarz that

$$\|1_A * 1_A\|_{\ell^2(G)} \geq |A|^{3/2}/K^{1/2}.$$

We now set

$$S := \left\{ g \in G : \|1_A * 1_A - \tau(g)1_A * 1_A\|_{\ell^2(G)} < \frac{1}{m} \frac{|A|^{3/2}}{K^{1/2}} \right\}.$$

Exercise 8.2.1. Show that S is symmetric, contains the origin, and that $S^m \subset A^4$.

To finish the proof of Sanders' lemma, it suffices to show that there are lots of almost periods of $1_A * 1_A$ in the sense that $|S| \gg_{K,m} |A|$.

The key observation here is that the translates $\tau(g)1_A * 1_A$ of $1_A * 1_A$, as g varies in A, range in a "totally bounded" set. This in turn comes from a "compactness" property of the convolution operator $f \mapsto f * 1_A$ when f is supported on A^2. Croot and Sisask establish this by using a random sampling argument to approximate this convolution operator by a bounded rank operator. More precisely, let $M \geq 1$ be an integer parameter to be chosen later, and select M sample points $y_1, \ldots, y_M$ from A^2 independently and uniformly at random (allowing repetitions). We will approximate the operator

$$Tf := f * 1_A = \sum_{y \in A^2} f(y)1_{yA}$$

for f supported on A^2 by the operator

$$Sf := \frac{|A^2|}{M} \sum_{i=1}^{M} f(y_i)1_{y_i A}.$$

The point is that as M gets larger, Sf becomes an increasingly good approximation to Tf:

Exercise 8.2.2. Let f be a function supported on A^2 that is bounded in magnitude by 1. Show that

$$\mathbf{E}Sf = Tf$$

and

$$\mathbf{E}\|Sf - Tf\|_{\ell^2(G)}^2 \ll_K |A|^3/M.$$

In particular, with probability at least $1/2$, one has

$$\|Sf - Tf\|_{\ell^2(G)} \ll_K |A|^{3/2}M^{-1/2}.$$

Thus, for any $g \in A$, we have

$$\|S1_{gA} - T1_{gA}\|_{\ell^2(G)} \ll_K |A|^{3/2}M^{-1/2}$$

with probability at least $1/2$. By the pigeonhole principle (or the first moment method), we thus conclude that there exists a choice of sample points $y_1, \ldots, y_M$ for which

$$(8.4) \qquad \|S1_{gA} - T1_{gA}\|_{\ell^2(G)} \ll_K |A|^{3/2} M^{-1/2}$$

for at least $|A|/2$ choices of $g \in A$.

Let A' denote the set of all $g \in A$ indicated above. For each $g \in A'$, the function $S1_{gA}$ is a linear combination of the M functions $\frac{|A^2|}{M} 1_{y_i A}$ with coefficients between 0 and 1. Each of these functions $\frac{|A^2|}{M} 1_{y_i A}$ has an $\ell^2(G)$ norm of $O_K(|A|^{3/2})$. Thus, by the pigeonhole principle, one can find a subset A'' of A' of size $|A''| \gg_{K,m,M} |A|$ such that the functions $S1_{gA}$ for $g \in A''$ all lie within $\frac{1}{2m} \frac{|A|^{3/2}}{K^{1/2}}$ in $\ell^2(G)$ norm of each other. If M is large enough depending on K, m, we then conclude from (8.4) and the triangle inequality that the functions $T1_{gA}$ for $g \in A''$ all lie within $\frac{1}{m} \frac{|A|^{3/2}}{K^{1/2}}$ in $\ell^2(G)$ norm of each other. Translating by some fixed element of A'', we obtain the claim.

Remark 8.2.3. For future reference, we observe that the above argument did not need the full strength of the hypothesis that A was a K-approximate group; it would have sufficed for A to be finite and nonempty with $|A^2| \leq K|A|$.

Remark 8.2.4. The above version of the Croot-Sisask argument gives a lower bound on S of the shape $|S| \geq \exp(-O(m^2 K \log K))|A|$. As was observed in a subsequent paper of Sanders [**Sa2010**], by optimising the argument (in particular, replacing ℓ^2 with ℓ^p for a large value of p, and replacing $1_A * 1_A$ by $1_A * 1_{A^2}$), one can improve this to $|S| \geq \exp(-O(m^2 \log^2 K))|A|$.

Remark 8.2.5. It is also possible to replace the random sampling argument above by a *singular value decomposition* of T (restricted to something like A^2) to split it as the sum of a bounded rank component and a small operator norm component, after computing the Frobenius norm of T in order to limit the number of large singular values. (In the abelian setting, this corresponds to a Fourier decomposition into large and small Fourier coefficients, which is a fundamental tool in additive combinatorics.) We will, however, not pursue this approach here.

As mentioned previously, the Sanders lemma will be useful in building topological group structure on ultra approximate groups A (and the group $\langle A \rangle$ that they generate). The connection can be seen from the simple observation that if U is a neighbourhood of the identity in a topological group and $m \geq 1$, then there exists another neighbourhood S of the identity such that $S^m \subset U$. However, in addition to this multiplicative structure, we will also need to impose some conjugacy structure as well. (Strangely, even though the conjugation operation $(g,h) \mapsto ghg^{-1}$ can be defined in terms of the more "primitive" operations of multiplication $(g,h) \mapsto gh$ and inversion $g \mapsto g^{-1}$, it almost seems to be an independent group operation in some ways, at least for the purposes of studying approximate groups, and the most powerful results tend to come from combining multiplicative structure and conjugation structure together. I do not know a fundamental reason for this "product-conjugation phenomenon" but it does seem to come up a lot in this subject.) Given two nonempty subsets A, B of a group G, define

$$A^B := \{a^b : a \in A, b \in B\}$$

where $a^b := b^{-1}ab$ is the conjugate of a by b. (Thus, for instance, $H^B = H$ whenever H is a subgroup normalised by B.)

We can now establish a stronger version of the Sanders lemma:

Lemma 8.2.6 (Normal Sanders lemma). *Let A be a finite K-approximate group in a (global) group G, let $m \geq 1$, and let B be a symmetric subset of A with $|B| \geq \delta|A|$ for some $\delta > 0$. Then there exists a symmetric subset N of A^4 with $|N| \gg_{K,m,\delta} |A|$ containing the identity such that $(N^{A^m})^m \subset B^4$.*

Roughly speaking, one can think of the set S in Lemma 8.0.8 as a "finite index subgroup" of A^4, while the set in Lemma 8.2.6 is a "finite index normal subgroup" of A^4. To get from the former to the latter, we will mimic the proof of the following well-known fact:

Lemma 8.2.7. *Suppose that H is a finite index subgroup of a group G. Then there is a finite index normal subgroup N of G that is contained in H.*

Proof. Consider all the conjugates H^g of the finite index subgroup H as g ranges over G. As all the elements g from the same right coset Hk give the same conjugate H^g, and H is finite index, we see that there are only finitely many such conjugates. Since the intersection of two finite index subgroups is again a finite index subgroup (why?), the intersection $N := \bigcap_{g \in G} H^g$ is thus a finite index subgroup of G. But we have $N = N^g$ for all $g \in G$, and so N is normal as desired. $\square$

Remark 8.2.8. An inspection of the argument reveals that if H had index k in G, then the normal subgroup N has index at most k^k. One can do slightly better than this by looking at the left action of G on the k-element quotient space G/H, which one can think of as a homomorphism from G to the symmetric group S_k of k elements. The kernel N of this homomorphism is then clearly a normal subgroup of G of index at most $|S_k| = k!$ that is contained in H. However, this slightly more efficient argument seems to be more "fragile" than the more robust proof given above, in that it is not obvious (to me, at least) how to adapt it to the approximate group setting of the Sanders lemma. (This illustrates the advantages of knowing multiple proofs for various basic facts in mathematics.)

Inspired by the above argument, we now prove Lemma 8.2.6. The main difficulty is to find an approximate version of the claim that the intersection of two finite index subgroups is again a finite index subgroup. This is provided by the following lemma:

Lemma 8.2.9. *Let A be a K-approximate group, and let $A_1, A_2 \subset A$ be such that $|A_1| \geq \delta_1 |A|$ and $|A_2| \geq \delta_2 |A|$. Then there exists a subset B of A with $BB^{-1} \subset A_1 A_1^{-1} \cap A_2 A_2^{-1}$ and $|B| \geq \delta_1 \delta_2 |A|/K$.*

Proof. Since $A_1^{-1} A_2 \subseteq A^2$, we have $|A_1^{-1} A_2| \leq K|A|$. It follows that there is some x with at least $\delta_1 \delta_2 |A|/K$ representations as $a_1^{-1} a_2$. Let B be the set of all values of a_2 that appear. Obviously $BB^{-1} \subseteq A_2 A_2^{-1}$. Suppose that $a_2, a_2' \in B$. Then there are a_1, a_1' such that $x = a_1^{-1} a_2 = (a_1')^{-1} a_2'$, and so $a_1' a_1^{-1} = a_2' a_2^{-1}$. Thus BB^{-1} lies in $A_1 A_1^{-1}$ as well. $\qquad\square$

Now we prove Lemma 8.2.6. By Lemma 8.0.8, we may find a symmetric set S of B^4 containing the identity such that S^{8m} (say) is contained in B^4 and $|S| \gg_{K,m,\delta} |A|$. We need the following simple covering lemma:

Exercise 8.2.3 (Ruzsa covering lemma). Let A, B be finite nonempty subsets of a group G such that $|AB| \leq K|B|$. Show that A can be covered by at most K left-translates aBB^{-1} of BB^{-1} for various $a \in A$. (*Hint:* Find a maximal disjoint family of aB with $a \in A$.)

From the above lemma we see that A can be covered by $O_{K,m}(1)$ left-translates of S^2. As A^m can be covered by $O_{K,m}(1)$ left-translates of A, we conclude that A^m can be covered by $O_{K,m}(1)$ left-translates of S^2. Since $S^2 \subset A^8$, we conclude that $A^m \subset \bigcap_{j=1}^{J} a_j S^2$ for some $J = O_{K,m}(1)$ and $a_1, \ldots, a_J \in A^{m+8}$.

The conjugates $a_j^{-1} S a_j^{-1}$ all lie in A^{2m+20}. By many applications of Lemma 8.2.9, we see that we may find a subset D of A^{2m+4} with $|D| \gg_{K,m} |A|$ such that
$$DD^{-1} \subset a_j S^4 a_j^{-1}$$
for all $j = 1, \ldots, J$. In particular, if $N := DD^{-1}$, then $N^{a_j} \subset S^4$ for all $j = 1, \ldots, J$, and thus $N^{A^m} \subset S^8$. Thus $(N^{A^m})^m \subset B^4$, and the claim follows.

Remark 8.2.10. The quantitative bounds on $|N|$ given by this argument are quite poor, being of triple exponential type (!) in K. It is likely that this can be improved by a more efficient argument.

8.3. Locally compact models of ultra approximate groups

We now use the normal Sanders lemma to place a topology on ultra approximate groups. We first build a "neighbourhood base" associated to an ultra approximate group:

Exercise 8.3.1. Let A be an ultra approximate group. Show that there exist a sequence of ultra approximate groups

$$A^4 = A_0 \supset A_1 \supset A_2 \supset \cdots \supset \ldots$$

such that

$$(A_m^{A^m})^2 \subset A_{m-1}$$

for all $m \geq 1$, and such that A^4 can be covered by finitely many left-translates of A_m for each m. (*Hint:* You will need Lemma 8.2.6, Lemma 8.2.9, Exercise 8.2.3, and some sort of recursive construction.)

Remark 8.3.1. A simple example to keep in mind here is the nonstandard interval $A = [-N, N]$ for some unbounded nonstandard natural number N, with $A_m := [-4N/2^m, 4N/2^m]$.

This gives a good topology on the group $\langle A \rangle = \bigcup_{m=1}^{\infty} A^m$ generated by A:

Exercise 8.3.2. Let A be an ultra approximate group, and let $\langle A \rangle$ be the group generated by A. Let $A_0, A_1, \ldots$ be as in the preceding exercise. Given a subset E of $\langle A \rangle$, call a point g in E an *interior point* of E if one has $g A_m \subset E$ for some (standard) m. Call E *open* if every element of E is an interior point.

(i) Show that this defines a topology on E.

(ii) Show that this topology makes $\langle A \rangle$ into a topological group (i.e., the group operations are continuous).

(iii) Show that this topology is first countable (and thus pseudo-metrisable by the Birkhoff-Kakutani theorem, Theorem 5.1.1).

(iv) Show that one can build a left-invariant pseudometric $d : \langle A \rangle \times \langle A \rangle \to \mathbf{R}^+$ on $\langle A \rangle$ which generates the topology on $\langle A \rangle$, with the property that

$$B(1, c2^{-m}) \subset A_m \subset B(1, C2^{-m})$$

for all (standard) $m \geq 0$ and some (standard) $C > c > 0$. (*Hint:* Inspect the proof of the Birkhoff-Kakutani theorem.)

(v) Show that $\langle A \rangle$ with this pseudometric is complete (i.e., every Cauchy sequence is convergent). (*Hint:* Use the countable saturation property.)

(vi) Show that A^m is totally bounded for each standard m (i.e., covered by a finite number of ε-balls for each $\varepsilon > 0$).

(vii) Show that $\langle A \rangle$ is locally compact.

With this topology, the group $\langle A \rangle$ becomes a locally compact topological group. However, in general, this group will not be Hausdorff, because the identity 1 need not be closed. Indeed, it is easy to see that the closure of 1 is the set $\bigcap_{m=1}^{\infty} A_m$, which is not necessarily trivial. For instance, using the example from Remark 8.3.1, $\langle A \rangle$ is the group $\{n \in {}^*\mathbf{Z} : n = O(N)\}$, and the closure of the identity is $\{n \in {}^*\mathbf{Z} : n = o(N)\}$. However, we may quotient out by this closure of the identity to obtain a locally compact *Hausdorff* group L, which is now metrisable (with the metric induced from the pseudometric d on $\langle A \rangle$). Let $\pi : \langle A \rangle \to L$ be the quotient homomorphism. This homomorphism obeys a number of good properties, which we formalise as a definition:

Definition 8.3.2. Let A be an ultra approximate group. A (global) *good model* for A is a homomorphism $\pi : \langle A \rangle \to L$ from $\langle A \rangle$ to a locally compact Hausdorff group L that obeys the following axioms:

(1) (Thick image) There exists a neighbourhood U_0 of the identity in L such that $\pi^{-1}(U_0) \subset A$ and $U_0 \subset \pi(A)$. (In particular, the kernel of π lies in A.)

(2) (Compact image) $\pi(A)$ is precompact.

(3) (Approximation by nonstandard sets) Suppose that $F \subset U \subset U_0$, where F is compact and U is open. Then there exists a nonstandard finite set B (i.e., an ultraproduct of finite sets) such that $\pi^{-1}(F) \subset B \subset \pi^{-1}(U)$.

We will often abuse notation and refer to L as the good model, rather than π. (Actually, to be truly pedantic, it is the ordered pair (π, L) which is the good model, but we will not use this notation often.)

Remark 8.3.3. In the next section we will also need to consider *local* good models, which only model (say) A^8 rather than all of $\langle A \rangle$, and in which L is a local group rather than a global one. However, for simplicity we will not discuss local good models for the moment.

Remark 8.3.4. The thick and compact image axioms imply that the geometry of L in some sense corresponds to the geometry of A. In particular, if a is an element of $\langle A \rangle$, then a will be "small" (in the sense that it lies in A) if $\pi(a)$ is close to the identity, and "large" (in the sense that it lies outside A) if $\pi(a)$ is far away from the identity. Thus π is "faithful" in some "coarse" or "macroscopic" sense. The inclusion $U_0 \subset \pi(A)$ is a sort of "local surjectivity" condition, and ensures that L does not contain any "excess" or "redundant" components. The approximation by nonstandard set axiom is a technical "measurability" axiom, that ensures that the model of the ultra approximate group actually has something nontrivial to say about the finite approximate groups that were used to build that ultra approximate group (as opposed to being some artefact of the ultrafilter itself).

Example 8.3.5. We continue the example from Remark 8.3.1. A good model for $A = [-N, N]$ is provided by the homomorphism $\pi : \langle A \rangle \to \mathbf{R}$ given by the formula $\pi(x) := \mathrm{st}(x/N)$. The thick image and compact image properties are clear. To illustrate the approximation by nonstandard set property, take $F = [-r, r]$ and $U = (-s, s)$ for some (standard) real numbers $0 < r < s$. The preimages $\pi^{-1}(F) = \{n \in {}^*\mathbf{Z} : |n| \leq rN + o(N)\}$ and $\pi^{-1}(U) = \{n \in {}^*\mathbf{Z} : |n| < (s - \varepsilon)N$ for some standard $\varepsilon > 0\}$ are not nonstandard finite sets (why? use the least upper bound axiom), but one can find a nonstandard integer M such that $r < \mathrm{st}(M/N) < s$, and $[-M, M]$ will be a nonstandard finite set between F and U.

The following fundamental observation is essentially due to Hrushovski [**Hr2012**]:

Exercise 8.3.3 (Existence of good models). Let A be an ultra approximate group. Show that A^4 has a good model by a locally compact Hausdorff metrisable group, given by the construction discussed previously.

Exercise 8.3.4. The purpose of this exercise is to show why it is necessary to model A^4, rather than A, in Exercise 8.3.3. Let F_2 be the field of two elements. For each positive integer n, let A_n be a random subset of F_2^n formed by selecting one element uniformly at random from the set $x, x + e_n$ for each $x \in F_2^{n-1} \times \{0\}$, and also selecting the identity 0. Clearly, A_n is a 2-approximate group, since A_n is contained in $F_2^n = A_n + \{0, e_n\}$.

  (i) Show that almost surely, for all but finitely many n, A_n does not contain any set of the form $B + B$, with $|B| \geq 2^{0.9n}$. (*Hint:* Use the *Borel-Cantelli lemma.*)

  (ii) Show that almost surely, the ultraproduct $\prod_{n \to \alpha} A_n$ cannot be modeled by any locally compact group.

Let us now give some further examples of good models, beyond that given by Example 8.3.5.

Example 8.3.6 (Nonstandard finite groups). Suppose that A_n is a sequence of (standard) finite groups; then the ultraproduct $A := \prod_{n \to \alpha} A_n$ is an ultra approximate group. In this case, A is in fact a genuine group, thus $A = \langle A \rangle$. In this case, the trivial homomorphism $\pi : \langle A \rangle \to L = \{1\}$ to the trivial group $\{1\}$ is a good model of A. Conversely, it is easy to see that this is the only case in which $\{1\}$ is a good model for A.

Example 8.3.7 (Generalised arithmetic progression). We still work in the integers $\mathbf{Z}$, but now take A_n to be the rank two generalised arithmetic progression

$$A_n := P(1, n^{10}; n, n) := \{a + bn^{10} : a, b \in \{-n, \ldots, n\}\}.$$

Then the ultraproduct $A := \prod_{n \to \alpha} A_n$ is the subset of the nonstandard integers $^*\mathbf{Z}$ of the form

$$A = P(1, N^{10}; N, N) = \{a + bN^{10} : a, b \in \{-N, \ldots, N\}\},$$

where N is the unbounded natural number $N := \lim_{n \to \alpha} n$. This is an ultra approximate group, with

$$\langle A \rangle = \{a + bN^{10} : a, b \in {}^*\mathbf{Z}; a, b = O(N)\}.$$

Then $\langle A \rangle$ can be modeled by the Euclidean plane $\mathbf{R}^2$, using the model maps $\pi : \langle A \rangle \to \mathbf{R}^2$ defined for each standard m by the formula

$$\pi(a + bN^{10}) := \left(\text{st}\, \frac{a}{N}, \text{st}\, \frac{b}{N} \right)$$

whenever $a, b = O(N)$. The image $\pi(A^m)$ is then the square $[-m, m]^2$ for any standard m. Note here that while A lives in a "one-dimensional" group $^*\mathbf{Z}$, the model $\mathbf{R}^2$ is "two-dimensional". This is also reflected in the volume growth of the powers A_n^m of A_n for small m and large n, which grow quadratically rather than linearly in m. Informally, A is "modeled" by the unit square in $\mathbf{R}^2$.

Exercise 8.3.5. With the notation of Example 8.3.7, show that A cannot be modeled by the one-dimensional Lie group $\mathbf{R}$. (*Hint:* If A was modeled by $\mathbf{R}$, conclude that A^m could be covered by $O(m)$ translates of A for each standard m.)

Exercise 8.3.6 (Heisenberg box, I). We take each A_n to be the "nilbox"

$$A_n := \left\{ \begin{pmatrix} 1 & x_n & z_n \\ 0 & 1 & y_n \\ 0 & 0 & 1 \end{pmatrix} \in \begin{pmatrix} 1 & \mathbf{Z} & \mathbf{Z} \\ 0 & 1 & \mathbf{Z} \\ 0 & 0 & 1 \end{pmatrix} : |x_n|, |y_n| \leq n, |z_n| \leq n^2 \right\}.$$

Consider the ultraproduct $A := \prod_{n \to \alpha} A_n$; this is a subset of the nilpotent (nonstandard) group $\begin{pmatrix} 1 & {}^*\mathbf{Z} & {}^*\mathbf{Z} \\ 0 & 1 & {}^*\mathbf{Z} \\ 0 & 0 & 1 \end{pmatrix}$, consisting of all elements $\begin{pmatrix} 1 & x & z \\ 0 & 1 & y \\ 0 & 0 & 1 \end{pmatrix}$ with $|x|, |y| \leq N$ and $|z| \leq N^2$, where $N := \lim_{n \to \alpha} n$. Thus

$$\langle A \rangle = \begin{pmatrix} 1 & O(N) & O(N^2) \\ 0 & 1 & O(N) \\ 0 & 0 & 1 \end{pmatrix}.$$

Consider now the map

$$\pi : \langle A \rangle \to \begin{pmatrix} 1 & \mathbf{R} & \mathbf{R} \\ 0 & 1 & \mathbf{R} \\ 0 & 0 & 1 \end{pmatrix}$$

defined by

(8.5)
$$\pi\left(\begin{pmatrix} 1 & x & z \\ 0 & 1 & y \\ 0 & 0 & 1 \end{pmatrix}\right) := \begin{pmatrix} 1 & \text{st } \frac{x}{N} & \text{st } \frac{z}{N^2} \\ 0 & 1 & \text{st } \frac{y}{N} \\ 0 & 0 & 1 \end{pmatrix}.$$

(i) Show that $A \cup A^{-1}$ is an ultra approximate group.

(ii) Show that π is a good model of $A \cup A^{-1}$.

Exercise 8.3.7 (Heisenberg box, II). This is a variant of the preceding exercise, in which the A_n is now defined as

(8.6)
$$A_n := \left\{ \begin{pmatrix} 1 & x_n & z_n \\ 0 & 1 & y_n \\ 0 & 0 & 1 \end{pmatrix} : |x_n|, |y_n| \leq n, |z_n| \leq n^{10} \right\}$$

so that the ultralimit $A := \prod_{n \to \alpha} A_n$ takes the form

$$A := \left\{ \begin{pmatrix} 1 & x & z \\ 0 & 1 & y \\ 0 & 0 & 1 \end{pmatrix} \in \begin{pmatrix} 1 & {}^*\mathbf{Z} & {}^*\mathbf{Z} \\ 0 & 1 & {}^*\mathbf{Z} \\ 0 & 0 & 1 \end{pmatrix} : |x|, |y| \leq N, |z| \leq N^{10} \right\}$$

and

$$\langle A \rangle := \begin{pmatrix} 1 & O(N) & O(N^{10}) \\ 0 & 1 & O(N) \\ 0 & 0 & 1 \end{pmatrix},$$

where $N := \lim_{n \to \alpha} n$. Now consider the map

$$\pi: \langle A \rangle^8 \to \mathbf{R}^3$$

defined by

$$\pi\left(\begin{pmatrix} 1 & x & z \\ 0 & 1 & y \\ 0 & 0 & 1 \end{pmatrix}\right) = \left(\text{st } \frac{x}{N}, \text{st } \frac{y}{N}, \text{st } \frac{z}{N^{10}}\right).$$

(i) Show that $A \cup A^{-1}$ is an ultra approximate group.

(ii) Show that π is a good model of $A \cup A^{-1}$.

(iii) Show that $A \cup A^{-1}$ cannot be modeled by $\mathbf{R}^2$, or by the Heisenberg group.

Remark 8.3.8. Note in the above exercise that the homomorphism π: $A^8 \to \mathbf{R}^3$ is not associated to any exact homomorphisms π_n from A_n^8 to $\mathbf{R}^3$. Instead, it is only associated to *approximate* homomorphisms

$$\pi_n\left(\begin{pmatrix} 1 & x_n & z_n \\ 0 & 1 & y_n \\ 0 & 0 & 1 \end{pmatrix}\right) := \left(\frac{x_n}{n}, \frac{y_n}{n}, \frac{z_n}{n^{10}}\right)$$

into $\mathbf{R}^3$. Such approximate homomorphisms are somewhat less pleasant to work with than genuine homomorphisms; one of the main reasons why we work in the ultraproduct setting is so that we can use genuine group homomorphisms.

We also note the sets A_n^m for small m and large n grow cubically in m in Exercise 8.3.6, and quartically in m in Exercise 8.3.7. This reflects the corresponding growth rates in $\mathbf{R}^3$ and in the Heisenberg group respectively.

Finally, we observe that the nonabelian structure of the ultra approximate group A is lost in the model group L, because the nonabelianness is "infinitesimal" at the scale of A. More generally, good models can capture the "macroscopic" structure of A, but do not directly see the "microscopic" structure.

The following exercise demonstrates that model groups need not be simply connected.

Exercise 8.3.8 (Models of Bohr sets). Let α be a standard irrational, let $0 < \varepsilon < 0.1$ be a standard real number, and let $A_n \subset \mathbf{Z}$ be the sets

$$A_n := \{m \in [-n, n] : \|\alpha m\|_{\mathbf{R}/\mathbf{Z}} \leq \varepsilon\}$$

where $\|x\|$ denotes the distance to the nearest integer. Set $A := \prod_{n \to \alpha} A_n$.

    (i) Show that A is an ultra approximate group.

    (ii) Show that $\mathbf{R}/\mathbf{Z} \times \mathbf{R}$ is a good model for A.

    (iii) Show that $\mathbf{R}^2$ is not a good model for A. (*Hint:* Consider the growth of A^m, as measured by the number of translates of A needed to cover this set.)

    (iv) Show that $\mathbf{R}$ is not a good model for A. (*Hint:* Consider the decay of the sets $\{g : g, g^2, \ldots, g^m \in A\}$, as measured by the number of translates of this set needed to cover A.)

Exercise 8.3.9 (Haar measure). Let $\pi : \langle A \rangle \to L$ be a good model for an ultra approximate group $A = \prod_{n \to \alpha} A_n$ by a locally compact group L. For any continuous compactly supported function $f : L \to \mathbf{R}$, we can define a functional $I(f)$ by the formula

$$I(f) = \inf \operatorname{st} \frac{\sum_{a \in A} F^+(a)}{|A|}$$

where $F^+ = \lim_{n \to \alpha} F_n^+$ is the ultralimit of functions $F_n^+ : A_n \to \mathbf{R}$, with the nonstandard real $\sum_{a \in A} F^+$ and nonstandard natural number $|A|$ defined in the usual fashion as

$$\sum_{a \in A} F^+(a) := \lim_{n \to \alpha} \sum_{a_n \in A_n} F_n^+(a_n)$$

and

$$|A| := \lim_{n \to \alpha} |A_n|,$$

and the infimum is over all F^+ for which $F^+(a) \geq f(\pi(a))$ for all $a \in A$.

    (i) Establish the equivalent formula

$$I(f) = \sup \operatorname{st} \frac{\sum_{a \in A} F^-(a)}{\sum_{a \in A} 1}$$

    where the supremum is over all F^- for which $F^-(a) \leq f(\pi(a))$ for all $a \in A$.

(ii) Show that there exists a bi-invariant Haar measure μ on G such that $I(f) = \int_G f \, d\mu$ for all continuous compactly supported $f: L \to \mathbf{R}$. (In particular, this shows that L is necessarily unimodular.)

(iii) Show that
$$\mu(F)|A| \leq |A'| \leq \mu(U)|A|$$
whenever $F \subseteq U \subseteq U_1$, F is compact, U is open, and A' is a nonstandard set with
$$\pi^{-1}(F) \subset A' \subseteq \pi^{-1}(U).$$

8.4. Lie models of ultra approximate groups

In the examples of good models in the previous section, the model group L was a Lie group. We give now give some examples to show that the model need not *initially* be of Lie type, but can then be replaced with a Lie model after some modification.

Example 8.4.1 (Nonstandard cyclic group, revisited). Consider the nonstandard cyclic group $A := \mathbf{Z}/2^N\mathbf{Z} = \prod_{n \to \alpha} \mathbf{Z}/2^n\mathbf{Z}$. This is a nonstandard finite group and can thus be modeled by the trivial group $\{1\}$ as discussed in Example 8.3.6. However, it can also be modeled by the compact abelian group $\mathbf{Z}_2$ of 2-adic integers using the model $\pi: A \to \mathbf{Z}_2$ defined by the formula
$$\pi(a) := \lim_{n \to \infty} a \bmod 2^n$$
where for each standard natural number n, $a \bmod 2^n \in \{0, \ldots, 2^{n-1}\}$ is the remainder of a modulo 2^n (this is well-defined in A) and the limit is in the 2-adic metric. Note that the image $\pi(A)$ of A is the entire group $\mathbf{Z}_2$, and conversely the preimage of $\mathbf{Z}_2$ in $A^8 = A$ is trivially all of $\mathbf{Z}/2^N\mathbf{Z}$; as such, one can quotient out $\mathbf{Z}_2$ in this model and recover the trivial model of A.

Example 8.4.2 (Nonstandard abelian 2-torsion group). In a similar spirit to the preceding example, the nonstandard 2-torsion group $A := (\mathbf{Z}/2\mathbf{Z})^N = \prod_{n \to \alpha} (\mathbf{Z}/2\mathbf{Z})^n$ can be modeled by the compact abelian group $(\mathbf{Z}/2\mathbf{Z})^{\mathbf{N}}$ by the formula
$$\pi(a) := \lim_{n \to \infty} \pi_n(a)$$
where $\pi_n: A \to (\mathbf{Z}/2\mathbf{Z})^n$ is the obvious projection, and the limit is in the product topology of $(\mathbf{Z}/2\mathbf{Z})^{\mathbf{N}}$. As before, we can quotient out $(\mathbf{Z}/2\mathbf{Z})^{\mathbf{N}}$ and model A instead by the trivial group.

Remark 8.4.3. The above two examples can be generalised to model any nonstandard finite group $G = \prod_{n \to \alpha} G_n$ equipped with surjective homomorphisms from G_{n+1} to G_n by the inverse limit of the G_n.

Exercise 8.4.1 (Lamplighter group). Let F_2 be the field of two elements. Let G be the *lamplighter group* $\mathbf{Z} \ltimes F_2^{\mathbf{Z}}$, where $\mathbf{Z}$ acts on $F_2^{\mathbf{Z}}$ by the shift $T \colon F_2^{\mathbf{Z}}$ defined by $T(a_k)_{k \in \mathbf{Z}} := (a_{k-1})_{k \in \mathbf{Z}}$. Thus the group law in G is given by

$$(i, x)(j, y) := (i + j, x + T^i y).$$

For each n, we then set $A_n \subseteq G$ to be the set

$$A_n := \{(i, x) \in G : i \in \{-1, 0, +1\}; x \in F_2^n\},$$

where we identify F_2^n with the space of elements $(a_k)_{k \in \mathbf{Z}}$ of $F_2^{\mathbf{Z}}$ such that $a_k \neq 0$ only for $k \in \{1, \dots, n\}$. Let $F_2((t))$ be the ring of *formal Laurent series* $\sum_{k \in \mathbf{Z}} a_k t^k$ in which all but finitely many of the a_k for negative k are zero. We let G_0 be the modified lamplighter group $\mathbf{Z} \ltimes F_2((t))$, where $\mathbf{Z}$ acts on $F_2^{\mathbf{Z}}$ by the shift $T \colon f \mapsto tf$. We give $F_2((t))$ a topology by assigning each nonzero Laurent series $\sum_{k \in \mathbf{Z}} a_k t^k$ a norm of 2^{-k}, where k is the least integer for which $a_k \neq 0$; this induces a topology on G_0 via the product topology construction.

We will model the ultraproduct $A := \prod_{n \to \alpha} A_n \subset \mathbf{Z} \ltimes {}^*(\mathbf{Z}/2\mathbf{Z})^{\mathbf{Z}}$ (or more precisely, the set $A \cup A^{-1}$, since A is not quite symmetric) by the group

$$G_0 \times_{\mathbf{Z}} G_0 := \{((i, x), (j, y)) \in G_0 \times G_0 : i = j\}$$

using the map

$$\pi((i, \lim_{n \to \alpha} (a_k^{(n)})_{k \in \mathbf{Z}})) := ((i, \sum_{k \in \mathbf{Z}} \lim_{n \to \alpha} a_k^{(n)} t^k), (i, \sum_{k \in \mathbf{Z}} \lim_{n \to \alpha} a_{n-k}^{(n)} t^k)).$$

Roughly speaking, $\pi(a)$ captures the behaviour of a at the two "ends" of F_2^N, where $N := \lim_{n \to \alpha} n$. We give $G_0 \times_{\mathbf{Z}} G_0$ the topology induced from the product topology on $G_0 \times G_0$.

   (i) Show that $A \cup A^{-1}$ is an ultra approximate group.

   (ii) Show that π is a good model of $A \cup A^{-1}$.

   (iii) Show that π is no longer a good model if one projects $G_0 \times_{\mathbf{Z}} G_0$ to the first or second copy of G_0, or to the base group $\mathbf{Z}$.

   (iv) Show that $A \cup A^{-1}$ does not have a good model by a Lie group L. (*Hint:* L does not contain arbitrarily small elements of order two, other than the identity.)

Remark 8.4.4. In the above exercise, one needs a moderately complicated (though still locally compact) group $G \times_{\mathbf{Z}} G$ to properly model A and its powers A^m. This can also be seen from volume growth considerations: A_n^m grows like 4^m for fixed (large n), which is also the rate of volume growth of $\pi(A)$ in $G \times_{\mathbf{Z}} G$, whereas the volume growth in a single factor G would only grow like 2^m, and the volume growth in $\mathbf{Z}$ is only linear in m. However, if we pass to the large subset A' of A defined by $A' := \prod_{n \to \alpha} A'_n$, where

$$A'_n := \{(i,x) \in G : i = 0; x \in (\mathbf{Z}/2\mathbf{Z})^n\},$$

then A' is now a nonstandard finite group (isomorphic to the group $(\mathbf{Z}/2\mathbf{Z})^N$ considered in Example 8.4.2) and can be modeled simply by the trivial group $\{1\}$. Thus we see that we can sometimes greatly simplify the modeling of an ultra approximate group by passing to a large ultra approximate subgroup.

By using the Gleason-Yamabe theorem, one can formalise these examples. Given two ultra approximate groups A', we say that A' is an *large ultra approximate subgroup* of A if $(A')^4 \subset A^4$ and A can be covered by finitely many left-translates of A'.

Theorem 8.4.5 (Hrushovski's Lie model theorem). *Let A be an ultra approximate group. Then there exists a large ultra approximate subgroup A' of A that can be modeled by a connected Lie group L.*

Proof. By Exercise 8.3.3, we have a good model $\pi_0 \colon \langle A \rangle \to L_0$ of A^4 by some locally compact group L_0. In particular, there is an open neighbourhood U_0 of the identity in L_0 such that $\pi_0^{-1}(U_0) \subset A^4$ and $U_0 \subset \pi_0(A^4)$.

Let U_1 be a symmetric precompact neighbourhood of the identity such that $U_1^{100} \subset U_0$. By the Gleason-Yamabe theorem (Theorem 1.1.13), there is an open subgroup L'_0 of L_0, and a compact normal subgroup N of L'_0 contained in U_1, such that L'_0/N is isomorphic to a connected Lie group L. Let $\pi_1 \colon L'_0 \to L$ be the quotient map.

Write $U_2 := U_1 \cap L'_0$. As π_0 is a good model, we can find a nonstandard finite set A' with

$$\pi^{-1}(U_2) \subset A' \subset \pi^{-1}(U_2^2).$$

By replacing A' with $A' \cap (A')^{-1}$ if necessary, we may take A' to be symmetric. As U_2^4 can be covered by finitely many left-translates of U_2, we see that A' is an ultra approximate group. Since

$$(A')^4 \subset \pi^{-1}(U_2^8) \subset \pi^{-1}(U_1^{100}) \subset \pi^{-1}(U_0) \subset A^4$$

and $\pi(A^4)$ can be covered by finitely many left-translates of U_2, we see that A' is a large ultra approximate subgroup of A. It is then a routine matter to verify that $\pi_1 \circ \pi_0 \colon \langle A' \rangle \to L$ is a good model for A'. $\qquad\square$

The Lie model L need not be unique. For instance, the nonstandard cyclic group $\mathbf{Z}/N\mathbf{Z}$ can be modeled both by the trivial group and by the unit circle $\mathbf{R}/\mathbf{Z}$. However, as observed by Hrushovski [**Hr2012**], it can be shown that after quotienting out the (unique) maximal compact normal subgroup from the Lie model L, the resulting quotient group (which is also a Lie group, and in some sense describes the "large scale" structure of L) is unique up to isomorphism. The following exercise fleshes out the details of this observation.

Exercise 8.4.2 (Large-scale uniqueness of the Lie model). Let L, L' be connected Lie groups, and let A be an ultra approximate group with good models $\pi\colon \langle A \rangle \to L$ and $\pi'\colon \langle A' \rangle \to L'$.

(i) Show that the centre $Z(L) := \{g \in L : gh = hg \text{ for all } h \in L\}$ is an abelian Lie group.

(ii) Show that the connected component $Z(L)^0$ of the identity in $Z(L)$ is isomorphic to $\mathbf{R}^d \times (\mathbf{R}/\mathbf{Z})^{d'}$ for some $d, d' \geq 0$.

(iii) Show that the quotient group $Z(L)/Z(L)^0$ is a finitely generated abelian group, and is isomorphic to $\mathbf{Z}^{d''} \times H$ for some finite group H.

(iv) Show that the torsion points of $Z(L)$ are contained in a compact subgroup of $Z(L)$ isomorphic to $(\mathbf{R}/\mathbf{Z})^{d'} \times H$.

(v) Show that any finite normal subgroup of L is central, and thus lies in the compact subgroup indicated above. (*Hint:* L will act continuously by conjugation on this finite normal subgroup.)

(vi) Show that given any increasing sequence $N_1 \subset N_2 \subset \dots$ of compact normal subgroups of L, the upper bound $\overline{\bigcup_n N_n}$ is also a compact normal subgroup of L. (*Hint:* The dimensions of N_n (which are well-defined by Cartan's theorem) are monotone increasing but bounded by the dimension of L. In particular, the connected components N_n^0 must eventually stabilise. Quotient them out and then use (v).)

(vii) Show that L contains a unique maximal compact normal subgroup N. Similarly, L' contains a unique maximal compact normal subgroup N'. Show that the quotient groups $L/N, L'/N'$ contain no nontrivial compact normal subgroups.

(viii) Show that $\pi^{-1}(N) = (\pi')^{-1}(N')$. (*Hint:* If $g \in L$, then $g \in N$ iff the group generated by g and its conjugates is bounded.)

(ix) Show that L/N is isomorphic to L'/N'.

(x) Show that for sufficiently large standard m, A^m can be modeled by a Lie group with no nontrivial compact normal subgroups, which is unique up to isomorphism.

To illustrate how this theorem is useful, let us apply it in the bounded torsion case.

Exercise 8.4.3. Let A be an ultra approximate group in an r-torsion nonstandard group G for some standard $r \geq 1$. Show that A^4 contains a nonstandard finite group H such that A can be covered by finitely many left-translates of H. (*Hint:* If a Lie group has positive dimension, then it contains elements arbitrarily close to the identity of arbitrarily large order.)

Combining this exercise with Proposition 7.5.5, we conclude a finitary consequence, first observed by Hrushovski [**Hr2012**]:

Corollary 8.4.6 (Freiman theorem, bounded torsion case). *Let $r, K \geq 1$, and let A be a finite K-approximate subgroup of an r-torsion group G. Then A^4 contains a finite subgroup H such that A can be covered by $O_{K,r}(1)$ left-translates of H.*

Exercise 8.4.4 (Commutator self-containment).

(1) Show that if A is an ultra approximate group, then there exists a large approximate subgroup B of A such that $[B, B] \subset B$, where we write $[A, B] := \{[a, b] : a \in A, b \in B\}$, with $[a, b] := a^{-1}b^{-1}ab$. (Note that this is slightly different from the group-theoretic convention, when H and K are subgroups, to define $[H, K]$ to be the *group generated by* the commutators $[h, k]$ with $h \in H$ and $k \in K$.)

(2) Show that if A is a finite K-approximate group, then there exists a symmetric set B containing the origin with $B^4 \subset A^4$, $|B| \gg_K |A|$, and $[B, B] \subset B$.

Remark 8.4.7. I do not know of any proof of Exercise 8.4.4 that does not go through the Gleason-Yamabe theorem (or some other comparably deep fragment of the theory of Hilbert's fifth problem).

The following result of Hrushovski is a significant strengthening of the preceding exercise:

Exercise 8.4.5 (Hrushovski's structure theorem). Let A be a finite K-approximate group, and let $F \colon \mathbf{N} \times \mathbf{N} \to \mathbf{N}$ be a function. Show that there exist natural numbers L, M, N with $N \geq F(L, M)$ and $L, M \ll_{K,F} 1$, and nested sets
$$\{1\} \subset A_N \subseteq \cdots \subseteq A_1 \subseteq A^4$$
with the following properties:

(i) For each $1 \leq n \leq N$, A_n is symmetric.

(ii) For each $1 \leq n < N$, $A_{n+1}^2 \subseteq A_n$.

(iii) For each $1 \leq n \leq N$, A_n is contained in M left-translates of A_{n+1}.

(iv) For $1 \leq n, m, k \leq N$ with $k < n + m$, one has $[A_n, A_m] \subset A_k$.

(v) A can be covered by L left-translates of A_1.

(*Hint:* First find and prove an analogous statement for ultra approximate groups, in which the function F is not present.)

There is a finitary formulation of Theorem 8.4.5, but it takes some effort to state. Let L be a connected Lie group, with Lie algebra $\mathfrak{l}$ which we identify using some coordinate basis with $\mathbf{R}^d$, thus giving a Euclidean norm $\|\|$ on $\mathfrak{l}$. We say that L with this basis has *complexity at most M* for some $M \geq 1$ if:

(1) the dimension d of L is at most M;

(2) the exponential map $\exp\colon \mathfrak{l} \to L$ is injective on the ball $\{x \in \mathfrak{l} : \|x\| \leq 1/M\}$;

(3) one has $\|[x, y]\| \leq M\|x\|\|y\|$ for all $x, y \in \mathfrak{l}$.

We then define "balls" B_R on L by the formula

$$B_R := \{\exp(x) : x \in \mathfrak{l}; \|x\| < R\}.$$

Exercise 8.4.6 (Hrushovski's Lie model theorem, finitary version). Let $F\colon \mathbf{N} \to \mathbf{N}$ be a function, and let A be a finite K-approximate group. Show that there exists a natural number $1 \leq M \ll_{K,F} 1$, a connected Lie group L of complexity at most M, a symmetric set A' containing the identity with $(A')^4 \subset A^4$, and a map $\pi\colon (A')^{F(M)} \to L$ obeying the following properties:

(i) (Large subgroup) A can be covered by M left-translates of A'.

(ii) (Approximate homomorphism) One has $\pi(1) = 1$, and for all $a, b \in (A')^{F(M)}$ with $ab \in (A')^{F(M)}$, one has

$$\pi(ab)\pi(b)^{-1}\pi(a)^{-1} \in B_{1/F(M)}.$$

(iii) (Thick image) If $a \in (A')^{F(M)}$ and $\pi(a) \in B_{1/M}$, then $a \in A'$. Conversely, if $g \in B_{1/M}$, then $\pi(A')$ intersects $gB_{1/F(M)}$.

(iv) (Compact image) One has $\pi(A') \subset B_M^M$.

(*Hint:* One needs to argue by compactness and contradiction, *carefully* negating all the quantifiers in the above claim, and then use Theorem 8.4.5.)

Exercise 8.4.7. In the converse direction, show that Theorem 8.4.5 can be deduced from Exercise 8.4.6. (*Hint:* To get started, one needs a statement to the effect that if an ultra approximate group A is unable to be modeled by some Lie group of complexity at most M, then there is also some $\varepsilon > 0$ for which A cannot be "approximately modeled up to error ε" in some sense by such a Lie group. Once one has such a statement (provable via a compactness or ultralimit argument), one can use this ε to build a suitable function $F()$ which which to apply Exercise 8.4.6.)

Chapter 9

The microscopic structure of approximate groups

A common theme in mathematical analysis (particularly in analysis of a "geometric" or "statistical" flavour) is the interplay between "macroscopic" and "microscopic" scales. These terms are somewhat vague and imprecise, and their interpretation depends on the context and also on one's choice of normalisations, but if one uses a "macroscopic" normalisation, "macroscopic" scales correspond to scales that are comparable to unit size (i.e., bounded above and below by absolute constants), while "microscopic" scales are much smaller, being the minimal scale at which nontrivial behaviour occurs. (Other normalisations are possible, e.g., making the microscopic scale a unit scale, and letting the macroscopic scale go off to infinity; for instance, such a normalisation is often used, at least initially, in the study of groups of polynomial growth. However, for the theory of approximate groups, a macroscopic scale normalisation is more convenient.)

One can also consider "mesoscopic" scales which are intermediate between microscopic and macroscopic scales, or large-scale behaviour at scales that go off to infinity (and in particular are larger than the macroscopic range of scales), although the behaviour of these scales will not be the main focus of this section. Finally, one can divide the macroscopic scales into "local" macroscopic scales (less than ε for some small but fixed $\varepsilon > 0$) and "global" macroscopic scales (scales that are allowed to be larger than a given large absolute constant C). For instance, given a finite approximate group A:

(1) Sets such as A^m for some fixed m (e.g., A^{10}) can be considered to be sets at a global macroscopic scale. Sending m to infinity, one enters the large-scale regime.

(2) Sets such as the sets S that appear in the Sanders lemma from the previous section (thus $S^m \subset A^4$ for some fixed m, e.g., $m = 100$) can be considered to be sets at a local macroscopic scale. Sending m to infinity, one enters the mesoscopic regime.

(3) The nonidentity element u of A that is "closest" to the identity in some suitable metric (cf. the proof of Jordan's theorem from Chapter 1) would be an element associated to the microscopic scale. The orbit $u, u^2, u^3, \ldots$ starts out at microscopic scales, and (assuming some suitable "escape" axioms) will pass through mesoscopic scales and finally entering the macroscopic regime. (Beyond this point, the orbit may exhibit a variety of behaviours, such as periodically returning back to the smaller scales, diverging off to ever larger scales, or filling out a dense subset of some macroscopic set; the escape axioms we will use do not exclude any of these possibilities.)

For comparison, in the theory of locally compact groups, properties about small neighbourhoods of the identity (e.g., local compactness, or the NSS property) would be properties at the local macroscopic scale, whereas the space $L(G)$ of one-parameter subgroups can be interpreted as an object at the microscopic scale. The exponential map then provides a bridge connecting the microscopic and macroscopic scales.

We return now to approximate groups. The macroscopic structure of these objects is well described by the *Hrushovski Lie model theorem* (Theorem 8.4.5), which informally asserts that the macroscopic structure of an (ultra) approximate group can be modeled by a Lie group. This is already an important piece of information about general approximate groups, but it does not directly reveal the full structure of such approximate groups, because these Lie models are unable to see the *microscopic* behaviour of these approximate groups.

To illustrate this, let us review one of the examples of a Lie model of an ultra approximate group, namely Exercise 8.3.7. In this example one studied a "nilbox" from a Heisenberg group, which we rewrite here in slightly different notation. Specifically, let G be the Heisenberg group

$$G := \{(a, b, c) : a, b, c \in \mathbf{Z}\}$$

with group law

(9.1) $$(a, b, c) * (a', b', c') := (a + a', b + b', c + c' + ab')$$

and let $A = \prod_{n \to \alpha} A_n$, where $A_n \subset G$ is the box

$$A_n := \{(a, b, c) \in G : |a|, |b| \leq n; |c| \leq n^{10}\};$$

thus A is the nonstandard box

$$A := \{(a, b, c) \in {}^*G : |a|, |b| \leq N; |c| \leq N^{10}\}$$

where $N := \lim_{n \to \alpha} n$. As the above exercise establishes, $A \cup A^{-1}$ is an ultra approximate group with a Lie model $\pi \colon \langle A \rangle \to \mathbf{R}^3$ given by the formula

$$\pi(a, b, c) := \left(\operatorname{st} \frac{a}{N}, \operatorname{st} \frac{b}{N}, \operatorname{st} \frac{c}{N^{10}} \right)$$

for $a, b = O(N)$ and $c = O(N^{10})$. Note how the nonabelian nature of G (arising from the ab' term in the group law (9.1)) has been lost in the model $\mathbf{R}^3$, because the effect of that nonabelian term on $\frac{c}{N^{10}}$ is only $O(\frac{N^2}{N^8})$ which is infinitesimal and thus does not contribute to the standard part. In particular, if we replace G with the abelian group $G' := \{(a, b, c) : a, b, c \in \mathbf{Z}\}$ with the additive group law

$$(a, b, c) *' (a', b', c') := (a + a', b + b', c + c')$$

and let A' and π' be defined exactly as with A and π, but placed inside the group structure of G' rather than G, then $A \cup A^{-1}$ and $A' \cup (A')^{-1}$ are essentially "indistinguishable" as far as their models by $\mathbf{R}^3$ are concerned, even though the latter approximate group is abelian and the former is not. The problem is that the nonabelianness in the former example is so microscopic that it falls entirely inside the kernel of π and is thus not detected at all by the model.

The problem of not being able to "see" the microscopic structure of a group (or approximate group) also was a key difficulty in the theory surrounding Hilbert's fifth problem that was discussed in Chapters 3, 5. A key tool in being able to resolve such structure was to build left-invariant metrics d (or equivalently, norms $\|\|$) on one's group, which obeyed useful "Gleason axioms" such as the commutator axiom

(9.2) $$\|[g, h]\| \ll \|g\| \|h\|$$

for sufficiently small g, h, or the escape axiom

(9.3) $$\|g^n\| \gg |n| \|g\|$$

when $|n| \|g\|$ was sufficiently small. Such axioms have important and nontrivial content even in the microscopic regime where g or h are extremely close to the identity. For instance, in the proof of Jordan's theorem (Theorem 1.0.2), a key step was to apply the commutator axiom (9.2) (for the distance to the identity in operator norm) to the most "microscopic" element of G, or more precisely a nonidentity element of G of minimal norm. The key point was that this microscopic element was virtually central in G, and as such it restricted much of G to a lower-dimensional subgroup of the unitary group, at which point one could argue using an induction-on-dimension argument. As we shall see, a similar argument can be used to place "virtually nilpotent" structure on finite approximate groups. For instance, in the Heisenberg-type approximate groups $A \cup A^{-1}$ and $A' \cup (A')^{-1}$ discussed earlier, the element $(0, 0, 1)$ will be "closest to the origin" in a suitable sense to be defined later, and is centralised by both approximate groups; quotienting out (the orbit of) that central element and iterating the process two more times, we shall see that one can express both $A \cup A^{-1}$ and $A' \cup (A')^{-1}$ as a tower of central cyclic extensions, which in particular establishes the nilpotency of both groups.

The escape axiom (9.3) is a particularly important axiom in connecting the microscopic structure of a group G to its macroscopic structure; for instance, as shown in Section 3, this axiom (in conjunction with the closely related commutator axiom) tends to imply dilation estimates such as $d(g^n, h^n) \sim n d(g, h)$ that allow one to understand the microscopic geometry of points g, h close to the identity in terms of the (local) macroscopic geometry of points g^n, h^n that are significantly further away from the identity.

It is thus of interest to build some notion of a norm (or left-invariant metrics) on an approximate group A that obeys the escape and commutator axioms (while being nondegenerate enough to adequately capture the geometry of A in some sense), in a fashion analogous to the Gleason metrics that played such a key role in the theory of Hilbert's fifth problem. It is tempting to use the Lie model theorem to do this, since Lie groups certainly come with Gleason metrics. However, if one does this, one ends up, roughly speaking, with a norm on A that only obeys the escape and commutator estimates *macroscopically*; roughly speaking, this means that one has a macroscopic commutator inequality

$$\|[g, h]\| \ll \|g\| \|h\| + o(1)$$

and a macroscopic escape property

$$\|g^n\| \gg |n| \|g\| - o(|n|),$$

but such axioms are too weak for analysis at the microscopic scale and, in particular, in establishing centrality of the element closest to the identity.

Another way to proceed is to build a norm that is specifically designed to obey the crucial escape property. Given an approximate group A in a group G, and an element g of G, we can define the *escape norm* $\|g\|_{e,A}$ of g

by the formula

$$\|g\|_{e,A} := \inf\left\{\frac{1}{n+1} : n \in \mathbf{N}; g, g^2, \ldots, g^n \in A\right\}.$$

Thus, $\|g\|_{e,A}$ equals 1 if g lies outside of A, equals $1/2$ if g lies in A but g^2 lies outside of A, and so forth. Such norms had already appeared in Section 5, in the context of analysing NSS groups.

As it turns out, this expression will obey an escape axiom, as long as we place some additional hypotheses on A which we will present shortly. However, it need not actually be a norm; in particular, the triangle inequality

$$\|gh\|_{e,A} \le \|g\|_{e,A} + \|h\|_{e,A}$$

is not necessarily true. Fortunately, it turns out that by a (slightly more complicated) version of the Gleason machinery from Chapter 5 we can establish a usable substitute for this inequality, namely the quasi-triangle inequality

$$\|g_1 \ldots g_k\|_{e,A} \le C(\|g_1\|_{e,A} + \cdots + \|g_k\|_{e,A}),$$

where C is a constant independent of k. As we shall see, these estimates can then be used to obtain a commutator estimate (9.2).

However, to do all this, it is not enough for A to be an approximate group; it must obey two additional "trapping" axioms that improve the properties of the escape norm. We formalise these axioms (somewhat arbitrarily) as follows:

Definition 9.0.8 (Strong approximate group). Let $K \ge 1$. A *strong K-approximate group* is a finite K-approximate group A in a group G with a symmetric subset S obeying the following axioms:

(1) (S small) One has

(9.4) $$(S^{A^4})^{1000K^3} \subset A.$$

(2) (First trapping condition) If $g, g^2, \ldots, g^{1000} \in A^{100}$, then $g \in A$.

(3) (Second trapping condition) If $g, g^2, \ldots, g^{10^6 K^3} \in A$, then $g \in S$.

An *ultra strong K-approximate group* is an ultraproduct $A = \prod_{n \to \alpha} A_n$ of strong K-approximate groups.

The first trapping condition can be rewritten as

$$\|g\|_{e,A} \le 1000\|g\|_{e,A^{100}}$$

and the second trapping condition can similarly be rewritten as

$$\|g\|_{e,S} \le 10^6 K^3 \|g\|_{e,A}.$$

This makes the escape norms of A, A^{100}, and S comparable to each other, which will be needed for a number of reasons (and in particular to close a certain bootstrap argument properly). Compare this with (5.11), which used the NSS hypothesis to obtain similar conclusions. Thus, one can view the strong approximate group axioms as being a sort of proxy for the NSS property.

Example 9.0.9. Let N be a large natural number. Then the interval $A = [-N, N]$ in the integers is a 2-approximate group, which is also a strong 2-approximate group (setting $S = [10^{-6}N, 10^{-6}N]$, for instance). On the other hand, if one places A in $\mathbf{Z}/5N\mathbf{Z}$ rather than in the integers, then the first trapping condition is lost and one is no longer a strong 2-approximate group. Also, if one remains in the integers, but deletes a few elements from A, e.g., deleting $\pm\lfloor 10^{-10}N \rfloor$ from A), then one is still a $O(1)$-approximate group, but is no longer a strong $O(1)$-approximate group, again because the first trapping condition is lost.

A key consequence of the Hrushovski Lie model theorem is that it allows one to replace approximate groups by strong approximate groups:

Exercise 9.0.8 (Finding strong approximate groups)**.**

    (i) Let A be an ultra approximate group with a good Lie model $\pi : \langle A \rangle \to L$, and let B be a symmetric convex body (i.e., a convex open bounded subset) in the Lie algebra $\mathfrak{l}$. Show that if $r > 0$ is a sufficiently small standard number, then there exists a strong ultra approximate group A' with

$$\pi^{-1}(\exp(rB)) \subset A' \subset \pi^{-1}(\exp(1.1rB)) \subset A,$$

and with A can be covered by finitely many left-translates of A'. Furthermore, π is also a good model for A'.

    (ii) If A is a finite K-approximate group, show that there is a strong $O_K(1)$-approximate group A' inside A^4 with the property that A can be covered by $O_K(1)$ left-translates of A'. (*Hint:* Use (i), Hrushovski's Lie model theorem, and a compactness and contradiction argument.)

The need to compare the strong approximate group to an exponentiated small ball $\exp(rB)$ will be convenient later, as it allows one to easily use the geometry of L to track various aspects of the strong approximate group.

As mentioned previously, strong approximate groups exhibit some of the features of NSS locally compact groups. In Chapter 5, we saw that the escape norm for NSS locally compact groups was comparable to a Gleason metric. The following theorem is an analogue of that result:

Theorem 9.0.10 (Gleason lemma)**.** *Let A be a strong K-approximate group in a group G.*

(i) *(Symmetry)* For any $g \in G$, one has $\|g^{-1}\|_{e,A} = \|g\|_{e,A}$.

(ii) *(Conjugacy bound)* For any $g, h \in A^{10}$, one has $\|g^h\|_{e,A} \ll \|g\|_{e,A}$.

(iii) *(Triangle inequality)* For any $g_1, \ldots, g_k \in G$, one has $\|g_1 \cdots g_k\|_{e,A} \ll_K (\|g_1\|_{e,A} + \cdots + \|g_k\|_{e,A})$.

(iv) *(Escape property)* One has $\|g^n\|_{e,A} \gg |n|\|g\|_{e,A}$ whenever $|n|\|g\|_{e,A} < 1$.

(v) *(Commutator inequality)* For any $g, h \in A^{10}$, one has $\|[g,h]\|_{e,A} \ll_K \|g\|_{e,A}\|h\|_{e,A}$.

The proof of this theorem will occupy a large part of this section. We then aim to use this theorem to classify strong approximate groups. The basic strategy (temporarily ignoring a key technical issue) follows the Bieberbach-Frobenius proof of Jordan's theorem, as given in Chapter 1, is as follows.

(1) Start with an (ultra) strong approximate group A.

(2) From the Gleason lemma, the elements with zero escape norm form a normal subgroup of A. Quotient these elements out. Show that all nonidentity elements will have positive escape norm.

(3) Find the nonidentity element g_1 in (the quotient of) A of minimal escape norm. Use the commutator estimate (assuming it is inherited by the quotient) to show that g_1 will centralise (most of) this quotient. In particular, the orbit $\langle g_1 \rangle$ is (essentially) a central subgroup of $\langle A \rangle$.

(4) Quotient this orbit out; then find the next nonidentity element g_2 in this new quotient of A. Again, show that $\langle g_2 \rangle$ is essentially a central subgroup of this quotient.

(5) Repeat this process until A becomes entirely trivial. Undoing all the quotients, this should demonstrate that $\langle A \rangle$ is virtually nilpotent, and that A is essentially a coset nilprogression.

There are two main technical issues to resolve to make this strategy work. The first is to show that the iterative step in the argument terminates in finite time. This we do by returning to the Lie model theorem. It turns out that each time one quotients out by an orbit of an element that escapes, the dimension of the Lie model drops by at least one. This will ensure termination of the argument in finite time.

The other technical issue is that while the quotienting out all the elements of zero escape norm eliminates all "torsion" from A (in the sense that the quotient of A has no nontrivial elements of zero escape norm), further quotienting operations can inadvertently re-introduce such torsion. This torsion can be re-eradicated by further quotienting, but the price one pays for this is that the final structural description of $\langle A \rangle$ is no longer as strong as "virtually nilpotent", but is instead a more complicated tower alternating between (ultra) finite extensions and central extensions.

Example 9.0.11. Consider the strong $O(1)$-approximate group

$$A := \{aN^{10} + 5b : |a| \leq N; |b| \leq N^2\}$$

in the integers, where N is a large natural number not divisible by 5. As $\mathbf{Z}$ is torsion-free, all nonzero elements of A have positive escape norm, and the nonzero element of minimal escape norm here is $g = 5$ (or $g = -5$). But if one quotients by $\langle g \rangle$, A projects down to $\mathbf{Z}/5\mathbf{Z}$, which now has torsion (and all elements in this quotient have zero escape norm). Thus torsion has been re-introduced by the quotienting operation. (A related observation is that the intersection of A with $\langle g \rangle = 5\mathbf{Z}$ is not a simple progression, but is a more complicated object, namely a generalised arithmetic progression of rank two.)

To deal with this issue, we will not quotient out by the entire cyclic group $\langle g \rangle = \{g^n : n \in \mathbf{Z}\}$ generated by the element g of minimal escape norm, but rather by an arithmetic progression $P = \{g^n : |n| \leq N\}$, where N is a natural number comparable to the reciprocal $1/\|g\|_{e,A}$ of the escape norm, as this will be enough to cut the dimension of the Lie model down by one without introducing any further torsion. Of course, this cannot be done in the category of global groups, since the arithmetic progression P will not, in general, be a group. However, it is still a *local* group, and it turns out that there is an analogue of the quotient space construction in local groups. This fixes the problem, but at a cost: in order to make the inductive portion of the argument work smoothly, it is now more natural to place the *entire* argument inside the category of local groups rather than global groups, even though the primary interest in approximate groups A is in the global case when A lies inside a global group. This necessitates some technical modification to some of the preceding discussion (for instance, the Gleason-Yamabe theorem must be replaced by the local version of this theorem, Theorem 5.6.9); details can be found in [**BrGrTa2011**], but will only be sketched here.

9.1. Gleason's lemma

Throughout this section, A is a strong K-approximate group in a global group G. We will prove the various estimates in Theorem 9.0.10. The arguments will be very close to those in Chapter 5; indeed, it is possible to unify the results here with the results in that section by a suitable modification of the notation, but we will not do so here.

We begin with the easy estimates. The symmetry property is immediate from the symmetry of A. Now we turn to the escape property. By symmetry, we may take n to be positive (the $n = 0$ case is trivial). We may of course assume that $\|g\|_{e,A}$ is strictly positive, say equal to $1/(m+1)$; thus $g, \ldots, g^m \in A$ and $g^{m+1} \notin A$, and $n \le m$. By the first trapping property, this implies that $g^{j(m+1)} \notin A^{100}$ for some $1 \le j \le 1000$.

Let kn be the first multiple of n larger than or equal to $j(m+1)$, then $kn \ll m+1$. Since $kn - j(m+1)$ is less than m, we have $g^{kn-j(m+1)} \in A$; since $g^{j(m+1)} \notin A^{100}$, we conclude that $g^{kn} \notin A^{99}$. In particular this shows that $\|g^n\|_{e,A} \gg 1/k \gg n/(m+1)$, and the claim follows.

The escape property implies the conjugacy bound:

Exercise 9.1.1. Establish the conjugacy bound. (*Hint:* One can mimic the arguments establishing a nearly identical bound in Section 5.2.)

Now we turn to the triangle inequality, which (as in Chapter 5) is the most difficult property to establish. Our arguments will closely resemble the proof of Proposition 5.3.4, with S^{A^4} and A playing the roles of U_1 and U_0 from that argument. As in that theorem, we will initially assume that we have an *a priori* bound of the form

$$(9.5) \qquad \|g_1 \ldots g_k\|_{e,A} \le M(\|g_1\|_{e,A} + \cdots + \|g_k\|_{e,A})$$

for all $g_1, \ldots, g_k$, and some (large) M independent of k, and remove this hypothesis later. We then introduce the norm

$$\|g\|_{*,A} := \inf\{\|g_1\|_{e,A} + \cdots + \|g_k\|_{e,A} : g = g_1 \ldots g_k\}.$$

Then $\|g\|_{*,A}$ is symmetric, obeys the triangle inequality, and is comparable to $\| \|_{e,A}$ in the sense that

$$(9.6) \qquad \frac{1}{M}\|g\|_{e,A} \le \|g\|_{*,A} \le \|g\|_{e,A}$$

for all $g \in G$.

We introduce the function $\psi \colon G \to \mathbf{R}^+$ by

$$\psi(x) := (1 - M\operatorname{dist}_{*,A}(x, A))_+,$$

where $\operatorname{dist}_{*,A}(x, A) := \inf_{y \in A} \|x^{-1}y\|_{*,A}$. Then ψ takes values between 0 and 1, equals 1 on A, is supported on A^2, and obeys the Lipschitz bound

$$(9.7) \qquad \|\partial_g \psi\|_{\ell^\infty(G)} \le M\|g\|_{e,A}$$

for all g, thanks to the triangle inequality for $\| \|_{*,A}$ and (9.6). We also introduce the function $\eta \colon G \to \mathbf{R}^+$ by

$$\eta(x) := \sup\{1 - \frac{j}{10^4 K^3} : x \in (S^{A^2})^j A\} \cup \{0\},$$

then η also takes values between 0 and 1, equals 1 on A, is supported on A^2, and obeys the Lipschitz bound

(9.8) $$\|\partial_{h^y}\eta\|_{\ell^\infty(G)} \leq \frac{1}{10^4 K^3}$$

for all $h \in S$ and $y \in A^4$.

Now we form the convolution $\phi: G \to \mathbf{R}^+$ by the formula

$$\phi(x) := \frac{1}{|A|} \sum_{y \in G} \psi(y)\eta(y^{-1}x)$$

$$= \frac{1}{|A|} \sum_{y \in G} \psi(xy)\eta(y^{-1}).$$

By construction, ϕ is supported on A^4 and at least 1 at the identity. As ψ or η is supported in A^2, which has cardinality at most $K|A|$, we have the uniform bound

(9.9) $$\|\phi\|_{\ell^\infty(G)} \leq K^2.$$

Similarly, from the identity

$$\partial_g \phi(x) = \frac{1}{|A|} \sum_{y \in G} \partial_g \psi(y)\eta(y^{-1}x)$$

and (9.7) we have the Lipschitz bound

(9.10) $$\|\partial_g \phi\|_{\ell^\infty(G)} \leq K^2 M \|g\|_{e,A}.$$

Finally, from the identity

$$\partial_h \partial_g \phi(x) = \frac{1}{|A|} \sum_{y \in G} \partial_g \psi(y)\partial_{h^y}\eta(y^{-1}x)$$

and restricting g to A (so that $\partial_g \psi$ is supported on A^4, which has cardinality at most $K^3|A|$) we see from (9.7), (9.8) that

$$\|\partial_h \partial_g \phi\|_{\ell^\infty(G)} \leq \frac{1}{10^4} M \|g\|_{e,A}$$

for $g \in A$ and $h \in S$.

We can use this to improve the bound (9.10). Indeed, using the telescoping identity

$$\partial_{g^n} = n\partial_g + \sum_{i=0}^{n-1} \partial_{g^i}\partial_g$$

we see that

$$\|\partial_g \phi\|_{\ell^\infty(G)} \leq \frac{1}{n}\|\partial_{g^n}\phi\|_{\ell^\infty(G)} + \frac{1}{n}\sum_{i=0}^{n-1}\|\partial_{g^i}\partial_g \phi\|_{\ell^\infty(G)}$$

and thus

$$\|\partial_g \phi\|_{\ell^\infty(G)} \le \frac{1}{n} + \frac{1}{10^4} M \|g\|_{e,A}$$

whenever $g, g^2, \ldots, g^{n-1} \in S$. Using the second trapping property, this implies that

$$\|\partial_g \phi\|_{\ell^\infty(G)} \le (\frac{1}{10^4} M + O_K(1)) \|g\|_{e,A}.$$

In the converse direction, if $\|\partial_g \phi\|_{\ell^\infty(G)} < \frac{1}{n}$, then

$$\|\partial_{g^i} \phi\|_{\ell^\infty(G)} < 1$$

and thus $g^i \in A^4$ from the support of ϕ, for all $1 \le i \le n$. But then by the first trapping property, this implies that $g^i \in A$ for all $1 \le i \le n/1000$. We conclude that

$$\|g\|_{e,A} \le 1000 \|\partial_g \phi\|_{\ell^\infty(G)}.$$

The triangle inequality for $\|\partial_g \phi\|_{\ell^\infty(G)}$ then implies a triangle inequality for $\|g\|_{e,A}$,

$$\|g_1 \ldots g_k\|_{e,A} \le \left(\frac{1}{10} M + O_K(1) \right) (\|g_1\|_{e,A} + \cdots + \|g_k\|_{e,A}),$$

which is (9.5) with M replaced by $\frac{1}{10} M + O_K(1)$. If we knew (9.5) for some large but finite M, we could iterate this argument and conclude that (9.5) held with M replaced by $O_K(1)$, which would give the triangle inequality. Now it is not immediate that (9.5) holds for any finite M, but we can avoid this problem with the usual regularisation trick of replacing $\|g\|_{e,A}$ with $\|g\|_{e,A} + \varepsilon$ throughout the argument for some small $\varepsilon > 0$, which makes (9.5) automatically true with $M = O(1/\varepsilon)$, run the above iteration argument, and then finally send ε to zero.

Exercise 9.1.2. Verify that the modifications to the above argument sketched above actually do establish the triangle inequality.

A final application of the Gleason convolution machinery then gives the final estimate in Gleason's lemma:

Exercise 9.1.3. Use the properties of the escape norm already established (and in particular, the escape property and the triangle inequality) to establish the commutator inequality. (*Hint:* Adapt the argument from Chapter 5.2.)

The proof of Theorem 9.0.10 is now complete.

Exercise 9.1.4. Generalise Theorem 9.0.10 to the setting where A is not necessarily finite, but is instead an open precompact subset of a locally compact group G. (*Hint:* Replace cardinality by left-invariant Haar measure and follow the arguments in Section 5 closely.) Note that this already gives most of one of the key results from that section, namely that any NSS group admits a Gleason metric, since it is not difficult to show that NSS groups contain open precompact strong approximate groups.

9.2. A cheap version of the structure theorem

In this section we use Theorem 9.0.10 to establish a "cheap" version of the structure theorem for ultra approximate groups. We begin by eliminating the elements of zero escape norm. Let us say that an approximate group A is *NSS* if it contains no nontrivial subgroups of the ambient group or, equivalently, if every nonidentity element of A has a positive escape norm. We say that an ultra approximate group is *NSS* if it is the ultralimit of NSS approximate groups.

Using the Gleason lemma, we can easily reduce to the NSS case:

Exercise 9.2.1 (Reduction to the NSS case). Let L be a connected Lie group with Lie algebra $\mathfrak{l}$, let B be a bounded symmetric convex body in $\mathfrak{l}$, let $r > 0$ be a sufficiently small standard real. Let $0 < r' < r/2$, and let A be an ultra strong approximate group which has a good model $\pi\colon \langle A \rangle \to L$ with
$$\pi^{-1}(\exp(rB)) \subset A \subset \pi^{-1}(\exp(1.1rB)).$$
Let H be the set of all elements in A of zero (nonstandard) escape norm. Show that H is a normal nonstandard finite subgroup of $\langle A \rangle$ that lies in $\ker(\pi)$. If $\eta\colon \langle A \rangle \to \langle A \rangle/H$ is the quotient map, and $\pi'\colon \langle A \rangle/H \to L$ is the map π factored through η, show that there exists an ultra strong NSS approximate group A' in $\eta(A)$ which has π' as a good model with
$$(\pi')^{-1}(\exp(r'B)) \subset A' \subset (\pi')^{-1}(\exp(1.1r'B)),$$
and such that A is covered by finitely many left-translates of $\pi^{-1}(A')$.

Let us now analyse the NSS case. Let L be a connected Lie group, with Lie algebra $\mathfrak{l}$, let B be a bounded symmetric convex body in $\mathfrak{l}$, let $r > 0$ be a sufficiently small standard real. Let A be an ultra strong NSS approximate group which has a good model $\pi\colon \langle A \rangle \to L$ with
$$\pi^{-1}(\exp(rB)) \subset A \subset \pi^{-1}(\exp(1.1rB)).$$

If L is zero-dimensional, then by connectedness it is trivial, and then (by the properties of a good model) A is a nonstandard finite group; since it is NSS, it is also trivial. Now suppose that L is not zero-dimensional. Then A contains nonidentity elements whose image under π is arbitrarily close to the identity of L; in particular, A does not consist solely of the identity element, and thus contains elements of positive escape norm by the NSS assumption. Let g be a nonidentity element of G with minimal escape norm $\|u\|_{e,A}$, then

$\pi(u)$ must be the identity (so in particular $\|u\|_{e,A}$ is infinitesimal). (Note that any nontrivial NSS finite approximate group will contain nonidentity elements of minimal escape norm, and the extension of this claim to the ultra approximate group case follows from Łos's theorem.) From Theorem 9.0.10 one has

$$\|[u,h]\|_{e,A} \ll \|u\|_{e,A}\|h\|_{e,A}$$

for all $h \in A$. (Here we are now using the nonstandard asymptotic notation, thus $X \ll Y$ means that $X \leq CY$ for some standard C.) In particular, from the minimality of $\|u\|_{e,A}$, we see that there is a standard $c > 0$ such that u commutes with all elements h with $\|h\|_{e,A} < c$. In particular, if $r' > 0$ is a sufficiently small standard real number, we can find an ultra approximate subgroup A' of A with

$$\pi^{-1}(\exp(r'B)) \subset A' \subset \pi^{-1}(\exp(1.1r'B))$$

which is centralised by u.

Now we show that u generates a one-parameter subgroup of the model Lie group L.

Exercise 9.2.2 (One-parameter subgroups from orbits). Let the notation be as above. Let $g \in A$ be such that $\|g\|_{e,A}$ is infinitesimal but nonzero.

(i) Show that $\pi(g^i) = 1$ whenever $i = o(1/\|g\|_{e,A})$.

(ii) Show that the map $t \mapsto \pi(g^{\lfloor t/\|g\|_{e,A}\rfloor})$ is a one-parameter subgroup in L (i.e., a continuous homomorphism from $\mathbf{R}$ to L).

(iii) Show that there exists an element X of $1.1rB\backslash rB$ such that $\pi(g^i) = \exp(\mathrm{st}(i\|g\|_{e,A})X)$ for all $i = O(1/\|g\|_{e,A})$.

Similar statements hold with A, r replaced by A', r'.

We can now quotient out by the centraliser of u and reduce the dimension of the Lie model:

Exercise 9.2.3. Let $Z(u) := \{h \in {}^*G : uh = hu\}$ be the centraliser of u in ${}^*G$, and let $H := \{u^n : n \in {}^*\mathbf{Z}\}$ be the nonstandard cyclic group generated by u. (Thus, by the preceding discussion, $Z(u)$ contains A', and H is a central subgroup of $Z(u)$ containing u. It will be important for us that $Z(u)$ and H are both nonstandard sets, i.e., ultraproducts of standard sets.)

(i) Show that $\pi(H \cap A^m)$ is a compact subset of L for each standard m.

(ii) Show that $\pi(H \cap \langle A\rangle)$ is a central subgroup of L that contains $\phi(\mathbf{R})$.

(iii) Show that $\overline{\pi(H \cap \langle A\rangle)}$ is a central subgroup of L that is a Lie group of dimension at least one, and so the quotient group $L' := L/\overline{\pi(H \cap \langle A\rangle)}$ is a Lie group of dimension strictly less than the dimension of L.

(iv) Let $\eta\colon Z(u) \to Z(u)/H$ be the quotient map, and let $\pi'\colon \eta(\langle A'\rangle) \to L'$ be the obvious quotient of π. Let B' be a convex symmetric body in the Lie algebra $\mathfrak{l}'$ of L. Show that for sufficiently small standard $r'' > 0$, there exists an ultra strong approximate group

$$(\pi')^{-1}(\exp(r''B)) \subset A'' \subset (\pi')^{-1}(\exp(1.1r''B))$$

with π' as a good model, with $A'' \subset \eta(A')$, and with $\pi'(A')$ covered by finitely many left-translates of A''.

Note that the quotient approximate group A'' obtained by the above procedure is not necessarily NSS. However, it can be made NSS by Exercise 9.2.1. As such, one can iterate the above exercise until the dimension of the Lie model shrinks all the way to zero, at which point the NSS approximate group one is working with becomes trivial. This leads to a "cheap" structure theorem for approximate groups:

Exercise 9.2.4 (Cheap structure theorem). Let A be an ultra approximate group in a nonstandard group G.

(i) Show that if A has a good model by a connected Lie group L, then L is nilpotent. (*Hint:* First use Exercise 9.0.8, and then induct on the dimension of L.)

(ii) Show that A is covered by finitely many left-translates of a nonstandard subgroup G' of G which admits a *normal series*

$$G' = G'_0 \rhd G'_1 \rhd G'_2 \rhd \cdots \rhd G'_k = \{1\}$$

for some standard k, where for every $0 \le i < k$, G'_{i+1} is a normal nonstandard subgroup of G' and of G'_i, and G'_i/G'_{i+1} is either a nonstandard finite group or a nonstandard central subgroup of G'/G'_{i+1}. Furthermore, if G'_i/G'_{i+1} is not central, then it is contained in the image of $A^4 \cap G'$ in G'/G'_{i+1}; and G' is generated (as a nonstandard subgroup) by $A^4 \cap G'$. (*Hint:* First use the Lie model theorem and Exercise 9.0.8, and then induct on the dimension of L.)

Exercise 9.2.5 (Cheap structure theorem, finite version). Let A be a finite K-approximate group in a group G. Show that A is covered by $O_K(1)$ left-translates of a subgroup G' of G which admits a normal series

$$G' = G'_0 \rhd G'_1 \rhd G'_2 \rhd \cdots \rhd G'_k = \{1\}$$

for some $k = O_K(1)$, where for every $0 \le i < k$, G'_{i+1} is a normal subgroup of G'_i and of G', and G'_i/G'_{i+1} is either finite or central in G'/G'_{i+1}. Furthermore, if G'_i/G'_{i+1} is not central, then it is contained in the image of $A^4 \cap G'$ in G'/G'_{i+1}, and G' is generated as a group by $A^4 \cap G$.

Exercise 9.2.6.

(i) Show that any finite extension G of a virtually nilpotent group H (thus there is a short exact sequence $0 \to K \to G \to H \to 0$ with K finite) is virtually nilpotent. (*Hint:* G acts on K by conjugation; look at the stabiliser of this action.)

(ii) Conclude that the group G' in the previous exercise is virtually nilpotent.

One can push the cheap structure theorem a bit further by controlling the dimension of the nilpotent Lie group in terms of the covering number K of the ultra approximate group, as laid out in the following exercise.

Exercise 9.2.7 (Nilpotent groups). A Lie algebra $\mathfrak{g}$ is said to be *nilpotent* if the derived series $\mathfrak{g}_1 := \mathfrak{g}$, $\mathfrak{g}_2 := [\mathfrak{g}_1, \mathfrak{g}]$, $\mathfrak{g}_3 := [\mathfrak{g}_2, \mathfrak{g}], \dots$ becomes trivial after a finite number of steps.

(i) Show that a connected Lie group is nilpotent if and only if its Lie algebra is nilpotent.

(ii) If $\mathfrak{g}$ is a finite-dimensional nilpotent Lie algebra, show that there is a simply connected Lie group G with Lie algebra $\mathfrak{g}$, for which the exponential map $\exp \colon \mathfrak{g} \to G$ is a (global) homeomorphism. Furthermore, any other connected Lie group with Lie algebra $\mathfrak{g}$ is a quotient of G by a discrete central subgroup of G.

(iii) If $\mathfrak{g}$ and G are as in (ii), show that the pushforward of a Haar measure (or Lebesgue measure) on $\mathfrak{g}$ is a bi-invariant Haar measure on G. (Recall from Exercise 4.1.7 that connected nilpotent Lie groups are unimodular.)

(iv) If $\mathfrak{g}$ and G are as in (ii), and μ is a bi-invariant Haar measure on G, show that $\mu(A^2) \geq 2^d \mu(A)$ for all open precompact $A \subset G$, where d is the dimension of G.

(v) If $\tilde{G}$ is a connected (but not necessarily simply connected) nilpotent Lie group, and N is the maximal compact normal subgroup of G (which exists by Exercise 8.4.2), show that N is central, and $\tilde{G}/N$ is simply connected. As a consequence, conclude that if $\tilde{\mu}$ is a left-Haar measure of $\tilde{G}$, then $\tilde{\mu}(\tilde{A}^2) \geq 2^d \tilde{\mu}(\tilde{A})$ for all open precompact $\tilde{A} \subset \tilde{G}$, where d is the dimension of G/N.

(vi) Show that if A is an ultra K-approximate group which has a Lie model L, and N is the maximal compact normal subgroup of L, then L/N has dimension at most $\log_2 K$.

(vii) Show that if A is an ultra K-approximate group, then there is an ultra $K^{O(1)}$-approximate group A' in A^4 that is modeled by a Lie group L, such that A is covered by finitely many left-translates of

A. (*Hint:* A^4 has a good model $\pi \colon A^4 \to G$ by a locally compact group G; by the Gleason-Yamabe theorem, G has an open subgroup G' and a normal subgroup N of G' inside $\pi(A^4)$ with G'/N a Lie group. Set $A' := A^4 \cap \pi^{-1}(G')$.)

(viii) Show that if A is an ultra K-approximate group, then there is an ultra $K^{O(1)}$-approximate group A'' in $A^{O(1)}$ that is modeled by a nilpotent group of dimension $O(\log K)$, such that A can be covered by finitely many left-translates of A.

9.3. Local groups

The main weakness of the cheap structure theorem in the preceding section is the continual reintroduction of torsion whenever one quotients out by the centraliser $Z(u)$, which can destroy the NSS property. We now address the issue of how to fix this, by moving to the context of local groups rather than global groups. We will omit some details, referring to [**BrGrTa2011**] for details.

We need to extend many of the notions we have been considering to the local group setting. We begin by generalising the concept of an approximate group.

Definition 9.3.1 (Approximate groups)**.** A (local) K-*approximate group* is a subset A of a local group G which is symmetric and contains the identity, such that A^{200} is well-defined in G, and for which A^2 is covered by K left-translates of A (by elements in A^3). An *ultra approximate group* is an ultraproduct $A = \prod_{n \to \alpha} A_n$ of K-approximate groups.

Note that we make no topological requirements on A or G in this definition; in particular, we may as well give the local group G the discrete topology. There are some minor technical advantages in requiring the local group to be symmetric (so that the inversion map is globally defined) and cancellative (so that $gh = gk$ or $hg = kg$ implies $h = k$), although these assumptions are essentially automatic in practice.

The exponent 200 here is not terribly important in practice, thanks to the following variant of the Sanders lemma:

Exercise 9.3.1 (local Sanders lemma)**.** Let A be a finite K-approximate group in a local group G, except with only A^8 known to be well-defined rather than A^{200}. Let $m \geq 1$. Show that there exists a finite $O_{K,m}(1)$-approximate subgroup A' in G with $(A')^m$ well-defined and contained in A^4, and with A covered by $O_{K,m}(1)$ left-translates of A' (by elements in A^5). (*Hint:* Adapt the proof of Lemma 8.0.8.)

Just as global approximate groups can be modeled by global locally compact groups (and in particular, global Lie groups), local approximate groups can be modeled by local locally compact groups:

Definition 9.3.2 (Good models). Let A be a (local) ultra approximate group. A (local) *good model* for A is a homomorphism $\pi \colon A^8 \to L$ from A^8 to a locally compact Hausdorff local group L that obeys the following axioms:

(1) (Thick image) There exists a neighbourhood U_0 of the identity in L such that $\pi^{-1}(U_0) \subset A$ and $U_0 \subset \pi(A)$.

(2) (Compact image) $\pi(A)$ is precompact.

(3) (Approximation by nonstandard sets) Suppose that $F \subset U \subset U_0$, where F is compact and U is open. Then there exists a nonstandard finite set B such that $\pi^{-1}(F) \subset B \subset \pi^{-1}(U)$.

We make the pedantic remark that with our conventions, a global good model $\pi \colon \langle A \rangle \to L$ of a global approximate group only becomes a local good model of A by L after restricting the domain of π to A^8. It is also convenient for minor technical reasons to assume that the local group L is symmetric (i.e., the inversion map is globally defined) but this hypothesis is not of major importance.

The Hrushovski Lie Model theorem can be localised:

Theorem 9.3.3 (Local Hrushovski Lie model theorem). *Let A be a (local) ultra approximate group. Then there is an ultra approximate subgroup A' of A (thus $(A')^4 \subset A^4$) with A covered by finitely many left-translates of A' (by elements in $A \cdot (A')^{-1}$), which has a good model by a connected local Lie group L.*

The proof of this theorem is basically a localisation of the proof of the global Lie model theorem from Chapter 8, and it is omitted (see **[BrGrTa2011]** for details). One key replacement is that if A is a local approximate group rather than a global one, then the global Gleason-Yamabe theorem (Theorem 1.1.13) must be replaced by the local Gleason-Yamabe theorem (Theorem 5.6.9).

One can define the notion of a strong K-approximate group and ultra strong approximate group in the local setting without much difficulty, since strong approximate groups only need to work inside A^{100}, which is well-defined. Using the local Lie model theorem, one can obtain a local version of Exercise 9.0.8. The Gleason lemma (Theorem 9.0.10) also localises without much difficulty to local strong approximate groups, as does the reduction to the NSS case in Exercise 9.2.1.

Now we once again analyse the NSS case. As before, let L be a connected (local) Lie group, with Lie algebra $\mathfrak{l}$, let B be a bounded symmetric convex body in $\mathfrak{l}$, let $r > 0$ be a sufficiently small standard real. Let A be a (local) ultra strong NSS approximate group which has a (local) good model $\pi \colon \langle A \rangle \to L$ with

$$\pi^{-1}(\exp(rB)) \subset A \subset \pi^{-1}(\exp(1.1rB)).$$

Again, we assume L has dimension at least 1, since A is trivial otherwise. We let u be a nonidentity element of minimal escape norm. As before, u will have an infinitesimal escape norm and lie in the kernel of π. If we set $N := \|u\|_{e,A}$, then N is an unbounded natural number, and the map $\phi \colon t \mapsto \pi(g^{\lfloor tN \rfloor})$ will be a local one-parameter subgroup, i.e., a continuous homomorphism from $[-1, 1]$ to L. This one-parameter subgroup will be nontrivial and centralised by a neighbourhood of the identity in L.

In the global setting, we quotiented (the group generated by a large portion of) A by the centraliser $Z(u)$ of u. In the local setting, we perform a more "gentle" quotienting, which roughly speaking arises by quotienting A by the geometric progression $P := \{u^n : -cN \le n \le cN\}$, where $c > 0$ is a sufficiently small standard quantity to be chosen later. However, P is only a local group rather than a global one, and so we will need the machinery of notion of quotients of local groups from Section 5.6; the reader is encouraged to review that section now.

We now begin the analysis of the NSS ultra strong approximate group A. We give the ambient local group G the discrete topology.

Exercise 9.3.2. If $r' > 0$ is a standard real that is sufficiently small depending on c, show that there exists an ultra approximate group A' with

$$\pi^{-1}(\exp(r'B)) \subset A' \subset \pi^{-1}(\exp(1.1r'B)),$$

such that P is a sub-local group of A with normalising neighbourhood $(A')^6 \cup P$, that is also centralised by A'.

By Exercise 5.6.2, we may now form the quotient set $A'' := A'/P$. Show that this is an ultra approximate group that is modeled by $U/\phi(-c, c)$, where U is an open neighbourhood of the identity in L and $\phi \colon [-1, 1] \to L$ is the local one-parameter subgroup of L introduced earlier. In particular, A'' is modeled by a local Lie group of dimension one less than the dimension of L.

Now we come to a key observation, which is the main reason why we work in the local groups category in the first place:

Lemma 9.3.4 (Preservation of the NSS property). *A'' is NSS.*

We will in fact prove a stronger claim:

Lemma 9.3.5 (Lifting lemma). *If $g \in A''$, then there exists $\tilde{g} \in A'$ such that $\kappa(\tilde{g}) = g$ and $\|\tilde{g}\|_{e,A'} \ll \|g\|_{e,A''}$, where $\kappa \colon A' \to A''$ is the projection map.*

Since A' is NSS, all nonidentity elements $\tilde{g}$ of A' have nonzero escape norm, and so by the lifting lemma, all nonidentity elements of A'' also have nonzero escape norm, giving Lemma 9.3.4.

Proof of Lemma 9.3.5. We choose $\tilde{g}$ to be a lift of g (i.e., an element of $\kappa^{-1}(\tilde{g})$ in A') that minimises the escape norm $\|\tilde{g}\|_{e,A'}$. (Such a minimum exists since A' is nonstandard finite, thanks to Los's theorem.) If $\tilde{g}$ is trivial, then so is g and there is nothing to prove. Therefore we may assume that $\tilde{g}$ is not the identity and hence, since A' is NSS, that it has positive escape norm. Suppose, by way of contradiction, that $\|g\|_{e,A'/P} = o(\|\tilde{g}\|_{e,A'})$. Our goal will be to reach a contradiction by finding another lift h of g with strictly smaller escape norm than $\tilde{g}$. We will do this by setting $h = \tilde{g}u^m$ for some suitably chosen m.

We may assume that $\|g\|_{e,A''/P}$ is infinitesimal, since otherwise there is nothing to prove; in particular, g lies in the kernel of the local model $\tilde{\pi} \colon A'/P \to U/\phi(-c,c)$. We may thus find a lift $\tilde{g}$ of g in the kernel of π. In particular, we may assume that $\tilde{g}$ has infinitesimal escape norm.

Set $M := 1/\|\tilde{g}\|_{e,A'}$, then M is unbounded. By hypothesis, $\|g\|_{e,A'/P} = o(1/M)$; thus $g^n \in A'/P$ whenever $n = O(M)$. In particular, for every (standard) integer $k \in \mathbf{N}$, $g^{kn} \in A'/P$. This implies that the group generated by g^n lies in A'/P. In particular, g^n lies in the kernel of $\tilde{\pi}$, and hence $\pi(\tilde{g}^n)$ lies in $\phi(-c,c)$ for all $-M \leq n \leq M$.

By (an appropriate local version of) Exercise 9.2.2, we can find $X \in 1.1B \backslash B$ such that

$$(9.11) \qquad \pi(\tilde{g}^n) = \exp(\mathrm{st}(n/M)r'X)$$

whenever $|n| \leq \frac{r}{r'}M$; since $\pi(\tilde{g}^n)$ lies in $\phi(-c,c)$ for $|n| \leq M$, we conclude that X must be parallel to the generator $\phi'(0)$ of ϕ. Similarly, we have

$$(9.12) \qquad \pi(u^n) = \exp(\mathrm{st}(n/N)rY)$$

whenever $|n| \leq 4N$ (say) for some $Y \in 1.1B \backslash B$ that is also parallel to $\phi'(0)$. In particular, $Y = \alpha X$ for some

$$\frac{1}{1.1} \leq \alpha \leq 1.1.$$

Since $1/N = \|u\|_{e,A}$ is the minimal escape norm of nonidentity elements of A, we have $\|\tilde{g}\|_{e,A} \geq 1/N$, and thus $\tilde{g}^i \in A^2 \backslash A$ for some $1 \leq i \leq N$; in particular, $\pi(\tilde{g}^i) \notin \exp(rB)$. Comparing this with (9.11) we see that

$$\mathrm{st}(i/M)r'X \notin rB$$

and thus

$$\operatorname{st}\left(\frac{N}{M}\right) \geq \frac{1}{1.1}\frac{r}{r'},$$

and hence

$$\frac{M\alpha}{N} \leq 1.3\frac{r'}{r}.$$

By the Euclidean algorithm, we can thus find a nonstandard integer number m such that the quantity

$$\theta := 1 + m\frac{M\alpha r}{Nr'}$$

lies in the interval $[-0.5, 0.5]$. In particular,

$$|m| \leq 2\frac{Nr'}{Mr}.$$

If we set $h := \tilde{g}u^m$ then (as u commutes with $\tilde{g}$) we see for all $|n| \leq M$ that

$$h^n = \tilde{g}^n u^{mn}$$

and thus by (9.11), (9.12)

$$\begin{aligned}
\pi(h^n) &= \exp((\operatorname{st}(n/M)r' + \operatorname{st}(mn/N)\alpha r)X) \\
&= \exp(\operatorname{st}(n/M)\theta r'X) \\
&\in \exp(r'B)
\end{aligned}$$

for all $|n| \leq M$. In particular, $\|h\|_{e,A'} < 1/M$. Since h is also a lift of g, this contradicts the minimality of $\|\tilde{g}\|_{e,A'} = 1/M$, and the claim follows.   □

Because the NSS property is preserved, it is possible to improve upon Exercise 9.2.4:

Exercise 9.3.3. Strengthen Exercise 9.2.4 by ensuring the final quotient $G'_{k-1}/G'_k = G'_{k-1}$ is nonstandard finite, and all the other quotients G'_i/G'_{i+1} are central in G'/G'_{i+1}.

As a consequence, one obtains a stronger structure theorem than Exercise 9.2.4. Call a symmetric subset U containing the identity in a local group *nilpotent of step at most s* if every iterated commutator in U of length $s + 1$ is well-defined and trivial.

Exercise 9.3.4 (Helfgott-Lindenstrauss conjecture).

(i) Let A be a (local) NSS ultra strong approximate group. Show that there is a symmetric subset A' of A containing the identity which is nilpotent of some finite step, such that A is covered by a finite number of left-translates of A'.

(ii) Let A be a global NSS ultra strong approximate group with ambient group G. Show that there is a nonstandard nilpotent subgroup G' of G such that A is covered by a finite number of left-translates of G'.

(iii) Let A be an NSS strong K-approximate group in a global group G. Show that there is a nilpotent subgroup G' of G of step $O_K(1)$ such that A can be covered by a finite number of left-translates of G'.

(iv) Let A be a K-approximate group in a global group G. Show that there exists a subgroup G' of G and a normal subgroup N of G' contained in A^4, such that A is covered by $O_K(1)$ left-translates of G', and G'/N is nilpotent of step $O_K(1)$.

In fact, a stronger statement is true, involving the nilprogressions defined in Chapter 7:

Proposition 9.3.6.

(i) *If A is an NSS ultra strong approximate group, then there is an ultra nilprogression Q in G such that A contains Q, and A can be covered by finitely many left-translates of Q.*

(ii) *If A is an ultra approximate group, then there is an ultra coset nilprogression Q in G such that A^4 contains Q, and A can be covered by finitely many left-translates of Q.*

(iii) *For all $K \geq 1$, there exists $C_K, s_K, r_K \geq 1$ such that, given a finite K-approximate group A in a group $G = (G, \cdot)$, one can find a coset nilprogression Q in G of rank at most r_K and step at most s_K such that A^4 contains Q, and A can be covered by at most C_K left-translates of Q.*

This proposition is established in [**BrGrTa2011**]. The key point is to use the lifting lemma to observe that if (with the notation of the preceding discussion) A'/P contains a large nilprogression, then A' also contains a large nilprogression. One consequence of this proposition is that there is essentially no difference between local and global approximate groups, at the qualitative level at least:

Corollary 9.3.7. *Let A be a local K-approximate group. Then there exists a $O_K(1)$-approximate subgroup A' of A, with A covered by $O_K(1)$ left-translates of A', such that A' is isomorphic to a global $O_K(1)$-approximate subgroup.*

This is because coset nilprogressions (or large fractions thereof) can be embedded into global groups; again, see [**BrGrTa2011**] for details.

For most applications, one does not need the full strength of Proposition 9.3.6; Exercise 9.3.4 will suffice. We will give some examples of this in the next section.

Chapter 10

Applications of the structural theory of approximate groups

In Chapter 9, we obtained the following structural theorem concerning approximate groups:

Theorem 10.0.8. *Let A be a finite K-approximate group. Then there exists a coset nilprogression P of rank and step $O_K(1)$ contained in A^4, such that A is covered by $O_K(1)$ left-translates of P (and hence also by $O_K(1)$ right-translates of P).*

Remark 10.0.9. Under some mild additional hypotheses (e.g., if the dimensions of P are sufficiently large, or if P is placed in a certain "normal form", details of which may be found in [**BrGrTa2011**]), a coset nilprogression P of rank and step $O_K(1)$ will be an $O_K(1)$-approximate group, thus giving a partial converse to Theorem 10.0.8. (It is not quite a full converse though, even if one works qualitatively and forgets how the constants depend on K: if A is covered by a bounded number of left- and right-translates gP, Pg of P, one needs the group elements g to "approximately normalise" P in some sense if one wants to then conclude that A is an approximate group.) The mild hypotheses alluded to above can be enforced in the statement of the theorem, but we will not discuss this technicality here, and refer the reader to the above-mentioned paper for details.

By placing the coset nilprogression in a virtually nilpotent group, we have the following corollary in the global case:

Corollary 10.0.10. *Let A be a finite K-approximate group in an ambient group G. Then A is covered by $O_K(1)$ left-cosets of a virtually nilpotent subgroup G' of G.*

In this section, we give some applications of the above results. The first application is to replace "K-approximate group" by "sets of bounded doubling":

Proposition 10.0.11. *Let A be a finite nonempty subset of a (global) group G such that $|A^2| \leq K|A|$. Then there exists a coset nilprogression P of rank and step $O_K(1)$ and cardinality $|P| \gg_K |A|$ such that A can be covered by $O_K(1)$ left-translates of P, and also by $O_K(1)$ right-translates of P.*

We will also establish (a strengthening of) a well-known theorem of Gromov [**Gr1981**] on groups of polynomial growth, as promised back in Chapter 1, as well as a variant result (of a type known as a "generalised Margulis lemma") controlling the almost stabilisers of discrete actions of isometries.

The material here is largely drawn from [**BrGrTa2011**].

10.1. Sets of bounded doubling

In this section we will deduce Proposition 10.0.11 from Theorem 10.0.8. This can be done using the general (nonabelian) additive combinatorics machinery from [**Ta2008b**], but we will give here an alternate argument relying on a version of the Croot-Sisask lemma used in Chapter 8 which is a little weaker with regards to quantitative bounds, but is slightly simpler technically (once one has the Croot-Sisask lemma).

We recall the Croot-Sisask lemma:

Lemma 10.1.1 (Croot-Sisask [**CrSi2010**]). *Let A be a nonempty finite subset of a group G such that $|A^2| \leq K|A|$. Then any $M \geq 1$, there is a symmetric set S containing the origin with $|S| \gg_{K,M} |A|$ such that*

$$(10.1) \qquad \|1_A * 1_A - \tau(g)1_A * 1_A\|_{\ell^2(G)} \leq \frac{1}{M}|A|^{3/2}$$

for all $g \in S$.

Let A be as in Proposition 10.0.11. We apply Lemma 10.1.1 with some large M depending on K to be chosen later. Then for any $g \in S^{100}$ one has

$$\|1_A * 1_A - \tau(g)1_A * 1_A\|_{\ell^2(G)} \leq \frac{100}{M}|A|^{3/2}.$$

Since $1_A * 1_A$ has an $\ell^1(G)$ norm of $|A|^2$ and is supported on the set A^2, which has cardinality at most $K|A|$, we see from Cauchy-Schwarz that

$$\|1_A * 1_A\|_{\ell^2(G)} \gg_K |A|^{3/2}$$

and hence (if M is large enough depending on K)

$$\|\tau(g)1_A * 1_A\|_{\ell^2(A^2)} \gg_K |A|^{3/2}.$$

In particular, we have $|gA^2 \cap A^2| \gg_K |A|$, thus every element of S^{100} has $\gg_K |A|$ representations of the form xy^{-1} with $x, y \in A^2$. As there are at most $K^2|A|^2$ pairs (x, y) with $x, y \in A^2$, we conclude that $|S^{100}| \ll_K |A|$. In particular, by the Ruzsa covering lemma (Exercise 8.2.3) we see that S^4 can be covered by $O_{K,M}(1)$ left-translates of S^2, and hence S^2 is a $O_{K,M}(1)$-approximate group.

In view of Theorem 10.0.8, we thus see that to conclude the proof of Proposition 10.0.11, it suffices to show that A can be covered by $O_{K,M}(1)$ left-translates (or right-translates) of S^2 if M is sufficiently large depending on K.

We will just prove the claim for left-translates, as the claim for right-translates is similar. We will need the following useful inequality:

Lemma 10.1.2 (Ruzsa triangle inequality). *Let A, B, C be finite nonempty subsets of a group G. Then $|A \cdot C^{-1}| \leq \frac{|A \cdot B^{-1}||B \cdot C^{-1}|}{|B|}$.*

Proof. Observe that if x is an element of $A \cdot C^{-1}$, so that $x = ac^{-1}$ for some $a \in A$ and $c \in C$, then x has at least $|B|$ representations of the form $x = yz$ with $y \in A \cdot B^{-1}$ and $z \in B \cdot C^{-1}$, since $ac^{-1} = (ab^{-1})(bc^{-1})$ for all $b \in B$. As there are only $|A \cdot B^{-1}||B \cdot C^{-1}|$ possible pairs (y, z) that could form such representations, the claim follows. $\square$

Now we return to the proof of Proposition 10.0.11. From (10.1) and Minkowski's inequality we see that

$$\left\| 1_A * 1_A - \frac{1}{|S|} 1_S * 1_A * 1_A \right\|_{\ell^2(G)} \leq \frac{1}{M} |A|^{3/2},$$

and thus (if M is sufficiently large depending on K)

$$\left\| \frac{1}{|S|} 1_S * 1_A * 1_A \right\|_{\ell^2(A^2)} \gg_K |A|^{3/2}$$

and, in particular,

$$\left\| \frac{1}{|S|} 1_S * 1_A * 1_A \right\|_{\ell^\infty(A^2)} \gg_K |A|$$

and thus by Young's inequality

$$\left\| \frac{1}{|S|} 1_S * 1_A \right\|_{\ell^\infty(G)} \gg_K 1,$$

and so

$$1_S * 1_A(x) \gg_{K,M} |A|$$

for some $x \in G$. In other words,

$$|S \cap xA^{-1}| \gg_{K,M} |A|.$$

Now if we set $B := S \cap xA^{-1}$, then

$$|B \cdot S^{-1}| \le |S^2| \ll_{K,M} |A|$$

and

$$|A \cdot B^{-1}| \le |A^2 x| = |A^2| \le K|A|$$

and hence by the Ruzsa triangle inequality

$$|A \cdot S| = |A \cdot S^{-1}| \ll_{K,M} |A|.$$

By the Ruzsa covering lemma (Exercise 8.2.3), this implies that A can be covered by $O_{K,M}(1)$ left-translates of S, as required. This proves Proposition 10.0.11. By placing the coset nilprogression in a virtually nilpotent group, we obtain a strengthening of Corollary 10.0.10:

Corollary 10.1.3. *Let A be a finite nonempty subset of an ambient group G such that $|A^2| \le K|A|$. Then A is covered by $O_K(1)$ left-cosets (and also by $O_K(1)$ right-cosets) of a virtually nilpotent subgroup G' of G.*

We remark that there is also an "off-diagonal" version of Proposition 10.0.11:

Proposition 10.1.4. *Let A, B be finite nonempty subsets of a (global) group G such that $|AB| \le K|A|^{1/2}|B|^{1/2}$. Then there exists a coset nilprogression P of rank and step $O_K(1)$ and cardinality $|P| \gg_K |A|$ such that A can be covered by $O_K(1)$ left-translates of P, and B can be covered by $O_K(1)$ right-translates of P.*

This is a consequence of Theorem 10.0.8 combined with [**Ta2008b**, Theorem 4.6]; we omit the details. There is also a "statistical" variant (using [**Ta2008b**, Theorem 5.4] instead), based on an additional tool, the (nonabelian) Balog-Szemerédi-Gowers theorem, which will not be discussed in detail here:

Proposition 10.1.5. *Let A, B be finite nonempty subsets of a (global) group G such that*

$$|\{(a, b, a', b') \in A \times B \times A \times B : ab = a'b'\}| \ge |A|^{3/2}|B|^{3/2}/K.$$

Then there exists a coset nilprogression P of rank and step $O_K(1)$ and cardinality $|P| \gg_K |A|$ such that A intersects a left-translate of P in a set of cardinality $\gg_K |A|$, and B intersects a right-translate of P in a set of cardinality $\gg_K |A|$.

10.2. Polynomial growth

The above results show that finite approximate groups A (as well as related objects, such as finite sets of bounded doubling) can be efficiently covered by virtually nilpotent groups. However, they do not place *all* of A inside a virtually nilpotent group. Indeed, this is not possible in general:

Exercise 10.2.1. Let G be the "$ax + b$ group", that is to say, the group of all affine transformations $x \mapsto ax + b$ on the real line, with $a \in \mathbf{R}\backslash\{0\}$ and $b \in \mathbf{R}$. Show that there exists an absolute constant K and arbitrarily large finite K-approximate groups A in G that are not contained in any virtually nilpotent group. (*Hint:* Build a set A which is very "long" in the b direction and very "thin" in the a direction.)

Such counterexamples have the feature of being "thin" in at least one of the directions of G. However, this can be fixed by adding a "thickness" assumption to the approximate group. In particular, we have the following result:

Theorem 10.2.1 (Thick sets of bounded doubling are virtually nilpotent). *For every $K \geq 1$ there exists $M \geq 1$ such that the following statement holds: whenever G is a group, S is a finite symmetric subset of G containing the identity, and A is a finite set containing S^M such that $|A^2| \leq K|A|$, then S generates a virtually nilpotent group.*

Proof. Fix K, and let M be a sufficiently large natural number depending on K to be chosen later. Let S, A, G be as in the theorem. By Proposition 10.0.11, there exists a virtually nilpotent group H such that A is covered by $O_K(1)$ left-cosets of H. In particular, $S^m H$ consists of $O_K(1)$ left-cosets of H for all $0 \leq m \leq M$. On the other hand, as S contains the identity, $S^m H$ is nondecreasing in m. If M is large enough, then by the pigeonhole principle we may thus find some $0 \leq m < M$ such that $S^{m+1} H = S^m H$. By induction this implies that $S^k H = S^m H$ for all $k \geq m$; we conclude that

$$\langle S \rangle \subset \langle S \rangle H = \bigcup_{k=m}^{\infty} S^k H = S^m H.$$

In particular, $H \cap \langle S \rangle$ has finite index in $\langle S \rangle$. Since H is virtually nilpotent, we conclude that $\langle S \rangle$ is virtually nilpotent also. $\square$

This theorem leads to the following Gromov-type theorem:

Exercise 10.2.2 (Gromov-type theorem). Show that for every $C, d > 0$ there exists $M \geq 1$ such that the following statement holds: whenever G is a group generated by a finite symmetric set S of generators containing the identity, and $|S^m| \leq Cm^d|S|$ for some $m \geq M$, then G is virtually nilpotent.

Note that this implies Gromov's original theorem (Theorem 1.3.1) as a corollary, but it is stronger because (a) one only needs a polynomial growth bound at a single scale m, rather than at all scales, and (b) the lower bound required on m does not depend on the size of the generating set S. (A previous result in this direction, which obtained (a) but not (b), was established in [**ShTa2010**], by a rather different argument based on an argument of Kleiner [**Kl2010**]. The original proof of Gromov [**Gr1981**] of his theorem had some features in common with the arguments given here, in particular, using the machinery of Gromov-Hausdorff limits as well as some of the theory surrounding Hilbert's fifth problem, and was also amenable to nonstandard analysis methods as demonstrated in [**vdDrWi1984**], but differed in a number of technical details.)

Remark 10.2.2. By inspecting the arguments carefully, one can obtain a slightly sharper description of the group G in Exercise 10.2.2, namely that G contains a normal subgroup G' of index $O_d(1)$ which is the extension of a finitely nilpotent group of step and rank $O(d)$ by a finite group contained in S^m. See [**BrGrTa2011**] for details.

Exercise 10.2.3 (Gap between polynomial and nonpolynomial growth). Show that there exists a function $f \colon \mathbf{N} \to \mathbf{R}^+$ which grows faster than any polynomial (i.e., $f(m)/m^d \to \infty$ as $m \to +\infty$ for any d), with the property that $|S^m| \geq f(m)|S|$ whenever G is any finitely generated group that is not virtually nilpotent, and S is any symmetric set of generators of G that contains the identity.

Remark 10.2.3. No effective bound for the function f in this exercise is explicitly known, though in principle one could eventually extract such a bound by painstakingly finitising the proof of the structure theorem for approximate groups. If one restricts the size of S to be bounded, then one can take $f(m)$ to be $m^{(\log \log m)^c}$ for some $c > 0$ and m sufficiently large depending on the size of S, by the result of my paper with Shalom, but this is unlikely to be best possible. (In the converse direction, Grigorchuk's construction [**Gr1984**] of a group of intermediate growth shows that $f(m)$ cannot grow faster than $\exp(m^\alpha)$ for some absolute constant $\alpha < 1$ (and it is believed that one can take $\alpha = 1/2$).)

Exercise 10.2.4 (Infinite groups have at least linear growth). If G is an infinite group generated by a finite symmetric set S containing the identity, show that $|S^m| \geq |S| + m - 1$ for all $m \geq 1$.

Exercise 10.2.5 (Linear growth implies virtually cyclic). Let G be an infinite group generated by a finite symmetric set S containing the identity. Suppose that G is of *linear growth*, in the sense that $|S^m| \leq Cm$ for all $m \geq 1$ and some finite C.

(i) Place a left-invariant metric d on G by defining $d(x, y)$ to be the least natural number m for which $x \in yS^m$. Define a geodesic to be a finite or infinite sequence $(g_n)_{n \in I}$ indexed by some discrete interval $I \subset \mathbf{Z}$ such that $d(g_n, g_m) = |n - m|$ for all $n, m \in I$. Show that there exist arbitrarily long finite geodesics.

(ii) Show that there exists a doubly infinite geodesic $(g_n)_{n \in \mathbf{Z}}$ with $g_0 = 1$. (*Hint:* Use (i) and a compactness argument.)

(iii) Show that $|S^{2m}| \geq 2|S^m| - 1$ for all $m \geq 1$. (*Hint:* Study the balls of radius m centred at g_m and g_{-m}.) More generally, show that $|S^{km}| \geq 2k|S^m| - 2k + 1$ for all $m, k \geq 1$.

(iv) Show that $|S^m|/m$ converges to a finite nonzero limit α as $m \to \infty$, thus $|S^m| = \alpha m + o(m)$ for all $m \geq 1$, where $o(m)$ denotes a quantity which, when divided by m, goes to zero as $m \to \infty$.

(v) Show that for all $m \geq 1$, then all elements of S^m lie within a distance at most $o(m)$ of the geodesic $(g_{-m}, \ldots, g_m)$. (*Hint:* First show that all but at most $o(m)$ elements of S^m lie within this distance, using (iv) and the argument used to prove (iii).)

(vi) Show that for sufficiently large m, g_m^{-1} lies within distance $o(m)$ of g_{-m}.

(vii) Show that for sufficiently large m, S^{m+1} lies within distance m of $\{1, g_m, g_m^{-1}\}$.

(viii) Show that G is virtually cyclic (i.e., it has a cyclic subgroup of finite index).

Exercise 10.2.6 (Nilpotent groups have polynomial growth). Let S be a finite symmetric subset of a nilpotent group G containing the identity.

(i) Let s be an element of S that is not the identity, and let S' be the minimal symmetric set containing $S \backslash \{s, s^{-1}\}$ that is closed under the operations $g \mapsto [g, s^{\pm 1}]^{\pm 1}$. Show that S' is also a finite symmetric subset of G containing the identity, and that every element of S^m can be written in the form $s^i h$ for some $|i| \leq m$ and $h \in (S')^{O(m^2)}$, where the implied constant can depend on S, G.

(ii) Show that $|S^m| \leq Cm^d$ for all $m \geq 1$ and some $C, d > 0$ depending on S, G (i.e., G is of polynomial growth).

(iii) Show that every virtually nilpotent group is of polynomial growth.

10.3. Fundamental groups of compact manifolds (optional)

This section presupposes some familiarity with Riemannian geometry. Throughout this text, Riemannian manifolds are always understood to be complete and without boundary.

We now apply the above theory to establish some relationships between the topology (and more precisely, the *fundamental group*) of a compact Riemannian manifold, and the curvature of such manifolds. A basic theme in this subject is that lower bounds on curvature tend to give somewhat restrictive control on the topology of a manifold. Consider, for instance, *Myers' theorem* [**My1941**], which among other things tells us that a connected Riemannian manifold M with a uniform positive lower bound on the *Ricci curvature* Ric is necessarily compact (with an explicit upper bound on the diameter). In a similar vein we have the *splitting theorem*, which asserts that if a connected Riemannian manifold M has everywhere nonnegative Ricci curvature, then it splits as the product of a Euclidean space and a manifold without straight lines (i.e., embedded copies of $\mathbf{R}$).

To analyse the fundamental group $\pi_1(M)$ of a connected Riemannian manifold M, it is convenient to work with its universal cover:

Exercise 10.3.1. Let M be a connected Riemannian manifold.

(i) (Existence of universal cover) Show that there exists a simply connected Riemannian manifold $\tilde{M}$ with the same dimension as M with a smooth surjective map $\pi\colon \tilde{M} \to M$ which is a local diffeomorphism and a Riemannian isometry (i.e., the metric tensors are preserved); such a manifold (or more precisely, the pair (M, π)) is known as a *universal cover* of M. (*Hint:* Take $\tilde{M}$ to be the space of all paths from a fixed base point p_0 in $\tilde{M}$, quotiented out by homotopies fixing the endpoints. Once π is constructed, pull back the smooth and Riemannian structures.)

(ii) (Universality) Show that if M' is any smooth connected manifold with a smooth surjective map $f\colon M' \to M$ that is a local diffeomorphism and Riemannian isometry, and $p' \in M', p \in M, \tilde{p} \in \tilde{M}$ are point such that $f(p') = \pi(\tilde{p}) = p$, then there exists a unique smooth map $\pi'\colon \tilde{M} \to M'$ with $\pi'(\tilde{p}) = p'$ that makes $\tilde{M}$ a universal cover of M' also.

(iii) (Uniqueness) Show that a universal cover is unique up to isometric isomorphism.

(iv) (Covering space) Show that for every $p \in M$ there exists a neighbourhood U of p such that $\pi^{-1}(U)$ is isometric (as a Riemannian manifold) to $\pi_1(M) \times U$, where we give the fundamental group $\pi_1(M)$ the discrete topology. In particular, the fibres $\pi^{-1}(\{p\})$ of a point $p \in M$ are discrete and can be placed in bijection with $\pi_1(M)$.

(v) (Deck transformations) Show that $\pi_1(M)$ acts freely and isometrically on M', in such a way that the orbits of $\pi_1(M)$ are the fibres of π. Conversely, show that every isometry on M' that preserves the fibres of π arises from an element of $\pi_1(M)$.

(vi) (Cocompactness) if M is compact, and $\pi^{-1}(\{p\})$ is a fibre of M, show that every element of M' lies a distance at most $\mathrm{diam}(M)$ from an element of the fibre $\pi^{-1}(\{p\})$.

(vii) (Finite generation) If M is compact and $p \in \tilde{M}$, show that the set $\{g \in \pi_1(M) \colon \mathrm{dist}(gp, p) \leq 2\mathrm{diam}(M)\}$ is finite and generates $\pi_1(M)$. (*Hint:* If gp is further than $2\mathrm{diam}(M)$ from p, use (vi) to find a factorisation $g = hk$ such that kp is closer to p or gp than p is to gp.)

(viii) (Polynomial growth) If M is compact, show that the group $\pi_1(M)$ is of polynomial growth (thus $|S^m| \leq Cm^d$ for some generating set S, some $C, d \geq 0$, and all $m \geq 1$) if and only if the universal cover $\tilde{M}$ is of polynomial growth (thus $\mathrm{Vol}(B(x_0, r)) \leq Cr^d$ for some base point $x_0 \in \tilde{M}$, some $C, d \geq 0$, and all $r \geq 1$).

The above exercise thus links polynomial growth of groups to polynomial growth of manifolds. To control the latter, a useful tool is the *Bishop-Gromov inequality*:

Proposition 10.3.1 (Bishop-Gromov inequality). *Let M be a Riemannian manifold whose Ricci curvature is everywhere bounded below by some constant $\rho \in \mathbf{R}$. Let $\tilde{M}$ be the simply connected Riemannian manifold of constant curvature ρ and the same dimension as M (this will be a Euclidean space for $\rho = 0$, a sphere for $\rho > 0$, and hyperbolic space for $\rho < 0$). Let p be a point in M and $\tilde{p}$ a point in $\tilde{M}$. Then the expression*

$$\frac{\mathrm{vol}(B(p, r))}{\mathrm{vol}(B(\tilde{p}, r))}$$

is monotone nonincreasing in r.

We will not prove this proposition here (as it requires, among other things, a definition of Ricci curvature, which would be beyond the scope of this text); but see for instance [**Pe2006**, Chapter 9] or [**Ta2009b**, §2.10] for a proof. This inequality is consistent with the geometric intuition that an increase in curvature on a manifold should correspond to a stunting of the growth of the volume of balls. For instance, in the positively curved spheres, the volume of balls eventually stabilises as a constant; in the zero curvature Euclidean spaces, the volume of balls grows polynomially; and in the negative curvature hyperbolic spaces, the volume of balls grows exponentially.

Informally, the balls in M cannot grow any faster than the balls in $\tilde{M}$. Setting $\rho = 0$, we conclude in particular that if M has nonnegative Ricci curvature and dimension d, then $\mathrm{vol}(B(p, r))/r^d$ is nondecreasing in r; in particular, any manifold of nonnegative Ricci curvature is of polynomial growth. Applying Exercise 10.3.1 and Gromov's theorem, we conclude that any manifold of nonnegative Ricci curvature has a fundamental group which is virtually nilpotent. (In fact, such fundametal groups can be shown, using the splitting theorem, to be virtually abelian; see [**ChGr1972**]. However, this improvement seems to be beyond the combinatorial methods used here.)

The above monotonicity also shows that whenever M has Ricci curvature at least zero, we have the doubling bound

$$\mathrm{vol}(B(p, 2r)) \leq 2^d \, \mathrm{vol}(B(p, r)).$$

A continuity argument then shows that for every $R > 0$ and $\varepsilon > 0$, there exists $\delta > 0$ such that if M has Ricci curvature at least $-\delta$, then one has

$$\mathrm{vol}(B(p, 2r)) \leq (2^d + \varepsilon) \, \mathrm{vol}(B(p, r))$$

for all $0 < r \leq R$.

Exercise 10.3.2. By using the above observation combined with Exercise 10.3.1 and Exercise 10.2.2, show that for every dimension d there exists $\delta > 0$ such that if M is any compact Riemannian manifold of diameter at most 1 and Ricci curvature at least $-\delta$ everywhere, then $\pi_1(M)$ is virtually nilpotent.

Remark 10.3.2. This result was first conjectured by Gromov, and was proven by Cheeger and Colding [**ChCo1996**] and Kapovitch and Wilking [**KaWi2011**] using deep Riemannian geometry tools (beyond just the Bishop-Gromov inequality). (See also the paper [**KaPeTu2010**], which established the analogous result assuming a lower bound on *sectional curvature* rather than Ricci curvature.)

The arguments in these papers in fact give more precise information on the fundamental group $\pi_1(M)$, namely that there is a nilpotent subgroup of step and rank $O_d(1)$ and index $O_d(1)$. The methods here (based purely on controlling the growth of balls in $\pi_1(M)$) can give the step and rank bounds, but appear to be insufficient to obtain the index bound. The exercise cannot be immediately obtained via a compactness-and-contradiction argument from the (easier) $\delta = 0$ case mentioned previously, because of the problem of *collapsing*: there is no lower bound assumed on the injectivity radius of M, and as such the space of all manifolds with the indicated diameter is noncompact even if one bounds the derivatives of the metric to all orders. (An equivalent way of phrasing the problem is that the orbits of $\pi_1(M)$ in the universal cover may be arbitrarily dense, and so a ball of bounded radius in $\tilde{M}$ may correspond to an arbitrarily large subset S of the fundamental group. For this application it is thus of importance that there is no upper bound on the size of the sets S or A assumed in Exercise 10.2.2.)

Remark 10.3.3. One way to view the above results is as an assertion that it is quite rare for a compact manifold to be equippable with a Riemannian metric with (almost) nonnegative Ricci curvature. Indeed, an application of van Kampen's theorem [**Se1931**], [**vKa1833**] shows that every fundamental group $\pi_1(M)$ of a compact manifold is finitely presented, and conversely a gluing argument for four-manifolds shows that every finitely presented group is the fundamental group of some (four-dimensional) manifold. Intuitively, "most" finitely presented groups are not virtually nilpotent, and so "most" compact manifolds cannot have metrics with almost nonnegative Ricci curvature.

10.4. A Margulis-type lemma

In Exercise 10.3.1 we saw that the fundamental group $\pi_1(M)$ of a connected Riemannian manifold can be viewed as a discrete group of isometries acting on a Riemannian manifold $\tilde{M}$. The curvature properties of $\tilde{M}$ then give doubling properties of the balls in $\tilde{M}$, and hence of $\pi_1(M)$, allowing one to use tools such as Exercise 10.2.2.

It turns out that one can abstract this process by replacing the universal cover $\tilde{M}$ by a more general metric space X:

Lemma 10.4.1 (Margulis-type lemma). *Let $K \geq 1$. Let $X = (X, d)$ be a metric space, with the property that every ball of radius 4 can be covered by K balls of radius 1. Let Γ be a group of isometries of X, which acts discretely in the sense that $\{g \in \Gamma : gx \in B\}$ is finite for every $x \in X$ and every bounded set $B \subset X$. Then if $x \in X$ and ε is sufficiently small depending on K, the set $S_\varepsilon := \{g \in \Gamma : d(gx, x) \leq \varepsilon\}$ generates a virtually nilpotent group.*

Proof. We can can cover $B(x, 4)$ by K balls $B(x_i, 1)$. If $g, h \in S_4$ and $gx, hx \in B(x_i, 1)$, then (by the isometric action of Γ) we see that $h^{-1}g \in S_2$. We conclude that S_4 can be covered by K right-translates of S_2. Since $S_2^2 \subset S_4$ and $S_{2/M}^M \subset S^2$ for all $M \geq 1$, the claim then follows from Exercise 10.2.2. $\square$

Roughly speaking, the above lemma asserts that for discrete actions of isometries on "spaces of bounded doubling", the "almost stabiliser" of a point is "virtually nilpotent". In the case when X is a Riemannian manifold with a lower bound on curvature, this result was established in [**ChCo1996**], [**KaWi2011**] (and, as mentioned in the previous section, stronger control on S_ε was established). The original lemma of Margulis addressed the case when X was a hyperbolic space, and relied on commutator estimates not unrelated to the commutator estimates of Gleason metrics and of strong approximate groups that were used in previous sections.

Part 2

Related Articles

Chapter 11

The Jordan-Schur theorem

We recall *Jordan's theorem*, proven in Exercise 1.0.3:

Theorem 11.0.2 (Jordan's theorem). *Let G be a finite subgroup of the general linear group $GL_d(\mathbf{C})$. Then there is an abelian subgroup G' of G of index $[G : G'] \leq C_d$, where C_d depends only on d.*

Informally, Jordan's theorem asserts that finite linear groups over the complex numbers are almost abelian. The theorem can be extended to other fields of characteristic zero, and also to fields of positive characteristic so long as the characteristic does not divide the order of G, but we will not consider these generalisations here.

The finiteness hypothesis on the group G in this theorem can be relaxed to the significantly weaker condition of *periodicity*. Recall that a group G is periodic if all elements are of finite order. Jordan's theorem with "finite" replaced by "periodic" is known as the *Jordan-Schur theorem*.

The Jordan-Schur theorem can be quickly deduced from Jordan's theorem, and the following result of Schur:

Theorem 11.0.3 (Schur's theorem). *Every finitely generated periodic subgroup of a general linear group $GL_d(\mathbf{C})$ is finite. (Equivalently, every periodic linear group is* locally finite.*)*

Remark 11.0.4. The question of whether *all* finitely generated periodic subgroups (not necessarily linear in nature) are finite was known as the *Burnside problem*; the answer was shown to be negative in [**GoSa1964**].

Let us see how Jordan's theorem and Schur's theorem combine via a compactness argument to form the Jordan-Schur theorem. Let G be a periodic subgroup of $GL_d(\mathbf{C})$. Then for every finite subset S of G, the group G_S generated by S is finite by Theorem 11.0.3. Applying Jordan's theorem, G_S contains an abelian subgroup G'_S of index at most C_d.

In particular, given any finite number $S_1, \ldots, S_m$ of finite subsets of G, we can find abelian subgroups $G'_{S_1}, \ldots, G'_{S_m}$ of $G_{S_1}, \ldots, G_{S_m}$, respectively, such that each G'_{S_j} has index at most C_d in G_{S_j}. We claim that we may furthermore impose the compatibility condition $G'_{S_i} = G'_{S_j} \cap G_{S_i}$ whenever $S_i \subset S_j$. To see this, we set $S := S_1 \cup \cdots \cup S_m$, locate an abelian subgroup G'_S of G_S of index at most C_d, and then set $G'_{S_i} := G'_S \cap G_{S_i}$. As G_S is covered by at most C_d cosets of G'_S, we see that G_{S_i} is covered by at most C_d cosets of G'_{S_i}, and the claim follows.

Note that for each S, the set of possible G'_S is finite, and so the product space of all configurations $(G'_S)_{S \subset G}$, as S ranges over finite subsets of G, is compact by *Tychonoff's theorem*. Using the *finite intersection property*, we may thus locate a subgroup G'_S of G_S of index at most C_d for *all* finite subsets S of G, obeying the compatibility condition $G'_T = G'_S \cap G_T$ whenever $T \subset S$. If we then set $G' := \bigcup_S G'_S$, where S ranges over all finite subsets of G, we then easily verify that G' is abelian and has index at most C_d in G, as required.

11.1. Proofs

We now give a proof of Schur's theorem, based on the proof in [**We1973**]. We begin with a lemma of Burnside. Given a vector space V, let $\mathrm{End}(V)$ denote the ring of linear transformations from V to itself.

Lemma 11.1.1. *Let V be a finite-dimensional complex vector space, and let A be a complex algebra with identity in $\mathrm{End}(V)$, i.e., a linear subspace of $\mathrm{End}(V)$ that is closed under multiplication and contains the identity operator. Then either $A = \mathrm{End}(V)$, or there exists some proper subspace $\{0\} \subsetneq W \subsetneq V$ which is A-invariant, i.e., $aW \subset W$ for all $a \in A$.*

Proof. Suppose that no such proper A-invariant subspace W exists. Then for any nonzero $v \in V$, the vector space Av must equal all of V, since it is a nontrivial A-invariant subspace. By duality, this implies that for any nonzero dual vector $u \in V^*$, the vector space A^*u must equal all of V^*.

Let u, v be linearly independent elements of V. We claim that there exists an element a of A such that $au \neq 0$ and $av = 0$. Suppose that this is not the case; then by the Hahn-Banach theorem, there exists $t \in \mathrm{End}(V)$ such that $au = tav$ for all $a \in A$. In particular, setting $a = 1$ we obtain

$u = tv$, and thus $atv = tav$. Replacing a by ab for some $b \in A$, we conclude that $tabv = abtv = atbv$, thus $ta - at$ annihilates all of Av. Since $Av = V$, we conclude that $at = ta$, thus A lies in the centraliser of t. But since $u = tv$ and u, v are linearly independent, t is not a multiple of the identity, and thus by the spectral theorem, t has at least one proper eigenspace. But this eigenspace is fixed by A, a contradiction.

Thus we can find $a \in A$ such that $au \neq 0$ and $av = 0$ for any linearly independent u, v. Iterating this, we see that for any nonzero $u \in V$ and any $0 \leq k \leq \dim(V) - 1$, we can find $a \in A$ of corank at least k that does not annihilate u. In particular, A contains a rank one transformation. Since $Av = V$ and $A^*u = V^*$ for all $v \in V$ and $u \in V^*$, this implies that A contains *all* rank one transformations, and hence contains all of $\text{End}(V)$ by linearity. $\qquad\square$

Corollary 11.1.2 (Burnside's theorem). *Let V be a complex vector space of some finite dimension d, and let G be a subgroup of $GL(V)$ where every element has order at most r. Then G is finite with cardinality $O_{d,r}(1)$.*

Proof. We induct on dimension, assuming the claim has already been proven for smaller values of d. Let $A \subset \text{End}(V)$ be the complex algebra generated by G (or equivalently, the complex linear span of G). Suppose first that there is a proper A-invariant subspace W. Then G projects down to $GL(W)$ and to $GL(V/W)$, and by induction hypothesis both of these projections are finite with cardinality $O_{d,r}(1)$. Thus there exists a subgroup G' of G of index $O_{d,r}(1)$ whose projections to $GL(W)$ and $GL(V/W)$ are trivial; in particular, all elements of G' are unipotent. But as the complex numbers have zero characteristic, the only unipotent element of finite order is the identity, and so G' is trivial, and the claim follows.

Hence we may assume that V has no proper A-invariant subspace. By Lemma 11.1.1, A must be all of $GL(V)$. In particular, one can find d^2 linearly independent elements $g_1, \ldots, g_{d^2}$ of G.

For any $g \in G$, the element gg_i has order at most r, and thus all the eigenvalues of gg_i are roots of unity of order at most r. This means that there are at most $O_{r,d}(1)$ possible values of the trace $\text{tr}(gg_i)$, which is a linear functional of g. Letting g_i vary among the basis $g_1, \ldots, g_{d^2}$ of $\text{End}(V)$, we conclude that there are at most $O_{r,d}(1)$ possible values of g, and the claim follows. $\qquad\square$

Remark 11.1.3. The question of whether any finite group with m generators in which all elements of order at most r is necessarily of order $O_{m,r}(1)$ is known as the *restricted Burnside problem*, and was famously solved affirmatively by Zelmanov [**Ze1990**]. (Note, however, that for certain values of m and r it is possible for the group to be infinite. Also, while any finite group is trivially embedded in some linear group, one does not have any obvious control on the dimension of that group in terms of m and r, so one cannot immediately solve this problem just from Corollary 11.1.2.)

To prove Schur's theorem (Theorem 11.0.3), it thus suffices to establish the following proposition:

Proposition 11.1.4. *Let k be a finitely generated extension of the field of rationals $\mathbf{Q}$. Then every periodic element of $GL_d(k)$ has order at most $O_{d,k}(1)$.*

Indeed, to obtain Schur's theorem, one applies Proposition 11.1.4 with k equal to the field generated by the coefficients of the generators of the finitely generated periodic group G, and then applies Corollary 11.1.2.

Proof. Suppose first that k is a finite extension of $\mathbf{Q}$. If $a \in GL_d(k)$ has period n, then the field generated by the eigenvalues of a contains a primitive n^{th} root of unity, and thus contains the cyclotomic field of that order. On the other hand, this field has degree $O_d(1)$ over k, and thus has degree $O_{d,k}(1)$ over the rationals. Thus $n = O_{d,k}(1)$, and the claim follows. Note that the bound on n depends only on the degree of k, and not on k itself.

Now we extend from the finite degree case to the finitely generated case. By using a *transcendence basis*, one can write k as a finite extension of $\mathbf{Q}(z_1, \ldots, z_m)$ for some algebraically independent $z_1, \ldots, z_m$ over $\mathbf{Q}$. By the *primitive element theorem*, one can then write $k = \mathbf{Q}(z_1, \ldots, z_m)(\alpha)$ where α is algebraic over $\mathbf{Q}(z_1, \ldots, z_m)$ of some degree D.

Now suppose we have an element $a \in GL_d(k)$ of period n, thus $a, \ldots, a^{n-1} \neq 1$ and $a^n = 1$. Let R be the ring in k generated by the coefficients of a. We can create a ring homomorphism $\phi \colon R \to \bar{k}$ to a finite extension $\bar{k}$ of the rationals by mapping each z_i to a rational number q_i, and then replace α with a root of the polynomial formed by replacing z_i with q_i in the minimal polynomial of α. As long as one chooses $(q_1, \ldots, q_m)$ generically (i.e., outside of a codimension one subset of $\mathbf{Q}^m$), this operation is well-defined on R (in that no division by zero issues arise in any of the coefficients of a). Furthermore, generically one has $\phi(a), \ldots, \phi(a)^{n-1} \neq 1$ and $\phi(a)^n = 1$, thus ϕa has period n. Furthermore, the degree of $\bar{k}$ over $\mathbf{Q}$ is at most the degree D of α over $\mathbf{Q}(z_1, \ldots, z_m)$ and is thus bounded uniformly in $q_1, \ldots, q_m$. The claim now follows from the finite extension case. $\qquad\square$

Chapter 12

Nilpotent groups and nilprogressions

In this chapter we present some of the algebra of nilpotent groups, and as an application establish that all nilprogressions are approximate groups (Proposition 1.2.11). This turns out to be a moderately lengthy computation; an easier partial result in this direction appears in [**BrGrTa2011**], in which an additional hypothesis is imposed that the nilprogression be in "C-normal form".

There are several ways to think about nilpotent groups; for instance, one can use the model example of the Heisenberg group

$$H(R) := \begin{pmatrix} 1 & R & R \\ 0 & 1 & R \\ 0 & 0 & 1 \end{pmatrix}$$

over an arbitrary ring R (which need not be commutative), or more generally any matrix group consisting of unipotent upper triangular matrices, and view a general nilpotent group as being an abstract generalisation of such concrete groups. (In the case of nilpotent Lie groups, at least, this is quite an accurate intuition, thanks to *Engel's theorem.*) Or, one can adopt a Lie-theoretic viewpoint and try to think of nilpotent groups as somehow arising from nilpotent Lie algebras; this intuition is rigorous when working with nilpotent Lie groups (at least when the characteristic is large, in order to avoid issues coming from the denominators in the Baker-Campbell-Hausdorff formula), but also retains some conceptual value in the non-Lie setting. In particular, nilpotent groups (particularly finitely generated ones) can be viewed in some sense as "nilpotent Lie groups over $\mathbf{Z}$", even though Lie theory does not quite work perfectly when the underlying scalars merely form an integral domain instead of a field.

Another point of view, which arises naturally both in analysis and in algebraic geometry, is to view nilpotent groups as modeling "infinitesimal" perturbations of the identity, where the infinitesimals have a certain finite order. For instance, given a (not necessarily commutative) ring R without identity (representing all the "small" elements of some larger ring or algebra), we can form the powers R^j for $j = 1, 2, \ldots$, defined as the ring generated by j-fold products $r_1 \ldots r_j$ of elements $r_1, \ldots, r_j$ in R; this is an ideal of R which represents the elements which are of "j^{th} order" in some sense. If one then formally adjoins an identity 1 onto the ring R, then for any $s \geq 1$, the multiplicative group $G := 1 + R \bmod R^{s+1}$ is a nilpotent group of step at most s. For instance, if R is the ring of strictly upper $s \times s$ matrices (over some base ring), then R^{s+1} vanishes and G becomes the group of unipotent upper triangular matrices over the same ring, thus recovering the previous matrix-based example. In analysis applications, R might be a ring of operators which are somehow of "order" $O(\varepsilon)$ or $O(\hbar)$ for some small parameter ε or $\hbar$, and one wishes to perform Taylor expansions up to order $O(\varepsilon^s)$ or $O(\hbar^s)$, thus discarding (i.e., quotienting out) all errors in R^{s+1}.

From a dynamical or group-theoretic perspective, one can also view nilpotent groups as towers of central extensions of a trivial group. Finitely generated nilpotent groups can also be profitably viewed as a special type of polycyclic group; this is the perspective taken for instance in [**Ta2010b**, §2.13]. Last, but not least, one can view nilpotent groups from a combinatorial group theory perspective, as being words from some set of generators of various "degrees" subject to some commutation relations, with commutators of two low-degree generators being expressed in terms of higher degree objects, and all commutators of a sufficiently high degree vanishing. In particular, generators of a given degree can be moved freely around a word, as long as one is willing to generate commutator errors of higher degree.

With this last perspective, in particular, one can start computing in nilpotent groups by adopting the philosophy that the lowest order terms should be attended to first, without much initial concern for the higher order errors generated in the process of organising the lower order terms. Only after the lower order terms are in place should attention then turn to higher order terms, working successively up the hierarchy of degrees until all terms are dealt with. This turns out to be a relatively straightforward philosophy to implement in many cases (particularly if one is not interested in explicit expressions and constants, being content instead with qualitative expansions of controlled complexity), but the arguments are necessarily recursive in nature and as such can become a bit messy, and require a fair amount of notation to express precisely. So, unfortunately, the arguments here will be somewhat cumbersome and notation-heavy, even if the underlying methods of proof are relatively simple.

12.1. Some elementary group theory

Let g, h be two elements of a group $G = (G, \cdot)$. We define the *conjugate* g^h and *commutator*[1] $[g, h]$ by the formulae

(12.1) $$g^h := h^{-1}gh$$

and

(12.2) $$[g, h] := g^{-1}h^{-1}gh.$$

Conjugation by a fixed element h is an automorphism of G, thus

$$(gk)^h = g^h k^h$$

and

$$(g^h)^{-1} = (g^{-1})^h$$

and

$$1^h = 1$$

and

$$[g, k]^h = [g^h, k^h]$$

for all $g, h, k \in G$. Conjugation is also an action, thus

$$(g^h)^k = g^{hk}$$

and

$$g^1 = g.$$

An automorphism of the form $g \mapsto g^h$ is called an *inner automorphism*.

Conjugation is related to multiplication by the identity

$$gh = hg^h,$$

thus one can pull g to the right of h at the cost of twisting (i.e., conjugating) it by h. Commutation is related to multiplication by the identity

$$gh = hg[g, h],$$

---

[1]Note that this convention for $[g, h]$ is not universal; for instance, the alternate convention $[g, h] = ghg^{-1}h^{-1}$ also appears in the literature. The distinctions between the two conventions, however, are quite minor; the conventions here are optimised for pulling group elements to the right of a word, whereas other conventions may be slightly better for pulling group elements to the left of a word.

thus one can pull g to the right of h at the cost of adding an additional commutator factor $[g, h]$ to the right. Finally, commutation is related to conjugation by the identity

$$g^h = g[g, h].$$

The commutator can be viewed as a nonlinear group-theoretic analogue of the *Lie bracket*. For instance, in a matrix group $G = GL_n(\mathbf{C})$, we observe that the commutator $[1 + \varepsilon A, 1 + \varepsilon B]$ of two elements $1 + \varepsilon A$ and $1 + \varepsilon B$ close to the identity is of the form $1 + \varepsilon^2(AB - BA) + O(\varepsilon^3)$, thus linking the group-theoretic commutator to the Lie bracket $A, B \mapsto AB - BA$.

Because of this link, we expect the group-theoretic commutator to obey some nonlinear analogues of the basic Lie bracket identities, and this is indeed the case. For instance, one easily observes that the commutator is antisymmetric in the sense that

$$(12.3) \qquad\qquad [h, g] = [g, h]^{-1}$$

and is approximately odd in the sense that

$$(12.4) \qquad\qquad [g^{-1}, h] = ([g, h]^{-1})^{g^{-1}}$$

and

$$(12.5) \qquad\qquad [g, h^{-1}] = ([g, h]^{-1})^{h^{-1}}$$

for any $g, h \in G$. We also have the easily verified approximate bilinearity identities

$$[g, hk] = [g, h][g, k]^h$$

and

$$[gh, k] = [g, k]^h [h, k]$$

for any $g, h, k \in G$. Finally, we have the approximate *Jacobi identity* (better known as the *Hall-Witt identity*)

$$(12.6) \qquad\qquad [[g, h^{-1}], k]^h [[h, k^{-1}], g]^k [[k, g^{-1}], h]^g = 1.$$

A subgroup H of G is said to be *normal* if it is preserved by all inner automorphisms, thus $H^g = H$ for all g (writing $H^g := \{h^g : h \in H\}$, of course), and *characteristic* if it is preserved by all automorphisms (not necessarily inner). Thus, all characteristic subgroups are normal, but the converse is not necessarily true. We write $H \leq G$ or $G \geq H$ if H is a subgroup of G, and $H \triangleright G$ or $G \triangleleft H$ if H is a normal subgroup of G.

If N is a normal subgroup of G, we write $g \mapsto g \bmod N$ for the quotient map from G to G/N, thus $g = h \bmod N$ if $g \in hN = Nh$. Given any other subgroup H of G, we write $H \bmod N$ for the image of H under the $\bmod N$ map, thus

$$H \bmod N \equiv HN/N = NH/N \equiv H/(H \cap N).$$

Given two subgroups H, K of a group G, we define the commutator group $[H, K]$ to be the group generated by the commutators $[h, k]$ with $h \in H, k \in K$. We say that H and K *commute* if $[H, K]$ is trivial, or equivalently if every element of H commutes with every element of K. Note that if H, K are normal, then $[H, K]$ is also normal. In this case, one can view $[H, K]$ as the smallest subgroup of G such that H, K commute modulo $[H, K]$ (or equivalently, that H mod $[H, K]$ and K mod $[H, K]$ commute). Similarly, if H, K are characteristic, then $[H, K]$ is also characteristic.

Observe from (12.3) that

$$[H, K] = [K, H]$$

for any subgroups H, K. Also observe that if N is a normal subgroup of G, then (as mod N is a homomorphism)

$$HK \text{ mod } N = (H \text{ mod } N)(K \text{ mod } N)$$

and

(12.7) $$[H, K] \text{ mod } N = [H \text{ mod } N, K \text{ mod } N].$$

Exercise 12.1.1.

(i) If H, K are normal subgroups of G generated by subsets $A \subset H$, $B \subset K$, respectively, show that $[H, K]$ is the normal subgroup of G generated (as a normal subgroup) by the commutators $[a, b]$ with $a \in A, b \in B$.

(ii) If N, H, K are normal subgroups of G, show that $[[N, H], K]$ lies in the normal subgroup generated by $[[H, K], N]$ and $[[K, N], H]$. (*Hint:* Use the Hall-Witt identity (12.6).)

Given an arbitrary group G, we define the *lower central series* $G = G_1 \lhd G_2 \ldots$ of G by setting $G_1 := G$ and $G_{i+1} := [G, G_i]$ for all $i \geq 1$. Observe that all of the groups G_i in this series are characteristic and thus normal. A group G is said to be *nilpotent of step at most s* if G_{s+1} is trivial; in particular, by (12.7), we see that G mod $G_{s+1} = G/G_{s+1}$ is nilpotent of step at most s for any group G. We have a basic inclusion:

Exercise 12.1.2 (Filtration property). We have $[G_i, G_j] \subset G_{i+j}$ for all $i, j \geq 1$. (*Hint:* Use Exercise 12.1.1.)

The commutator structure can be clarified by passing to the "top order" components of this commutator, as follows. (Strictly speaking, this analysis is not needed to study nilprogressions, but it is still conceptually useful nonetheless.) Consider the subquotients $V_i := G_i$ mod G_{i+1} of the group G for $i = 1, 2, \ldots$. As G_{i+1} contains $[G_i, G_i]$, we see that each group V_i is abelian. To emphasise this, we will write V_i additively instead of multiplicatively, thus we shall denote the group operation on V_i as $+$.

Lemma 12.1.1. *For each* $i, j \geq 1$*, the commutator map* $[,] : G_i \times G_j \to$ $[G_i, G_j]$ *descends to a map* $[,]_{i,j} : V_i \times V_j \to V_{i+j}$*, thus*

(12.8) $[g, h] \mod G_{i+j+1} = [g \mod G_{i+1}, h \mod G_{j+1}]_{i,j}$

for $g \in G_i, h \in G_j$.

Proof. We can quotient out by G_{i+j+1}, and assume that G_{i+j+1} is trivial. In particular, by Exercise 12.1.2, G_{i+1} commutes with G_j, and G_i commutes with G_{j+1}. As such, it is easy to see that if $g \in G_i, h \in G_j$, then $[g, h]$ is unchanged if one multiplies g (on the left or right) by an element of G_{i+1}, and similarly for h and G_{j+1}. This defines the map $[,]_{i,j}$ as required. $\square$

We refer to the maps $[,]_{i,j}$ as *quotiented commutator maps*. The identity (12.8) asserts that these maps $[,]_{i,j}$ capture the "top order" nonlinear behaviour of the group G. As the following exercise shows, these maps behave like (graded components of) a Lie bracket.

Exercise 12.1.3. Let G be a group with lower central series $G = G_1 \lhd G_2 \lhd \ldots$, subquotients V_i, and quotiented commutator maps $[,]_{i,j} : V_i \times V_j \to V_{i+j}$. Let $i, j, k \geq 1$, let $x, x' \in V_i$, $y, y' \in V_j$, and $z \in V_k$. Establish the antisymmetry

$$[x, y]_{i,j} = -[y, x]_{j,i},$$

the bihomomorphism properties

$$[x + x', y]_{i,j} = [x, y]_{i,j} + [x', y]_{i,j},$$
$$[-x, y]_{i,j} = -[x, y]_{i,j},$$
$$[0, y]_{i,j} = 0$$

and

$$[x, y + y']_{i,j} = [x, y]_{i,j} + [x, y']_{i,j},$$
$$[x, -y]_{i,j} = -[x, y]_{i,j},$$
$$[x, 0]_{i,j} = 0,$$

and the Jacobi identity

$$[[x, y]_{i,j}, z]_{i+j,k} + [[y, z]_{j,k}, x]_{j+k,i} + [[z, x]_{k,i}, y]_{k+i,j} = 0.$$

Remark 12.1.2. If one wished, one could combine all of these subquotients V_i and quotiented commutator maps $[,]_{i,j}$ into a single *graded* object, namely the additive group $V := \bigoplus_i V_i$ consisting of tuples $(v_i)_{i=1}^\infty$ with $v_i \in V_i$ (and all but finitely many of the v_i trivial), with a bracket $[,] : V \times V \to V$ defined by

$$[(v_i)_{i=1}^\infty, (w_j)_{j=1}^\infty] := \left(\sum_{i+j=k} [v_i, w_j]_{i,j} \right)_{k=1}^\infty.$$

Then this bracket is an antisymmetric bihomomorphism obeying the Jacobi identity, thus V is a Lie algebra over the integers $\mathbf{Z}$. One could view this object as the "top order" or "Carnot" component of the "Lie algebra" of G. We will, however, not need this object here.

We can iterate the "bilinear" commutator operations to create "multilinear" operations. Given a nonempty finite set $\mathcal{A}$ of formal group elements $\mathbf{g}$, we can define a *formal commutator word* w on $\mathcal{A}$ to be a word that can be generated by the following rules:

(1) If $\mathbf{g}$ is a formal group element, then $\mathbf{g}$ is a formal commutator word on $\{\mathbf{g}\}$.

(2) If $\mathcal{A}, \mathcal{B}$ are disjoint finite nonempty sets of formal group elements, and u, v are formal commutator words on $\mathcal{A}, \mathcal{B}$, respectively, then $[u, v]$ is a formal commutator word on $\mathcal{A} \cup \mathcal{B}$.

We refer to $|\mathcal{A}|$ as the *length* of the formal commutator word. Thus, for instance, $[[\mathbf{g}_1, \mathbf{g}_2], [\mathbf{g}_3, \mathbf{g}_4]]$ is a formal commutator word on $\{\mathbf{g}_1, \mathbf{g}_2, \mathbf{g}_3, \mathbf{g}_4\}$ of length 4. Given a formal commutator word w on $\mathcal{A}$ and an assignment $g \colon \mathbf{g} \mapsto g(\mathbf{g})$ of group elements $g(\mathbf{g}) \in G$ to each formal group element $\mathbf{g} \in \mathcal{A}$, we can define the group element $w(g) \in G$ by substituting $g(\mathbf{g})$ for $\mathbf{g}$ in w for each $\mathbf{g} \in G$. This gives a *commutator word map* $w \colon G^{\mathcal{A}} \to G$. Given an assignment $i \colon \mathbf{g} \mapsto i(\mathbf{g})$ of a positive integer (or "degree") $i(\mathbf{g})$ to each formal group element $\mathcal{A}$, Lemma 12.1.1 and induction then gives a map $w_i \colon \prod_{\mathbf{g} \in \mathcal{A}} V_{i(\mathbf{g})} \to V_{\sum_{\mathbf{g} \in \mathcal{A}} i(\mathbf{g})}$ which is a multihomomorphism (i.e., a homomorphism in each variable). From (12.8) one has that

$$w(g) \bmod G_{\sum_{\mathbf{g} \in \mathcal{A}} i(\mathbf{g})+1} = w_i(g \bmod G_{i+1})$$

where $g \bmod G_{i+1}$ is shorthand for the assignment $\mathbf{g} \mapsto g(\mathbf{g}) \bmod G_{i(\mathbf{g})+1}$.

In particular, using the degree map $i(\mathbf{g}) := 1$, we obtain a multihomomorphism

$$w_1 \colon V_1^l \to V_l$$

for any commutator word w of length l, such that

$$w(g) \bmod G_{l+1} = w_1(g \bmod G_2).$$

From the multihomomorphism nature of w_1 we conclude, in particular, that

(12.9) $$w(g_1^{n_1}, \ldots, g_l^{n_l}) = w(g_1, \ldots, g_l)^{n_1 \cdots n_l} \bmod G_{l+1}$$

for any $g_1, \ldots, g_l \in G$ and integers $n_1, \ldots, n_l$. A variant of this approximate identity will be key in understanding nilprogressions.

One can use commutator words to generate a nilpotent group G and its lower central series:

Exercise 12.1.4. Let G be a nilpotent group that is generated by a set S of generators.

(i) Show that for every positive integer i, G_i is generated by the words $w(g)$, where w ranges over formal commutator words of length l at least i, and g is a collection of l generators from S (possibly with repetition).

(ii) Suppose further that S is finite (so that there are only finitely many possible choices for $w(g)$ for any given length l, and let $e_1, \ldots, e_n$ be all the values of $w(g)$ as w varies over formal commutator words of length at most s, and g ranges over collections of generators from S, arranged in nondecreasing length of w (this arrangement is not unique, and may contain repeated values). Show that for every positive integer i, any element in G_i can be expressed in the form $e_j^{a_j} \ldots e_n^{a_n}$, where $e_j, \ldots, e_n$ are the elements of $e_1, \ldots, e_n$ arising from words w of length at least i, and $a_j, \ldots, a_n$ are integers.

12.2. Nilprogressions

Recall from Section 1 that a *noncommutative progression* $P(a_1, \ldots, a_r; N_1, \ldots, N_r)$ in a group G with generators $a_1, \ldots, a_r \in G$ and dimensions $N_1, \ldots, N_r > 0$ is the set of all words with alphabet $a_1^{\pm 1}, \ldots, a_r^{\pm 1}$, such that for each $1 \le i \le r$, the symbols a_i, a_i^{-1} are used a total of at most N_i times. A *nilprogression* is a noncommutative progression in a nilpotent group. The objective of this section is to prove Proposition 1.2.11, restated here:

Proposition 12.2.1. *Suppose that $a_1, \ldots, a_r \in G$ generate a nilpotent group of step s, and suppose that $N_1, \ldots, N_r$ are all sufficiently large depending on r, s. Then $P := P(a_1, \ldots, a_r; N_1, \ldots, N_r)$ is an $O_{r,s}(1)$-approximate group.*

Recall that a subset A of a group G is a K-approximate group if it is symmetric, contains the origin, and if A^2 can be covered by K left-translates (or equivalently, right-translates) of A. The first two properties are clear, so it suffices to show that P^2 can be covered by $O_{r,s}(1)$ right-translates of P.

We allow all implied constants to depend on r, s. We pick a small constant $\varepsilon > 0$ (depending on r, s). It will suffice to show the inclusion

$$P(a_1, \ldots, a_r; 2N_1, \ldots, 2N_r) \subset P(a_1, \ldots, a_r; O(\varepsilon N_1 + 1), \ldots, O(\varepsilon N_r + 1))X$$

for some set X of cardinality $|X| \ll_\varepsilon 1$.

We will need some auxiliary objects. For $i \ge 1$ and $t > 0$, let

(12.10) $Q_i^t(a_1, \ldots, a_r; N_1, \ldots, N_r) := P(a_1', \ldots, a_R'; tN_1', \ldots, tN_R'),$

where $a_1', \ldots, a_R'$ consists of all group elements of the form $w(a_{i_1}, \ldots, a_{i_l})$ where w is a formal commutator word of length $l \ge i$, $1 \le i_1, \ldots, i_l \le r$, and each $a'_j = w(a_{i_1}, \ldots, a_{i_l})$ is associated to the dimension $N'_j := N_{i_1} \ldots N_{i_l}$ (note that we allow some of the a'_i to be equal, or to degenerate to the identity). Thus the $Q_i(a_1, \ldots, a_r; N_1, \ldots, N_r)$ are nonincreasing in i and become trivial for $i \geq s$. We will usually abbreviate $Q_i^t(a_1, \ldots, a_r; N_1, \ldots, N_r)$ as Q_i^t. Trivially we also have

$$P(a_1, \ldots, a_r; N_1, \ldots, N_r) \subset Q_1^t.$$

Observe also that the Q_i^t are symmetric, contain the identity, and $Q_i^t Q_i^{t'} \subset Q_i^{t+t'}$.

Exercise 12.2.1 (Approximate filtration property). If $g \in Q_i^{O(1)}$ and $h \in Q_j(a_1, \ldots, a_r; N_1, \ldots, N_r)$, show that

$$[g, h] \in Q_{i+j}^{O(1)}$$

and

$$g^h \in Q_i^{O(1)}.$$

Next, we need the following variant of (12.9).

Lemma 12.2.2. *Let w be a formal commutator word of length $l \geq 1$, and let $1 \leq i_1, \ldots, i_l \leq r$. For each $1 \leq j \leq r$, let n_j be an integer with $n_j = O(N_{i_j})$. Then*

$$w(a_{i_1}^{n_1}, \ldots, a_{i_l}^{n_l}) \in w(a_{i_1}, \ldots, a_{i_l})^{n_1 \ldots n_l} Q_{l+1}^{O(1)}.$$

Proof. When $l > s$, the expressions involving w collapse to the identity, and the claim follows, so we may assume that $1 \leq l \leq s$. We induct on l. The claim $l = 1$ is trivial, so suppose that $1 < l \leq s$ and that the claim has been proven for smaller values of l.

We first establish the key case $l = 2$. In this case, it suffices to show that

$$(12.11) \qquad [a^n, b^m] \in [a, b]^{nm} Q_3^{O(1)}(a, b; |n|, |m|)$$

for any $a, b \in G$ and $n, m \in \mathbf{Z}$. By using commutator identities (12.4), (12.5) we may assume that n, m are positive. It suffices to show that

$$(a^n)^{b^m} \in a^n [a, b]^{nm} Q_3^{O(1)}(a, b; n, m).$$

We can write $(a^n)^{b^m} = (a^{b^m})^n$. It is then not difficult to see (and we leave as an exercise to the reader) that the claim will follow if we can show

$$(a)^{b^m} \in a[a, b]^m Q_3^{O(1)}(a, b; 1, m),$$

or equivalently,

$$[a, b^m] \in [a, b]^m Q_3^{O(1)}(a, b; 1, m).$$

Thus we have effectively reduced the problem to the case $n = 1$. A similar argument allows us to also obtain the additional reduction $m = 1$, at which point the claim is trivial. This completes the treatment of the $l = 2$ case.

Now we handle the general case. After some relabeling, we may write $w = [u, v]$, where u, v are words of length l_1, l_2, respectively, for some $l_1, l_2 < l$ adding up to l, with

$$w(a_{i_1}^{n_1}, \ldots, a_{i_l}^{n_l}) = [u(a_{i_1}^{n_1}, \ldots, a_{i_{l_1}}^{n_{l_1}}), v(a_{i_{l_1+1}}^{n_{l_1+1}}, \ldots, a_{i_l}^{n_l})]$$

and similarly

$$w(a_{i_1}, \ldots, a_{i_l}) = [u(a_{i_1}, \ldots, a_{i_{l_1}}, v(a_{i_{l_1+1}}, \ldots, a_{i_l})].$$

By induction hypothesis, one has

$$u(a_{i_1}^{n_1}, \ldots, a_{i_{l_1}}^{n_{l_1}}) \in u(a_{i_1}, \ldots, a_{i_{l_1}})^{n_1 \ldots n_{l_1}} Q_{l_1+1}^{O(1)}.$$

In particular, we have

$$u(a_{i_1}^{n_1}, \ldots, a_{i_{l_1}}^{n_{l_1}}) \in Q_{l_1}^{O(1)}.$$

Similarly we have

$$v(a_{i_{l_1+1}}^{n_{l_1+1}}, \ldots, a_{i_l}^{n_l}) \in v(a_{i_{l_1+1}}, \ldots, a_{i_l})^{n_{l_i+1} \ldots n_l} Q_{l_2+1}^{O(1)} \subset Q_{l_2}^{O(1)}.$$

Using (12.2.1) and the approximate homomorphism properties of commutators, we conclude that

$$[u(a_{i_1}^{n_1}, \ldots, a_{i_{l_1}}^{n_{l_1}}), v(a_{i_{l_1+1}}^{n_{l_1+1}}, \ldots, a_{i_l}^{n_l})]$$
$$\in [u(a_{i_1}, \ldots, a_{i_{l_1}})^{n_1 \ldots n_{l_1}}, v(a_{i_{l_1+1}}, \ldots, a_{i_l})^{n_{l_i+1} \ldots n_l}] Q_{l+1}^{O(1)},$$

but from (12.11) we have

$$[u(a_{i_1}, \ldots, a_{i_{l_1}})^{n_1 \ldots n_{l_1}}, v(a_{i_{l_1+1}}, \ldots, a_{i_l})^{n_{l_i+1} \ldots n_l}]$$
$$\subset [u(a_{i_1}, \ldots, a_{i_{l_1}}), v(a_{i_{l_1+1}}, \ldots, a_{i_l})]^{n_1 \ldots n_l} Q_{l+1}^{O(1)}.$$

The claim follows.   □

Exercise 12.2.2. For positive integers a, b, show that

$$g^n \in a^n [a, b]^{nm} Q_3^{O(1)}(a, b; n, m)$$

whenever

$$g \in a[a, b]^m Q_3^{O(1)}(a, b; 1, m).$$

Next, we need the following elementary number-theoretic lemma:

Exercise 12.2.3. Let $M_1, \ldots, M_l \geq 1$, and let n be an integer with $n = O(M_1 \ldots M_l)$. Show that n can be expressed as the sum of $O_l(1)$ terms of the form $n_1 \ldots n_l$, where the $n_1, \ldots, n_l$ are integers with $n_i = O(M_i)$ for each $i = 1, \ldots, l$. (*Hint:* Despite the superficial similarity here with nontrivial number-theoretic questions such as the *Waring problem*, this is actually a very elementary fact which can be proven by induction on l.)

Exercise 12.2.4. Let w be a formal commutator word of length $l \geq 1$, and let $1 \leq i_1, \ldots, i_l \leq r$. Let $n = O(N_{i_1} \ldots N_{i_r})$. Show that

$$w(a_{i_1}, \ldots, a_{i_l})^n \in P^{O(1)} Q_{l+1}^{O(1)}.$$

Conclude that

$$Q_l^C \subset P^{O(1)} Q_{l+1}^{O(1)}$$

whenever $C = O(1)$, and on iteration conclude that

$$Q_1^C \subset P^{O(1)}$$

whenever $C = O(1)$.

A similar argument shows that $Q_1^\varepsilon \subset P$ for a sufficiently small $\varepsilon > 0$ depending only on r, s, assuming that $N_1, \ldots, N_r$ are sufficiently large depending on r, s. Since $P^2 \subset Q_1^2$, it thus suffices to show that for any $\delta > 0$, that Q_1^2 can be covered by $O_\delta(1)$ right-translates of $Q_1^{O(\delta)}$. But this then follows by iterating the following exercise:

Exercise 12.2.5. Let $\delta > 0$, $l \geq 1$, and $C = O(1)$. Show that Q_l^C can be covered by $O_\delta(1)$ right-translates of $Q_l^{O(\delta)} \cdot Q_{l+1}^{O(1)}$. (*Hint:* This is similar to Exercise 12.2.1. Factor an element of Q_l^C into words $w(a_{i_1}, \ldots, a_{i_l})$ of length l, together with words of higher length. Gather all the words of length l into monomials $w(a_{i_1}, \ldots, a_{i_l})^n$ with $n = O(N_{i_1} \ldots N_{i_l})$, times a factor in $Q_{l+1}^{O(1)}$. Split these monomials into a monomial with exponent $O(\delta N_{i_1} \ldots N_{i_l})$, times a monomial which can take at most $O_\delta(1)$ possible values. Then push the latter monomials (and the $Q_{l+1}^{O(1)}$ factor) to the right.)

Exercise 12.2.6 (Polynomial growth). Suppose that $a_1, \ldots, a_r \in G$ generate a nilpotent group of step s, and suppose that $N_1, \ldots, N_r \geq 1$. Show that

$$|P(a_1, \ldots, a_r; N_1, \ldots, N_r)| \ll_{r,s} (N_1 \ldots N_r)^{O_{r,s}(1)}.$$

Exercise 12.2.7. Suppose that $a_1, \ldots, a_r \in G$ generate a nilpotent group of step s, and suppose that $N_1, \ldots, N_r$ are sufficiently large depending on r, s. Write $P := |P(a_1, \ldots, a_r; N_1, \ldots, N_r)|$. Let $a'_1, \ldots, a'_R, N'_1, \ldots, N'_R$ be as in (12.10), arranged in nondecreasing order of the length of the associated formal commutator words.

   (i) Show that every element of P can be represented in the form

$$(a'_1)^{n_1} \ldots (a'_R)^{n_R}$$

    for some integers $n_i = O_{r,s}(N'_i)$. (We do not claim however that this representation is unique, and indeed the generators $a'_1, \ldots, a'_R$ are likely to contain quite a lot of redundancy.)

(ii) Conversely, show that there exists an $\varepsilon > 0$ depending only on r, s such that any expression of the form

$$(a_1')^{n_1} \ldots (a_R')^{n_R}$$

with integers n_i with $|n_i| \leq \varepsilon N_i'$, lies in P.

Chapter 13

Ado's theorem

Recall from Definition 2.2.6 that a (complex) *abstract Lie algebra* is a complex vector space $\mathfrak{g}$ (either finite or infinite dimensional) equipped with a bilinear antisymmetric form $[] \colon \mathfrak{g} \times \mathfrak{g} \to \mathfrak{g}$ that obeys the *Jacobi identity*

$$(13.1) \qquad [[X, Y], Z] + [[Y, Z], X] + [[Z, X], Y] = 0.$$

One can of course define Lie algebras over other fields than the complex numbers $\mathbf{C}$, but it will be convenient in this section to work over an algebraically closed field. For most of this text it is preferable to work over the reals $\mathbf{R}$ instead, but in many cases one can pass from the complex theory to the real theory by complexifying the Lie algebra.

An important special case of the abstract Lie algebras are the *concrete Lie algebras*, in which $\mathfrak{g} \subset \operatorname{End}(V)$ is a (complex) vector space of linear transformations $X \colon V \to V$ on a vector space V (which again can be either finite or infinite-dimensional), and the bilinear form is given by the usual Lie bracket

$$[X, Y] := XY - YX.$$

It is easy to verify that every concrete Lie algebra is an abstract Lie algebra. In the converse direction, we have

Theorem 13.0.3. *Every abstract Lie algebra is isomorphic to a concrete Lie algebra.*

To prove this theorem, we introduce the useful algebraic tool of the *universal enveloping algebra* $U(\mathfrak{g})$ of the abstract Lie algebra $\mathfrak{g}$. This is the free (associative, complex) algebra generated by $\mathfrak{g}$ (viewed as a complex vector space), subject to the constraints

$$(13.2) \qquad [X, Y] = XY - YX.$$

This algebra is described by the *Poincaré-Birkhoff-Witt theorem*, which asserts that given an ordered basis $(X_i)_{i \in I}$ of $\mathfrak{g}$ as a vector space, that a basis of $U(\mathfrak{g})$ is given by "monomials" of the form

(13.3) $$X_{i_1}^{a_1} \ldots X_{i_m}^{a_m}$$

where m is a natural number, the $i_1 < \cdots < i_m$ are an increasing sequence of indices in I, and the $a_1, \ldots, a_m$ are positive integers. Indeed, given two such monomials, one can express their product as a finite linear combination of further monomials of the form (13.3) after repeatedly applying (13.2) (which we rewrite as $XY = YX + [X, Y]$) to reorder the terms in this product modulo lower order terms until all monomials have their indices in the required increasing order. It is then a routine exercise in basic abstract algebra (using all the axioms of an abstract Lie algebra) to verify that this multiplication rule on monomials does indeed define a complex associative algebra which has the universal properties required of the universal enveloping algebra.

The abstract Lie algebra $\mathfrak{g}$ acts on its universal enveloping algebra $U(\mathfrak{g})$ by left-multiplication: $X \colon M \mapsto XM$, thus giving a map from $\mathfrak{g}$ to $\mathrm{End}(U(\mathfrak{g}))$. It is easy to verify that this map is a Lie algebra homomorphism (so this is indeed an action (or representation) of the Lie algebra), and this action is clearly faithful (i.e., the map from $\mathfrak{g}$ to $\mathrm{End}(U\mathfrak{g})$ is injective), since each element X of $\mathfrak{g}$ maps the identity element 1 of $U(\mathfrak{g})$ to a different element of $U(\mathfrak{g})$, namely X. Thus $\mathfrak{g}$ is isomorphic to its image in $\mathrm{End}(U(\mathfrak{g}))$, proving Theorem 13.0.3.

In the converse direction, every representation $\rho \colon \mathfrak{g} \to \mathrm{End}(V)$ of a Lie algebra "factors through" the universal enveloping algebra, in that it extends to an algebra homomorphism from $U(\mathfrak{g})$ to $\mathrm{End}(V)$, which by abuse of notation we shall also call ρ.

One drawback of Theorem 13.0.3 is that the space $U(\mathfrak{g})$ that the concrete Lie algebra acts on will almost always be infinite-dimensional, even when the original Lie algebra $\mathfrak{g}$ is finite-dimensional. However, there is a useful *theorem of Ado* that rectifies this:

Theorem 13.0.4 (Ado's theorem). *Every finite-dimensional abstract Lie algebra is isomorphic to a concrete Lie algebra over a finite-dimensional vector space V.*

Among other things, this theorem can be used (in conjunction with the Baker-Campbell-Hausdorff formula) to show that every abstract (finite-dimensional) Lie group (or abstract *local Lie group*) is locally isomorphic to a linear group, a result also known as *Lie's third theorem*; see Section 2. It is well-known, though, that abstract Lie groups are not necessarily *globally* isomorphic to a linear group; see for instance Exercise 1.1.1 for a counterexample.

Ado's theorem is surprisingly tricky to prove in general, but some special cases are easy. For instance, one can try using the *adjoint representa-
tion* ad: $\mathfrak{g} \to \mathrm{End}(\mathfrak{g})$ of $\mathfrak{g}$ on itself, defined by the action $X: Y \mapsto [X, Y]$;
the Jacobi identity (2.3) ensures that this is indeed a representation of $\mathfrak{g}$.
The kernel of this representation is the centre $Z(\mathfrak{g}) := \{X \in \mathfrak{g} : [X, Y] =
0 \text{ for all } Y \in \mathfrak{g}\}$. This already gives Ado's theorem in the case when $\mathfrak{g}$ is
semisimple, in which case the center is trivial.

The adjoint representation does not suffice, by itself, to prove Ado's
theorem in the non-semisimple case. However, it does provide an important reduction in the proof, namely it reduces matters to showing that every finite-dimensional Lie algebra $\mathfrak{g}$ has a finite-dimensional representation
$\rho: \mathfrak{g} \to \mathrm{End}(V)$ which is faithful on the centre $Z(\mathfrak{g})$. Indeed, if one has such
a representation, one can then take the direct sum of that representation
with the adjoint representation to obtain a new finite-dimensional representation which is now faithful on all of $\mathfrak{g}$, which then gives Ado's theorem for
$\mathfrak{g}$.

It remains to find a finite-dimensional representation of $\mathfrak{g}$ which is faithful on the centre $Z(\mathfrak{g})$. In the case when $\mathfrak{g}$ is abelian, so that the centre
$Z(\mathfrak{g})$ is all of $\mathfrak{g}$, this is again easy, because $\mathfrak{g}$ then acts faithfully on $\mathfrak{g} \times \mathbf{C}$
by the infinitesimal shear maps $X: (Y, t) \mapsto (tX, 0)$. In matrix form, this
representation identifies each X in this abelian Lie algebra with an "upper-
triangular" matrix:

$$X \equiv \begin{pmatrix} 0 & X \\ 0 & 0 \end{pmatrix}.$$

This construction gives a faithful finite-dimensional representation of
the centre $Z(\mathfrak{g})$ of any finite-dimensional Lie algebra. The standard proof of
Ado's theorem (which I believe dates back to work of Harish-Chandra) then
proceeds by gradually "extending" this representation of the centre $Z(\mathfrak{g})$ to
larger and larger sub-algebras of $\mathfrak{g}$, while preserving the finite-dimensionality
of the representation and the faithfulness on $Z(\mathfrak{g})$, until one obtains a representation on the entire Lie algebra $\mathfrak{g}$ with the required properties. (For
technical inductive reasons, one also needs to carry along an additional property of the representation, namely that it maps the *nilradical* to nilpotent
elements, but we will discuss this technicality later.)

This procedure is a little tricky to execute in general, but becomes sim-
pler in the *nilpotent case*, in which the lower central series $\mathfrak{g}_1 := \mathfrak{g}; \mathfrak{g}_{n+1} :=
[\mathfrak{g}, \mathfrak{g}_n]$ becomes trivial for sufficiently large n:

Theorem 13.0.5 (Ado's theorem for nilpotent Lie algebras). *Let $\mathfrak{n}$ be a
finite-dimensional nilpotent Lie algebra. Then there exists a finite-dimen-
sional faithful representation $\rho: \mathfrak{n} \to \mathrm{End}(V)$ of $\mathfrak{n}$. Furthermore, there exists*

a natural number k such that $\rho(\mathfrak{n})^k = \{0\}$, i.e., one has $\rho(X_1) \ldots \rho(X_k) = 0$ for all $X_1, \ldots, X_k \in \mathfrak{n}$.

The second conclusion of Ado's theorem here is useful for induction purposes. (By *Engel's theorem*, this conclusion is also equivalent to the assertion that every element of $\rho(\mathfrak{n})$ is nilpotent, but we can prove Theorem 13.0.5 without explicitly invoking Engel's theorem.)

In this chapter, we give a proof of Theorem 13.0.5, and then extend the argument to cover the full strength of Ado's theorem. The presentation here is based on the one in [**FuHa1991**].

13.1. The nilpotent case

We first prove Theorem 13.0.5. We achieve this by an induction on the dimension of $\mathfrak{n}$. The claim is trivial for dimension zero, so we assume inductively that $\mathfrak{n}$ has positive dimension, and that Theorem 13.0.5 has already been proven for all lower-dimensional nilpotent Lie algebras.

As noted earlier, the adjoint representation already verifies the claim except for the fact that it is not faithful on the centre $Z(\mathfrak{n})$. (The nilpotency of $\mathfrak{n}$ ensures the existence of some k for which $\mathrm{ad}(\mathfrak{n})^k = 0$.) Thus, it will suffice to find a finite-dimensional representation $\rho \colon \mathfrak{n} \to \mathrm{End}(V)$ that is faithful on the center $Z(\mathfrak{n})$, and for which $\rho(\mathfrak{n})^k = \{0\}$ for some k.

We have already verified the claim when $\mathfrak{n}$ is abelian (and can take $k = 1$ in this case), so suppose that $\mathfrak{n}$ is not abelian; in particular, the centre $Z(\mathfrak{n})$ has strictly smaller dimension. We then observe that $\mathfrak{n}$ must contain a codimension one ideal $\mathfrak{a}$ containing $Z(\mathfrak{n})$, which will of course again be a nilpotent Lie algebra. This can be seen by passing to the quotient $\mathfrak{n}' := \mathfrak{n}/Z(\mathfrak{n})$ (which has positive dimension) and then to the abelianisation $\mathfrak{n}'/[\mathfrak{n}', \mathfrak{n}']$ (which still has positive dimension, by nilpotency), arbitrarily selecting a codimension one subspace of that abelianisation, and then passing back to $\mathfrak{n}$.

If we let $\mathfrak{h}$ be a one-dimensional complementary subspace of $\mathfrak{a}$, then $\mathfrak{h}$ is automatically an abelian Lie algebra, and we have the decomposition

$$\mathfrak{n} = \mathfrak{a} \oplus \mathfrak{h}$$

as vector spaces. This is not necessarily a direct sum of Lie algebras, though, because $\mathfrak{a}$ and $\mathfrak{h}$ need not commute. However, as $\mathfrak{a}$ is an ideal, we do know that $[\mathfrak{h}, \mathfrak{a}] \subset \mathfrak{a}$, thus there is an adjoint action $\mathrm{ad} \colon \mathfrak{h} \to \mathrm{End}(\mathfrak{a})$ of $\mathfrak{h}$ on $\mathfrak{a}$.

By induction hypothesis, we know that there is a faithful representation $\rho_0 \colon \mathfrak{a} \to \mathrm{End}(V_0)$ on some finite-dimensional space V_0 with $\rho_0(\mathfrak{a})^{k_0} = \{0\}$ for some k_0. We would like to somehow "extend" this representation to a finite-dimensional representation $\rho \colon \mathfrak{n} \to \mathrm{End}(V)$ which is still faithful on $\mathfrak{a}$ (and hence on $Z(\mathfrak{n})$), and with $\rho(\mathfrak{n})^k = \{0\}$ for some k.

To motivate the construction, let us begin not with ρ_0, but with the universal enveloping algebra representation $A \colon M \mapsto AM$ of $\mathfrak{a}$ on $U(\mathfrak{a})$. The idea is to somehow combine this action $\mathfrak{a}$ with an action of $\mathfrak{h}$ on the same space $U(\mathfrak{a})$ to obtain an action of the full Lie algebra $\mathfrak{n}$ on $U(\mathfrak{a})$. To do this, recall that we have the adjoint action $\mathrm{ad}(H) \colon A \mapsto [H, A]$ of $\mathfrak{h}$ on $\mathfrak{a}$. This extends to an adjoint action of $\mathfrak{h}$ on $U(\mathfrak{a})$, defined by the Leibniz rule

$$(13.4) \qquad [H, A_1 \ldots A_m] := \sum_{i=1}^{m} A_1 \ldots A_{i-1}[H, A_i]A_{i+1} \ldots A_m;$$

one can easily verify that this is indeed an action (and furthermore that the map $\mathrm{ad}(H) \colon M \mapsto [H, M]$ is a *derivation* on $U(\mathfrak{a})$). One can then combine the multiplicative action $A \colon M \mapsto AM$ of $\mathfrak{a}$ and the adjoint action $H \colon M \mapsto [H, M]$ of $\mathfrak{h}$ into a joint action

$$(13.5) \qquad (A + H) \colon M \mapsto AM + [H, M]$$

of $\mathfrak{a} + \mathfrak{h} = \mathfrak{n}$ on $U(\mathfrak{a})$. A minor algebraic miracle occurs here, namely that this combined action is still a genuine action of the Lie algebra $\mathfrak{a} + \mathfrak{h}$. In particular, one has

$$[A + H, A' + H']M = (A + H)(A' + H')M - (A' + H')(A + H)M$$

for all $A + H, A' + H' \in \mathfrak{a} + \mathfrak{h}$, as can be verified by a brief computation. Also, because the action of $\mathfrak{a}$ was already faithful on $U(\mathfrak{a})$, this enlarged action of $\mathfrak{n}$ will still be faithful on $\mathfrak{a}$.

Remark 13.1.1. One can explain this "miracle" by working first in the larger universal algebra $U(\mathfrak{n})$. This space has a left multiplicative action $X \colon M \mapsto XM$ of $\mathfrak{n}$, but also has a right multiplicative action $X \colon M \mapsto -MX$ which commutes with the left multiplicative action. Furthermore, the splitting $\mathfrak{n} = \mathfrak{a} + \mathfrak{h}$ induces a projection map $\pi \colon \mathfrak{n} \to \mathfrak{h}$ which will be a Lie algebra homomorphism because $\mathfrak{a}$ is an ideal. This gives a projected right multiplicative action $X \colon M \mapsto -M\pi(X)$ which still commutes with the left multiplicative action. In particular, the combined action $X \colon M \mapsto XM - M\pi(X)$ is still an action of $\mathfrak{n}$. One then observes that this action preserves $U(\mathfrak{a})$, and when restricted to that space, becomes precisely (13.5).

Now we need to project the above construction down to a finite-dimensional representation. It turns out to be convenient not to use the original representation ρ_0 directly, but only indirectly via the property $\rho_0(\mathfrak{a})^{k_0} = \{0\}$. Specifically, we consider the (two-sided) ideal $I := \langle (\mathfrak{a})^{k_0} \rangle$ of $U(\mathfrak{a})$ generated by k_0-fold products of elements in $\mathfrak{a}$, and then consider the quotient algebra $U(\mathfrak{a})/I$. Because $\rho_0(\mathfrak{a})^{k_0} = \{0\}$, this algebra still maps to $\mathrm{End}(V_0)$; since $\mathfrak{a}$ was faithful in ρ_0, it must also be faithful in $U(\mathfrak{a})/I$.

The point of doing this quotienting is that whereas $U(\mathfrak{a})$ is likely to be infinite-dimensional, the space $U(\mathfrak{a})/I$ is only finite-dimensional. Indeed, it is generated by those monomials (13.3) whose total degree $a_1 + \cdots + a_m$ is less than k_0; since $\mathfrak{a}$ is finite-dimensional, there are only finitely many such monomials.

From the Leibniz rule (13.4) we see that the adjoint action of $\mathfrak{h}$ on $U(\mathfrak{a})$ preserves I, and thus descends to an action on $U(\mathfrak{a})/I$. We can combine this action with the multiplicative action of $\mathfrak{a}$ as before using (13.5) to create an action of $\mathfrak{n}$ on $U(\mathfrak{a})/I$, which is now a finite-dimensional representation which (as observed previously) is faithful on $\mathfrak{a}$.

The only remaining thing to show is that for some sufficiently large k, that $\mathfrak{n}^k$ annihilates $U(\mathfrak{a})/I$. By linearity, it suffices to show that

$$X_1 \ldots X_k A_1 \ldots A_m = 0 \bmod I$$

whenever $X_1, \ldots, X_k$ lie in either $\mathfrak{a}$ or $\mathfrak{h}$, and $A_1 \ldots A_m$ is a monomial in $U(\mathfrak{a})$. But observe that if one multiplies a monomial $A_1 \ldots A_m$ by an element A of $\mathfrak{a}$, then one gets a monomial $A A_1 \ldots A_m$ of one higher degree; and if instead one multiplies a monomial $A_1 \ldots A_m$ by an element H of $\mathfrak{h}$, then from (13.4) one gets a sum of monomials in which one of the terms A_i has been replaced with the "higher order" term $[H, A_i]$. Using the nilpotency of $\mathfrak{n}$ (which implies that all sufficiently long iterated commutators must vanish), we thus see that if k is large enough, then $X_1 \ldots X_k A_1 \ldots A_m$ will consist only of terms of degree at least k_0, which automatically lie in I, and the claim follows.

Remark 13.1.2. The above argument gives an effective (though not particularly efficient) bound on the dimension of V and on the order k of $\rho(\mathfrak{n})$ in terms of the dimension of $\mathfrak{n}$. Similar remarks can be made for the arguments we give below to prove more general versions of Ado's theorem.

13.2. The solvable case

To go beyond the nilpotent case one needs to use more of the structural theory of Lie algebras. In particular, we will use *Engel's theorem*:

Theorem 13.2.1 (Engel's theorem). *Let $\mathfrak{g} \subset \mathrm{End}(V)$ be a concrete Lie algebra on a finite-dimensional space V which is* concretely nilpotent *in the sense that every element of $\mathfrak{g}$ is a nilpotent operator on V. Then there exists a basis of V such that $\mathfrak{g}$ consists entirely of upper-triangular matrices. In particular, if d is the dimension of V, then $\mathfrak{g}^d = \{0\}$ (i.e., $X_1 \ldots X_d = 0$ for all $X_1, \ldots, X_d \in \mathfrak{g}$).*

The proof of Engel's theorem is not too difficult, but we omit it here as we have nothing to add to the textbook proofs of this theorem (such as the one in [**FuHa1991**]). Note that this theorem, combined with Theorem 13.0.5, gives a correspondence between concretely nilpotent Lie algebras and abstractly nilpotent Lie algebras.

Engel's theorem has the following consequence:

Corollary 13.2.2 (Nilpotent ideals are null). *Let $\mathfrak{g} \subset \mathrm{End}(V)$ be a concrete Lie algebra on a finite-dimensional space V, and let $\mathfrak{a}$ be a concretely nilpotent Lie algebra ideal of $\mathfrak{g}$. Then for any $X_1, \ldots, X_m \in \mathfrak{g}$ with at least one of the X_i in $\mathfrak{a}$, one has $\mathrm{tr}(X_1 \ldots X_m) = 0$.*

Proof. Set $V_i := \mathfrak{a}^i V$ to be the vector space spanned by vectors of the form $A_1 \ldots A_i v$ for $A_1, \ldots, A_i \in \mathfrak{a}$ and $v \in V$, then $V = V_0 \supset \cdots \supset V_d = \{0\}$ is a flag, with $\mathfrak{a}$ mapping V_i to V_{i+1} for each i. As $\mathfrak{a}$ is an ideal of $\mathfrak{g}$, we see that each of the spaces V_i is invariant with respect to $\mathfrak{g}$. Thus $X_1 \ldots X_m$ also maps V_i to V_{i+1} for each i, and so must have zero trace. $\square$

This leads to the following curious algebraic fact:

Corollary 13.2.3. *Let $\mathfrak{g} \subset \mathrm{End}(V)$ be a concrete Lie algebra on a finite-dimensional space V, and let $\mathfrak{a}, \mathfrak{b}$ be ideals of $\mathfrak{g}$ such that $\mathfrak{b} \in [\mathfrak{a}, \mathfrak{g}]$. If $[\mathfrak{a}, \mathfrak{b}]$ is concretely nilpotent, then $\mathfrak{b}$ is also concretely nilpotent.*

Proof. Let $B \in \mathfrak{b}$. We need to show that B is nilpotent. By the *Cayley-Hamilton theorem*, it suffices to show that $\mathrm{tr}(B^k) = 0$ for each $k \geq 1$. Since $B \in \mathfrak{b} \subset [\mathfrak{a}, \mathfrak{g}]$, it suffices to show that $\mathrm{tr}([A, X]B^{k-1}) = 0$ for each $k \geq 1$, $A \in \mathfrak{a}$, and $X \in \mathfrak{g}$. But we can use the cyclic property of trace to rearrange

$$\mathrm{tr}([A, X]B^{k-1}) = -\mathrm{tr}(X[A, B^{k-1}]).$$

We can then use the Leibniz rule to expand

$$[A, B^{k-1}] = \sum_{i=1}^{k-1} B^{i-1}[A, B]B^{k-i-1}.$$

But $[A, B]$ lies in $[\mathfrak{a}, \mathfrak{b}]$, and the claim now follows from Corollary 13.2.2. $\square$

We can apply this corollary to the *radical* $\mathfrak{r}$ of an (abstract) finite-dimensional Lie algebra $\mathfrak{g}$, defined as the maximal solvable ideal of the Lie algebra. (One can easily verify that the vector space sum of two solvable ideals is again a solvable ideal, which implies that the radical is well-defined.)

Theorem 13.2.4. *Let $\mathfrak{g}$ be a finite-dimensional (abstract) Lie algebra, and let $\mathfrak{r}$ be a solvable ideal in $\mathfrak{g}$. Then $[\mathfrak{g}, \mathfrak{g}] \cap \mathfrak{r}$ is an (abstractly) nilpotent ideal of $\mathfrak{g}$.*

Proof. Without loss of generality we may take $\mathfrak{r}$ to be the radical of $\mathfrak{g}$.

Let us first verify the theorem in the case when $\mathfrak{g}$ is a concrete Lie algebra over a finite-dimensional vector space V. We let $\mathfrak{r}^{(i+1)} := [\mathfrak{r}^{(i)}, \mathfrak{r}^{(i)}]$ be the derived series of $\mathfrak{r}$. One easily verifies that each of the $\mathfrak{r}^{(i)}$ are ideals of $\mathfrak{g}$. We will show by downward induction on i that the ideals $[\mathfrak{g}, \mathfrak{g}] \cap \mathfrak{r}^{(i)}$ are concretely nilpotent. As $\mathfrak{r}$ is solvable, this claim is trivial for i large enough. Now suppose that i is such that the ideal $[\mathfrak{g}, \mathfrak{g}] \cap \mathfrak{r}^{(i+1)}$ is concretely nilpotent. This ideal contains $[\mathfrak{r}^{(i)}, [\mathfrak{r}^{(i)}, \mathfrak{g}]]$, which is then also concretely nilpotent. By Corollary 13.2.3, this implies that the ideal $[\mathfrak{r}^{(i)}, \mathfrak{g}]$ is concretely nilpotent, and hence the smaller ideal $[\mathfrak{g}, [\mathfrak{g}, \mathfrak{g}] \cap \mathfrak{r}^{(i)}]$ is also concretely nilpotent. By a second application of Corollary 13.2.3, we conclude that $[\mathfrak{g}, \mathfrak{g}] \cap \mathfrak{r}^{(i)}$ is also concretely nilpotent, closing the induction. This proves the theorem in the concrete case.

To establish the claim in the abstract case, we simply use the adjoint representation, which effectively quotients out the centre $Z(\mathfrak{g})$ of $\mathfrak{g}$ (which will also be an ideal of $\mathfrak{r}$) to convert an finite-dimensional abstract Lie algebra into a concrete Lie algebra over a finite-dimensional space. We conclude that the quotient of $[\mathfrak{g}, \mathfrak{g}] \cap \mathfrak{r}$ by the central ideal $Z(\mathfrak{g})$ is nilpotent, which implies that $[\mathfrak{g}, \mathfrak{g}] \cap \mathfrak{r}$ is nilpotent as required. $\square$

This has the following corollary. Recall that a *derivation* $D \colon \mathfrak{g} \to \mathfrak{g}$ on an abstract Lie algebra $\mathfrak{g}$ is a linear map such that $D[X, Y] = [DX, Y] + [X, DY]$ for each $X, Y \in \mathfrak{g}$. Examples of derivations include the *inner derivations* $DX := [A, X]$ for some fixed $A \in \mathfrak{g}$, but not all derivations are inner.

Corollary 13.2.5. *Let $\mathfrak{g}$ be a finite-dimensional (abstract) Lie algebra, and let $\mathfrak{r}$ be a solvable ideal in $\mathfrak{g}$. Then for any derivation $D \colon \mathfrak{g} \to \mathfrak{g}$, $D\mathfrak{r}$ is nilpotent.*

Proof. If D were an inner derivation, the claim would follow easily from Theorem 13.2.4. In the noninner case, the trick is simply to view D as an inner derivation coming from an extension of $\mathfrak{g}$. Namely, we work in the enlarged Lie algebra $\mathfrak{g} \rtimes_D \mathbf{C}$, which is the Cartesian product $\mathfrak{g} \times \mathbf{C}$ with Lie bracket

$$[(X, s), (Y, t)] := ([X, Y] + sDy - tDx, 0).$$

One easily verifies that this is an (abstract) Lie algebra which contains $\mathfrak{g} \equiv \mathfrak{g} \times \{0\}$ as a subalgebra. The derivation D on $\mathfrak{g}$ then arises from the inner derivation on $\mathfrak{g} \rtimes_D \mathbf{C}$ coming from $(0, 1)$. The claim then easily follows by applying Theorem 13.2.4 to this enlarged algebra. $\square$

Remark 13.2.6. I would be curious to know if there is a more direct proof of Corollary 13.2.5 (which in particular did not need the full strength of Engel's theorem), as this would give a simpler proof of Ado's theorem in the solvable case (indeed, it seems almost equivalent to that theorem, especially in view of Lie's third theorem, Exercise 2.5.6).

Define the *nilradical* of an (abstract) finite-dimensional Lie algebra $\mathfrak{g}$ to be the maximal nilpotent ideal in $\mathfrak{g}$. One can verify that the sum of two nilpotent ideals is again a nilpotent ideal, so the nilradical is well-defined. One can verify that the radical $\mathfrak{r}$ is characteristic (i.e., preserved by all derivations), and so $D\mathfrak{r}$ is also characteristic; in particular, it is an ideal. We conclude that $D\mathfrak{r}$ is in fact contained in the nilradical of $\mathfrak{g}$.

We can now prove Ado's theorem in the solvable case:

Theorem 13.2.7 (Ado's theorem for solvable Lie algebras)**.** *Let $\mathfrak{r}$ be a finite-dimensional solvable Lie algebra. Then there exists a finite-dimensional faithful representation $\rho \colon \mathfrak{r} \to \mathrm{End}(V)$ of $\mathfrak{r}$. Furthermore, if $\mathfrak{n}$ is the nilradical of $\mathfrak{r}$, one can ensure that $\rho(\mathfrak{n})$ is nilpotent.*

Proof. This is similar to the proof of Theorem 13.0.5. We induct on the relative dimension $\mathfrak{r}/\mathfrak{n}$ of $\mathfrak{r}$ relative to its nilradical. When this dimension is zero, the claim follows from Theorem 13.0.5, so we assume that this dimension is positive, and the claim has already been proven for lesser dimensions.

Using the adjoint representation as before, it suffices to establish a representation which is nilpotent on $\mathfrak{n}$ and faithful on the centre $Z(\mathfrak{r})$. As the centre is contained in the nilradical, it will suffice to establish nilpotency and faithfulness on $\mathfrak{n}$.

As in the proof of Theorem 13.0.5, we can find a codimension one ideal $\mathfrak{a}$ of $\mathfrak{r}$ that contains $\mathfrak{n}$, and so we can split $\mathfrak{r} = \mathfrak{a} \oplus \mathfrak{h}$ as vector spaces, for some abelian one-dimensional Lie algebra $\mathfrak{h}$. Note that the nilradical of $\mathfrak{a}$ is characteristic in $\mathfrak{a}$, and hence a nilpotent ideal in $\mathfrak{r}$; as a consequence, it must be identical to the nilradical $\mathfrak{n}$ of $\mathfrak{r}$.

By induction hypothesis, we have a finite-dimensional faithful representation $\rho_0 \colon \mathfrak{a} \to \mathrm{End}(V_0)$ which is nilpotent on $\mathfrak{n}$; we extend this representation to the universal enveloping algebra $U(\mathfrak{a})$. In particular (by Engel's theorem), there exists a natural number k such that $\rho_0(\mathfrak{n}^k) = 0$.

Let I be the two-sided ideal in the universal enveloping algebra $U(\mathfrak{a})$ generated by the kernel $\ker(\rho_0)$ (in the universal enveloping algebra $U(\mathfrak{a})$, not the Lie algebra) and $\mathfrak{n}$. Then the ideal I^k is contained in the kernel of ρ_0. Since ρ_0 is faithful on $\mathfrak{a}$, we see that the quotient map from $U(\mathfrak{a})$ to $U(\mathfrak{a})/I$ is also faithful on $\mathfrak{a}$.

By the Cayley-Hamilton theorem, for every element A of $\mathfrak{a}$ there is a monic polynomial $P(A)$ of A that is annihilated by ρ_0 and is thus in I. The monic polynomial $P(A)^k$ is thus in I^k. Using this, we see that we can express any monomial (13.3) with at least one sufficiently large exponent a_i

in terms of monomials of lower degree modulo I^k. From this we see that $U(\mathfrak{a})/I^k$ can be generated by just finitely many such monomials and is, in particular, finite-dimensional.

The Lie algebra $\mathfrak{h}$ has an adjoint action on $\mathfrak{a}$. These are derivations on $\mathfrak{a}$, so by Corollary 13.2.5, they take values in $\mathfrak{n}$ and in particular in I. If we extend these derivations to the universal enveloping algebra $U(\mathfrak{a})$ by the Leibniz rule (cf. (13.4)), we thus see that these derivations preserve I, and thus (by the Leibniz rule) also preserve I^k. Thus this also gives an action of $\mathfrak{h}$ on $U(\mathfrak{a})/I^k$ by derivations.

In analogy with (13.5), we may now combine the actions of $\mathfrak{a}$ and $\mathfrak{h}$ on $U(\mathfrak{a})/I^k$ to an action of $\mathfrak{r}$ which will remain faithful and nilpotent on $\mathfrak{n}$, and the claim follows. $\square$

13.3. The general case

Finally, we handle the general case. Actually we can use basically the same argument as in preceding cases, but we need one additional ingredient, namely *Levi's theorem*:

Theorem 13.3.1 (Levi's theorem). *Let $\mathfrak{g}$ be a finite-dimensional (abstract) Lie algebra. Then there exists a splitting $\mathfrak{g} = \mathfrak{r} \oplus \mathfrak{h}$ as vector spaces, where $\mathfrak{r}$ is the radical of $\mathfrak{g}$ and $\mathfrak{h}$ is another Lie algebra.*

We remark that $\mathfrak{h}$ necessarily has trivial radical and is thus *semisimple*. It is easy to see that the quotient $\mathfrak{g}/\mathfrak{r}$ is semisimple; the entire difficulty of Levi's theorem is to ensure that one can lift this quotient back up into the original space $\mathfrak{g}$. This requires some knowledge of the structural theory of semisimple Lie algebras. The proof will be omitted, as I have nothing to add to the textbook proofs of this result (such as the one in [**FuHa1991**]).

Using Levi's theorem and Theorem 13.2.7, one obtains the full Ado theorem:

Theorem 13.3.2 (Ado's theorem for general Lie algebras). *Let $\mathfrak{g}$ be a finite-dimensional Lie algebra. Then there exists a finite-dimensional faithful representation $\rho\colon \mathfrak{g} \to \mathrm{End}(V)$ of $\mathfrak{g}$. Furthermore, if $\mathfrak{n}$ is the nilradical of $\mathfrak{r}$, one can ensure that $\rho(\mathfrak{n})$ is nilpotent.*

Exercise 13.3.1. Prove Theorem 13.3.2 (and thus also Theorem 13.0.4). (*Hint:* Deduce this theorem from Theorem 13.2.7 and Theorem 13.3.1 by using the same argument used to deduce Theorem 13.2.7 from the induction hypothesis, with the key point again being Corollary 13.2.5 that places the adjoint action of $\mathfrak{h}$ on $\mathfrak{r}$ inside $\mathfrak{n}$.)

Chapter 14

Associativity of the Baker-Campbell-Hausdorff-Dynkin law

Let $\mathfrak{g}$ be a finite-dimensional Lie algebra (over the reals). Given two sufficiently small elements x, y of $\mathfrak{g}$, define the *right Baker-Campbell-Hausdorff-Dynkin law*

$$(14.1) \qquad R_y(x) := x + \int_0^1 F_R(\mathrm{Ad}_x \, \mathrm{Ad}_{ty}) y \; dt$$

where $\mathrm{Ad}_x := \exp(\mathrm{ad}_x)$, $\mathrm{ad}_x \colon \mathfrak{g} \to \mathfrak{g}$ is the adjoint map $\mathrm{ad}_x(y) := [x, y]$, and F_R is the function $F_R(z) := \frac{z \log z}{z - 1}$, which is analytic for z near 1. Similarly, define the *left Baker-Campbell-Hausdorff-Dynkin law*

$$(14.2) \qquad L_x(y) := y + \int_0^1 F_L(\mathrm{Ad}_{tx} \, \mathrm{Ad}_y) x \; dt$$

where $F_L(z) := \frac{\log z}{z - 1}$. One easily verifies that these expressions are well-defined (and depend smoothly on x and y) when x and y are sufficiently small.

We have the famous *Baker-Campbell-Hausdoff-Dynkin formula*:

Theorem 14.0.3 (BCH formula). *Let G be a finite-dimensional Lie group over the reals with Lie algebra $\mathfrak{g}$. Let $\log$ be a local inverse of the exponential map $\exp \colon \mathfrak{g} \to G$, defined in a neighbourhood of the identity. Then for* *sufficiently small* $x, y \in \mathfrak{g}$, *one has*

$$\log(\exp(x)\exp(y)) = R_y(x) = L_x(y).$$

See Chapter 2 for a proof of this formula. In particular, one can give a neighbourhood of the identity in $\mathfrak{g}$ the structure of a local Lie group by defining the group operation $*$ as

(14.3) $$x * y := R_y(x) = L_x(y)$$

for sufficiently small x, y, and the inverse operation by $x^{-1} := -x$ (one easily verifies that $R_x(-x) = L_x(-x) = 0$ for all small x).

It is tempting to reverse the BCH formula and conclude (the local form of) *Lie's third theorem*, that every finite-dimensional Lie algebra is isomorphic to the Lie algebra of some local Lie group, by using (14.3) to define a smooth local group structure on a neighbourhood of the identity; see Exercise 2.5.6. The main difficulty in doing so is in verifying that the definition (14.3) is well-defined (i.e., that $R_y(x)$ is always equal to $L_x(y)$) and locally associative. The well-definedness issue can be trivially disposed of by using just one of the expressions $R_y(x)$ or $L_x(y)$ as the definition of $*$ (though, as we shall see, it will be very convenient to use both of them simultaneously). However, the associativity is not obvious at all.

With the assistance of *Ado's theorem* (Theorem 13.0.4), which places $\mathfrak{g}$ inside the general linear Lie algebra $\mathfrak{gl}_n(\mathbf{R})$ for some n, one can deduce both the well-definedness and associativity of (14.3) from the Baker-Campbell-Hausdorff formula for $\mathfrak{gl}_n(\mathbf{R})$. However, Ado's theorem is rather difficult to prove, and it is natural to ask whether there is a way to establish these facts without Ado's theorem, thus giving an independent proof of the local version of Lie's third theorem. This will be the purpose of this section.

The key is to observe that the right and left BCH laws commute with each other:

Proposition 14.0.4 (Commutativity). *Let $\mathfrak{g}$ be a finite-dimensional Lie algebra. Then for sufficiently small x, y, z, one has*

(14.4) $$L_y(R_z(x)) = R_z(L_y(x)).$$

Note that this commutativity has to hold if (14.3) is to be both well-defined and associative. Assuming Proposition 14.0.4, we can set $x = 0$ in (14.4) and use the easily verified identities $R_z(0) = z$, $L_y(0) = y$ to conclude that $L_y(z) = R_z(y)$ for small y, z, ensuring that (14.3) is well-defined; and then inserting (14.3) into (14.4) we obtain the desired (local) associativity.

It remains to prove Proposition 14.0.4. We first make a convenient observation. Thanks to the Jacobi identity, the adjoint representation ad: $x \mapsto \mathrm{ad}_x$ is a Lie algebra homomorphism from $\mathfrak{g}$ to the Lie algebra $\mathfrak{gl}(\mathfrak{g})$. As this latter Lie algebra is the Lie algebra of a Lie group, namely $GL(\mathfrak{g})$, the Baker-Campbell-Hausdorff formula is valid for that Lie algebra. In particular, one has

$$\log(\exp(\mathrm{ad}_x)\exp(\mathrm{ad}_y)) = R_{\mathrm{ad}_y}(\mathrm{ad}_x) = L_{\mathrm{ad}_x}(\mathrm{ad}_y)$$

for sufficiently small x, y. But as ad is a Lie algebra homomorphism, one has

$$R_{\mathrm{ad}_y}(\mathrm{ad}_x) = \mathrm{ad}_{R_y(x)}$$

and similarly

$$L_{\mathrm{ad}_x}(\mathrm{ad}_y) = \mathrm{ad}_{L_x(y)}.$$

Exponentiating, we conclude that

(14.5) $$\mathrm{Ad}_x\,\mathrm{Ad}_y = \mathrm{Ad}_{R_y(x)} = \mathrm{Ad}_{L_x(y)}.$$

This would already give what we want if the adjoint representation was faithful. Unfortunately, we cannot assume this (and this is the main reason, by the way, why Ado's theorem is so difficult), but we can at least use (14.5) to rewrite the formulae (14.1), (14.2) as

$$R_y(x) = x + \int_0^1 F_R(\mathrm{Ad}_{R_{ty}(x)})\,dt$$

and

$$L_x(y) := y + \int_0^1 F_L(\mathrm{Ad}_{L_{tx}(y)})\,dt.$$

This leads to the important *radial homogeneity identities*

$$R_{(s+t)y}(x) = R_{sy}(R_{ty}(x))$$

and

$$L_{(s+t)x}(y) = L_{sx}(L_{tx}(y))$$

for all sufficiently small $x, y \in \mathfrak{g}$ and $0 \le s, t \le 1$, as can be seen by a short computation.

Because of these radial homogeneity identities (together with the smoothness of the right and left BCH laws), it will now suffice to prove the *approximate* commutativity law

(14.6) $$L_y(R_z(x)) = R_z(L_y(x)) + O(|y|^2|z|) + O(|y||z|^2)$$

for all small x, y, z. Indeed, this law implies that

(14.7) $$L_{y/n} \circ R_{z/n} = R_{z/n} \circ L_{y/n} + O(1/n^3)$$

for fixed small y, z, a large natural number n, and with the understanding that the operations are only applied to sufficiently small elements x. From radial homogeneity we have $L_y = L_{y/n}^n$ and $R_z = R_{z/n}^n$, and so a large number ($O(n^2)$, to be more precise) of iterations of (14.7) (using uniform smoothness to control all errors) gives

$$L_y \circ R_z = R_z \circ L_y + O(1/n),$$

and the claim (14.4) then follows by sending $n \to \infty$.

It remains to prove (14.6). When $y = 0$, then L_y is the identity map and the claim is trivial; similarly if $z = 0$. By Taylor expansion, it thus suffices to establish the *infinitesimal* commutativity law

$$\frac{\partial}{\partial a}\frac{\partial}{\partial b}L_{ay}(R_{bz}(x))|_{a=b=0} = \frac{\partial}{\partial a}\frac{\partial}{\partial b}R_{bz}(L_{ay}(x))|_{a=b=0}.$$

(One can interpret this infinitesimal commutativity as a commutativity of the vector fields corresponding to the infinitesimal generators of the left and right BCH laws, although we will not explicitly adopt that perspective here.) This is a simplification, because the infinitesimal versions of (14.1), (14.2) are simpler than the noninfinitesimal versions. Indeed, from the fundamental theorem of calculus one has

$$\frac{\partial}{\partial a}L_{ay}(w)|_{a=0} = F_L(\mathrm{Ad}_w)y$$

for any fixed y, w, and similarly

$$\frac{\partial}{\partial b}R_{bz}(v)|_{b=0} = F_R(\mathrm{Ad}_v)z.$$

Thus it suffices (by *Clairaut's theorem*) to show that

$$(14.8) \qquad \frac{\partial}{\partial b}F_L(\mathrm{Ad}_{R_{bz}(x)})y|_{b=0} = \frac{\partial}{\partial a}F_R(\mathrm{Ad}_{L_{az}(x)})z|_{a=0}.$$

It will be more convenient to work with the reciprocals F_L^{-1}, F_R^{-1} of the functions F_L, F_R. Recall the general matrix identity

$$\frac{d}{dt}A^{-1}(t) = -A^{-1}(t)A'(t)A^{-1}(t)$$

for any smoothly varying invertible matrix function $A(t)$ of a real parameter t. Using this identity, we can write the left-hand side of (14.8) as

$$-F_L(\mathrm{Ad}_x)(\frac{\partial}{\partial b}F_L^{-1}(\mathrm{Ad}_{R_{bz}(x)})|_{b=0})F_L(\mathrm{Ad}_x)y.$$

If we write $Y := F_L(\mathrm{Ad}_x)y$ and $Z := F_R(\mathrm{Ad}_x)z$, then from Taylor expansion we have

$$R_{bz}(x) = x + bZ + O(|b|^2)$$

and so we can simplify the above expression as

$$-F_L(\mathrm{Ad}_x)(\frac{\partial}{\partial b}F_L^{-1}(\mathrm{Ad}_{x+bZ})|_{b=0})Y.$$

Similarly, the right-hand side of (14.8) is

$$-F_R(\mathrm{Ad}_x)(\frac{\partial}{\partial a}F_R^{-1}(\mathrm{Ad}_{x+aY})|_{a=0})Z.$$

Since $F_R(\mathrm{Ad}_x) = \mathrm{Ad}_x F_L(\mathrm{Ad}_x)$, it thus suffices to show that

(14.9) $(\frac{\partial}{\partial b}F_L^{-1}(\mathrm{Ad}_{x+bZ})|_{b=0})Y = \mathrm{Ad}_x(\frac{\partial}{\partial a}F_R^{-1}(\mathrm{Ad}_{x+aY})|_{a=0})Z.$

Now, we write

$$F_L^{-1}(\mathrm{Ad}_x) = \int_0^1 \mathrm{Ad}_{tx}\ dt$$

and

$$F_R^{-1}(\mathrm{Ad}_x) = \int_0^1 \mathrm{Ad}_{-tx}\ dt$$

and thus expand (14.9) as

(14.10) $\int_0^1 (\frac{\partial}{\partial b}\mathrm{Ad}_{tx+tbZ})|_{b=0}Y\ dt = \mathrm{Ad}_x \int_0^1 (\frac{\partial}{\partial b}\mathrm{Ad}_{-tx-taY})|_{a=0}Z\ dt.$

We write Ad as the exponential of ad. Using the Duhamel matrix identity

$$\frac{d}{dt}\exp(A(t)) = \int_0^1 \exp(sA(t))A'(t)\exp((1-s)A(t))\ dt$$

for any smoothly varying matrix function $A(t)$ of a real variable t, together with the linearity of ad, we see that

$$(\frac{\partial}{\partial b}\mathrm{Ad}_{tx+tbZ})|_{b=0} = \int_0^1 \mathrm{Ad}_{stx}\, t\, \mathrm{ad}_Z\, \mathrm{Ad}_{(1-s)tx}\ ds$$

and similarly

$$(\frac{\partial}{\partial b}\mathrm{Ad}_{-tx-taY})|_{a=0} = -\int_0^1 \mathrm{Ad}_{-stx}\, t\, \mathrm{ad}_Y\, \mathrm{Ad}_{-(1-s)tx}\ ds.$$

Collecting terms, our task is now to show that

(14.11)
$$\int_0^1 \int_0^1 \mathrm{Ad}_{stx}\, \mathrm{ad}_Z\, \mathrm{Ad}_{(1-s)tx}\, Y\ tdsdt$$
$$= -\int_0^1 \int_0^1 \mathrm{Ad}_{(1-st)x}\, \mathrm{ad}_Y\, \mathrm{Ad}_{-(1-s)tx}\, Z\ tdsdt.$$

For any $x \in \mathfrak{g}$, the adjoint map $\mathrm{ad}_x : \mathfrak{g} \to \mathfrak{g}$ is a derivation in the sense that

$$\mathrm{ad}_x[y,z] = [\mathrm{ad}_x\, y, z] + [y, \mathrm{ad}_x\, z],$$

thanks to the Jacobi identity. Exponentiating, we conclude that

$$\mathrm{Ad}_x[y,z] = [\mathrm{Ad}_x\, y, \mathrm{Ad}_x\, z]$$

(thus each Ad_x is a Lie algebra homomorphism) and thus

$$\mathrm{Ad}_x\, \mathrm{ad}_y\, \mathrm{Ad}_x^{-1} = \mathrm{ad}_{\mathrm{Ad}_x\, y}.$$

Using this, we can simplify (14.11) as

$$\int_0^1 \int_0^1 \mathrm{ad}_{\mathrm{Ad}_{stx} Z} \, \mathrm{Ad}_{tx} \, Y \; tdsdt = -\int_0^1 \int_0^1 \mathrm{ad}_{\mathrm{Ad}_{(1-st)x} Y} \, \mathrm{Ad}_{(1-t)x} \, Y \; tdsdt$$

which we can rewrite as

$$\int_0^1 \int_0^1 [\mathrm{Ad}_{stx} \, Z, \mathrm{Ad}_{tx} \, Y] \; tdsdt = -\int_0^1 \int_0^1 [\mathrm{Ad}_{(1-st)x} \, Y, \mathrm{Ad}_{(1-t)x} \, Y] \; tdsdt.$$

But by an appropriate change of variables (and the anti-symmetry of the Lie bracket), both sides of this equation can be written as

$$\int_{0 \le a \le b \le 1} [\mathrm{Ad}_{ax} \, Z, \mathrm{Ad}_{bx} \, Y] \; dadb,$$

and the claim follows.

Remark 14.0.5. The above argument shows that every finite-dimensional Lie algebra $\mathfrak{g}$ can be viewed as arising from a *local* Lie group G. It is natural to then ask if that local Lie group (or a sufficiently small piece thereof) can in turn be extended to a global Lie group $\tilde{G}$. The answer to this is affirmative, as was first shown by Cartan. I have been unable, however, to find a proof of this result that does not either use Ado's theorem, the proof method of Ado's theorem (in particular, the structural decomposition of Lie algebras into semisimple and solvable factors), or some facts about group cohomology (particularly with regards to central extensions of Lie groups) which are closely related to the structural decompositions just mentioned. (As noted by Serre [**Se1964**], though, a certain amount of this sort of difficulty in the proof may in fact be necessary, given that the global form of Lie's third theorem is known to fail in the infinite-dimensional case.)

Chapter 15

Local groups

The material here is based in part on [**Oll996**], [**Go2010**].

One of the fundamental structures in modern mathematics is that of a *group*. Formally, a group is a set $G = (G, 1, \cdot, ()^{-1})$ equipped with an identity element $1 = 1_G \in G$, a multiplication operation $\cdot \colon G \times G \to G$, and an inversion operation $()^{-1} \colon G \to G$ obeying the following axioms:

(1) (Closure) If $g, h \in G$, then $g \cdot h$ and g^{-1} are well-defined and lie in G. (This axiom is redundant from the above description, but we include it for emphasis.)

(2) (Associativity) If $g, h, k \in G$, then $(g \cdot h) \cdot k = g \cdot (h \cdot k)$.

(3) (Identity) If $g \in G$, then $g \cdot 1 = 1 \cdot g = g$.

(4) (Inverse) If $g \in G$, then $g \cdot g^{-1} = g^{-1} \cdot g = 1$.

One can also consider additive groups $G = (G, 0, +, -)$ instead of multiplicative groups, with the obvious changes of notation. By convention, additive groups are always understood to be abelian, so it is convenient to use additive notation when one wishes to emphasise the abelian nature of the group structure. As usual, we often abbreviate $g \cdot h$ by gh (and 1_G by 1) when there is no chance of confusion.

If, furthermore, G is equipped with a topology, and the group operations $\cdot, ()^{-1}$ are continuous in this topology, then G is a *topological group*. Any group can be made into a topological group by imposing the *discrete topology*, but there are many more interesting examples of topological groups, such as *Lie groups*, in which G is not just a topological space, but is in fact a smooth manifold (and the group operations are not merely continuous, but also smooth).

There are many naturally occuring group-like objects that obey some, but not all, of the axioms. For instance, *monoids* are required to obey the closure, associativity, and identity axioms, but not the inverse axiom. If we also drop the identity axiom, we end up with a *semigroup*. *Groupoids* do not necessarily obey the closure axiom, but obey (versions of) the associativity, identity, and inverse axioms, and so forth.

Another group-like concept is that of a *local topological group* (or *local group*, for short), defined in Section 2.1. A prime example of a local group can be formed by *restricting* any global topological group G to an open neighbourhood $U \subset G$ of the identity, with the domains

$$\Omega := \{(g, h) \in U : g \cdot h \in U\}$$

and

$$\Lambda := \{g \in U : g^{-1} \in U\};$$

one easily verifies that this gives U the structure of a local group (which we will sometimes call $G \downharpoonright_U$ to emphasise the original group G). If U is symmetric (i.e., $U^{-1} = U$), then we in fact have a symmetric local group. One can also restrict *local* groups G to open neighbourhoods U to obtain a smaller local group $G \downharpoonright_U$ by the same procedure (adopting the convention that statements such as $g \cdot h \in U$ or $g^{-1} \in U$ are considered false if the left-hand side is undefined). (Note though that if one restricts to nonopen neighbourhoods of the identity, then one usually does not get a local group; for instance, $[-1, 1]$ is not a local group (why?).)

Finite subsets of (Hausdorff) groups containing the identity can be viewed as local groups. This point of view turns out to be particularly useful for studying *approximate groups* in additive combinatorics, a point which I hope to expound more on later. Thus, for instance, the discrete interval $\{-9, \ldots, 9\} \subset \mathbf{Z}$ is an additive symmetric local group, which informally might model an adding machine that can only handle (signed) one-digit numbers. More generally, one can view a local group as an object that behaves like a group near the identity, but for which the group laws (and, in particular, the closure axiom) can start breaking down once one moves far enough away from the identity.

In many situations (such as when one is investigating the local structure of a global group) one is only interested in the *local* properties of a (local or global) group. We can formalise this by the following definition. Let us call two local groups $G = (G, \Omega, \Lambda, 1_G, \cdot, ()^{-1})$ and $G' = (G', \Omega', \Lambda', 1_{G'}, \cdot, ()^{-1})$ *locally identical* if they have a common restriction, thus there exists a set $U \subset G \cap G'$ such that $G \downharpoonright_U = G' \downharpoonright_U$ (thus, $1_G = 1_{G'}$, and the topology and group operations of G and G' agree on U). This is easily seen to be an equivalence relation. We call an equivalence class $[G]$ of local groups a *group germ*.

Let $\mathcal{P}$ be a property of a local group (e.g., abelianness, connectedness, compactness, etc.). We call a group germ *locally* $\mathcal{P}$ if every local group in that germ has a restriction that obeys $\mathcal{P}$; we call a local or global group G *locally* $\mathcal{P}$ if its germ is locally $\mathcal{P}$ (or equivalently, every open neighbourhood of the identity in G contains a further neighbourhood that obeys $\mathcal{P}$). Thus, the study of local properties of (local or global) groups is subsumed by the study of group germs.

Exercise 15.0.2.

(i) Show that the above general definition is consistent with the usual definitions of the properties "*connected*" and "*locally connected*" from point-set topology.

(ii) Strictly speaking, the above definition is *not* consistent with the usual definitions of the properties "*compact*" and "*local compact*" from point-set topology because in the definition of local compactness, the compact neighbourhoods are certainly not required to be open. Show, however, that the point-set topology notion of "locally compact" is equivalent, using the above conventions, to the notion of "locally *precompact* inside of an ambient local group". Of course, this is a much more clumsy terminology, and so we shall abuse notation slightly and continue to use the standard terminology "locally compact" even though it is, strictly speaking, not compatible with the above general convention.

(iii) Show that a local group is locally discrete if and only if it is locally trivial.

(iv) Show that a connected global group is abelian if and only if it is locally abelian. (*Hint:* In a connected global group, the only open subgroup is the whole group.)

(v) Show that a global topological group is *first-countable* if and only if it is locally first countable. (By the Birkhoff-Kakutani theorem (Theorem 5.1.1), this implies that such groups are metrisable if and only if they are locally metrisable.)

- Let p be a prime. Show that the *solenoid group* $\mathbf{Z}_p \times \mathbf{R}/\mathbf{Z}^\Delta$, where $\mathbf{Z}_p$ is the *p-adic integers* and $\mathbf{Z}^\Delta := \{(n,n) : n \in \mathbf{Z}\}$ is the diagonal embedding of $\mathbf{Z}$ inside $\mathbf{Z}_p \times \mathbf{R}$, is connected but not locally connected.

Remark 15.0.6. One can also study the local properties of groups using *nonstandard analysis*. Instead of group germs, one works (at least in the case when G is first countable) with the *monad* $o(G)$ of the identity element

1_G of G, defined as the nonstandard group elements $g = \lim_{n \to \alpha} g_n$ in $^*G$ that are infinitesimally close to the origin in the sense that they lie in every standard neighbourhood of the identity. The monad $o(G)$ is closely related to the group germ $[G]$, but has the advantage of being a genuine (global) group, as opposed to an equivalence class of local groups. It is possible to recast most of the results here in this nonstandard formulation; see e.g. [**Ro1966**]. However, we will not adopt this perspective here.

A useful fact to know is that Lie structure is local; see Exercise 2.4.7.

As with so many other basic classes of objects in mathematics, it is of fundamental importance to specify and study the *morphisms* between local groups (and group germs). Given two local groups G, G', we can define the notion of a *(continuous) homomorphism* $\phi : G \to G'$ between them, defined as a continuous map with

$$\phi(1_G) = 1_{G'}$$

such that whenever $g, h \in G$ are such that gh is well-defined, then $\phi(g)\phi(h)$ is well-defined and equal to $\phi(gh)$; similarly, whenever $g \in G$ is such that g^{-1} is well-defined, then $\phi(g)^{-1}$ is well-defined and equal to $\phi(g^{-1})$. (In abstract algebra, the continuity requirement is omitted from the definition of a homomorphism; we will call such maps *discrete* homomorphisms to distinguish them from the continuous ones which will be the ones studied here.)

It is often more convenient to work locally: Define a *local (continuous) homomorphism* $\phi : U \to G'$ from G to G' to be a homomorphism from an open neighbourhood U of the identity to G'. Given two local homomorphisms $\phi : U \to G'$, $\tilde{\phi} : \tilde{U} \to \tilde{G}'$ from one pair of locally identical groups $G, \tilde{G}$ to another pair $G', \tilde{G}'$, we say that ϕ, ϕ' are *locally identical* if they agree on some open neighbourhood of the identity in $U \cap \tilde{U}'$ (note that it does not matter here whether we require openness in G, in $\tilde{G}$, or both). An equivalence class $[\phi]$ of local homomorphisms will be called a *germ homomorphism* (or *morphism* for short) from the group germ $[G]$ to the group germ $[G']$.

Exercise 15.0.3. Show that the class of group germs, equipped with the germ homomorphisms, becomes a *category*. (Strictly speaking, because group germs are themselves *classes* rather than sets, the collection of all group germs is a second-order class rather than a class, but this set-theoretic technicality can be resolved in a number of ways (e.g., by restricting all global and local groups under consideration to some fixed "universe") and should be ignored for this exercise.)

As is usual in category theory, once we have a notion of a morphism, we have a notion of an *isomorphism*: Two group germs $[G], [G']$ are isomorphic if there are germ homomorphisms $\phi\colon [G] \to [G']$, $\psi\colon [G'] \to [G]$ that invert each other. Lifting back to local groups, the associated notion is that of *local isomorphism*: Two local groups G, G' are locally isomorphic if there exist local isomorphisms $\phi\colon U \to G'$ and $\psi\colon U' \to G$ from G to G' and from G' to G that locally invert each other, thus $\psi(\phi(g)) = g$ for $g \in G$ sufficiently close to 1_G, and $\phi(\psi(g))$ for $g' \in G'$ sufficiently close to $1_{G'}$. Note that all local properties of (global or local) groups that can be defined purely in terms of the group and topological structures will be preserved under local isomorphism. Thus, for instance, if G, G' are locally isomorphic local groups, then G is locally connected iff G' is, G is locally compact iff G' is, and (by Exercise 2.4.7) G is Lie iff G' is.

Exercise 15.0.4.

   (i) Show that the additive global groups $\mathbf{R}/\mathbf{Z}$ and $\mathbf{R}$ are locally isomorphic.

   (ii) Show that every locally path-connected group G is locally isomorphic to a path-connected, simply connected group.

15.1. Lie's third theorem

Lie's fundamental theorems of Lie theory link the Lie group germs to Lie algebras. Observe that if $[G]$ is a locally Lie group germ, then the tangent space $\mathfrak{g} := T_1 G$ at the identity of this germ is well-defined, and is a finite-dimensional vector space. If we choose G to be symmetric, then $\mathfrak{g}$ can also be identified with the left-invariant (say) vector fields on G, which are first-order differential operators on $C^\infty(M)$. The *Lie bracket for vector fields* then endows $\mathfrak{g}$ with the structure of a *Lie algebra*. It is easy to check that every morphism $\phi\colon [G] \to [H]$ of locally Lie germs gives rise (via the derivative map at the identity) to a morphism $D\phi(1)\colon \mathfrak{g} \to \mathfrak{h}$ of the associated Lie algebras. From the Baker-Campbell-Hausdorff formula (which is valid for local Lie groups, as discussed in this previous post) we conversely see that $D\phi(1)$ uniquely determines the germ homomorphism ϕ. Thus the derivative map provides a *covariant functor* from the category of locally Lie group germs to the category of (finite-dimensional) Lie algebras. In fact, this functor is an isomorphism:

Exercise 15.1.1 (Lie's third theorem). For this exercise, all Lie algebras are understood to be finite-dimensional (and over the reals).

   (i) Show that every Lie algebra $\mathfrak{g}$ is the Lie algebra of a local Lie group germ $[G]$, which is unique up to germ isomorphism (fixing $\mathfrak{g}$).

   (ii) Show that every Lie algebra $\mathfrak{g}$ is the Lie algebra of some global connected, simply connected Lie group G, which is unique up to Lie group isomorphism (fixing $\mathfrak{g}$).

(iii) Show that every homomorphism $\Phi\colon \mathfrak{g} \to \mathfrak{h}$ between Lie algebras is the derivative of a unique germ homomorphism $\phi\colon [G] \to [H]$ between the associated local Lie group germs.

(iv) Show that every homomorphism $\Phi\colon \mathfrak{g} \to \mathfrak{h}$ between Lie algebras is the derivative of a unique Lie group homomorphism $\phi\colon G \to H$ between the associated global connected, simply connected, Lie groups.

(v) Show that every local Lie group germ is the germ of a global connected, simply connected Lie group G, which is unique up to Lie group isomorphism. In particular, every local Lie group is locally isomorphic to a global Lie group.

(*Hint:* Use Exercise 2.5.6.)

Lie's third theorem (which, actually, was proven in full generality by Cartan) demonstrates the equivalence of three categories: the category of finite-dimensonal Lie algebras, the category of local Lie group germs, and the category of connected, simply connected Lie groups.

15.2. Globalising a local group

Many properties of a local group improve after passing to a smaller neighbourhood of the identity. Here are some simple examples:

Exercise 15.2.1. Let G be a local group.

(i) Give an example to show that G does not necessarily obey the cancellation laws

(15.1) $$gk = hk \implies g = h; \quad kg = kh \implies g = h$$

for $g, h, k \in G$ (with the convention that statements such as $gk = hk$ are false if either side is undefined). However, show that there exists an open neighbourhood U of G within which the cancellation law holds.

(ii) Repeat the previous part, but with the cancellation law (15.1) replaced by the inversion law

(15.2) $$(gh)^{-1} = h^{-1}g^{-1}$$

for any $g, h \in G$ for which both sides are well-defined.

(iii) Repeat the previous part, but with the inversion law replaced by the involution law

(15.3) $$(g^{-1})^{-1} = g$$

for any g for which the left-hand side is well-defined.

Note that the counterexamples in the above exercise demonstrate that not every local group is the restriction of a global group, because global groups (and hence, their restrictions) always obey the cancellation law (15.1), the inversion law (15.2), and the involution law (15.3). Another way in which a local group can fail to come from a global group is if it contains relations which can interact in a "global' way to cause trouble, in a fashion which is invisible at the local level. For instance, consider the open unit cube $(-1, 1)^3$, and consider four points a_1, a_2, a_3, a_4 in this cube that are close to the upper four corners $(1, 1, 1), (1, 1, -1), (1, -1, 1), (1, -1, -1)$ of this cube, respectively. Define an equivalence relation $\sim$ on this cube by setting $x \sim y$ if $x, y \in (-1, 1)^3$ and $x - y$ is equal to either 0 or $\pm 2a_i$ for some $i = 1, \ldots, 4$. Note that this is indeed an equivalence relation if a_1, a_2, a_3, a_4 are close enough to the corners (as this forces all nontrivial combinations $\pm 2a_i \pm 2a_j$ to lie outside the doubled cube $(-2, 2)^3$). The quotient space $(-1, 1)^3 / \sim$ (which is a cube with bits around opposite corners identified together) can then be seen to be a symmetric additive local Lie group, but will usually not come from a global group. Indeed, it is not hard to see that if $(-1, 1)^3 / \sim$ is the restriction of a global group G, then G must be a Lie group with Lie algebra $\mathbf{R}^3$ (by Exercise 2.4.7), and so the connected component G° of G containing the identity is isomorphic to $\mathbf{R}^3 / \Gamma$ for some sublattice Γ of $\mathbf{R}^3$ that contains a_1, a_2, a_3, a_4; but for generic a_1, a_2, a_3, a_4, there is no such lattice, as the a_i will generate a dense subset of $\mathbf{R}^3$. (The situation here is somewhat analogous to a number of famous Escher prints, such as *Ascending and Descending*, in which the geometry is locally consistent but globally inconsistent.) We will give this sort of argument in more detail later, when we prove Proposition 15.3.1.

Nevertheless, the space $(-1, 1)^3 / \sim$ is still *locally* isomorphic to a global Lie group, namely $\mathbf{R}^3$; for instance, the open neighbourhood $(-0.5, 0.5)^3 / \sim$ is isomorphic to $(-0.5, 0.5)^3$, which is an open neighbourhood of $\mathbf{R}^3$. More generally, Lie's third theorem tells us that any local Lie group is locally isomorphic to a global Lie group.

Let us call a local group *globalisable* if it is locally isomorphic to a global group; thus Lie's third theorem tells us that every local Lie group is globalisable. Thanks to Goldbring's solution [**Go2010**] to the local version of Hilbert's fifth problem (Theorem 5.6.9), we also know that locally Euclidean local groups are globalisable. A modification of this argument [**vdDrGo2010**] shows in fact that every locally compact local group is globalisable.

In view of these results, it is tempting to conjecture that *all* local groups are globalisable; among other things, this would simplify the proof of Lie's third theorem (and of the local version of Hilbert's fifth problem). Unfortunately, this claim as stated is false:

Theorem 15.2.1. *There exists local groups G which are not globalisable.*

The counterexamples used to establish Theorem 15.2.1 are remarkably delicate; the first example I know of appears in [**vEsKo1964**]. One reason for this, of course, is that the previous results prevents one from using any local Lie group, or even a locally compact group as a counterexample. We will present a (somewhat complicated) example shortly, based on the unit ball in the infinite-dimensional Banach space $\ell^\infty(\mathbf{N}^2)$.

Despite such counterexamples, there are certainly many situations in which we can globalise a local group. For instance, this is the case if one has a locally faithful representation of that local group inside a global group:

Lemma 15.2.2 (Faithful representation implies globalisability)**.** *Let G be a local group, and suppose there exists an injective local homomorphism $\phi \colon U \to H$ from G into a global topological group H with U symmetric. Then U is isomorphic to the restriction of a global topological group to an open neighbourhood of the identity; in particular, G is globalisable.*

We now prove Lemma 15.2.2. Let G, ϕ, U, H be as in that lemma. The set $\phi(U)$ generates a subgroup $\langle \phi(U) \rangle$ of H, which contains an embedded copy $\phi(U)$ of U. It is then tempting to restrict the topology of H to that of $\langle \phi(U) \rangle$ to give $\langle \phi(U) \rangle$ the structure of a global topological group and then declare victory, but the difficulty is that $\phi(U)$ need not be an open subset of $\langle \phi(U) \rangle$, as the following key example demonstrates.

Example 15.2.3. Take $G = U = (-1, 1)$, $H = (\mathbf{R}/\mathbf{Z})^2$, and $\phi(t) := (t, \alpha t) \bmod \mathbf{Z}^2$, where α is an irrational number (e.g., $\alpha = \sqrt{2}$). Then $\langle \phi(U) \rangle$ is the dense subgroup $\{(t, \alpha t) \bmod \mathbf{Z}^2 : t \in \mathbf{R}\}$ of $(\mathbf{R}/\mathbf{Z})^2$, which is not locally isomorphic to G if endowed with the topology inherited from $(\mathbf{R}/\mathbf{Z})^2$ (for instance, $\langle \phi(U) \rangle$ is not locally connected in this topology, whereas G is). Also, $\phi(U)$, while homeomorphic to U, is not an open subset of $\langle \phi(U) \rangle$. Thus we see that the "global" behaviour of $\phi(U)$, as captured by the group $\langle \phi(U) \rangle$, can be rather different from the "local" structure of $\phi(U)$.

However, the problem can be easily resolved by giving $\langle \phi(U) \rangle$ a different topology, as follows. We use the sets $\{\phi(W) : 1 \in W \subset U, W \text{ open}\}$ as a neighbourhood base for the identity in $\langle \phi(U) \rangle$, and their left-translates $\{g\phi(W) : 1 \in W \subset U, W \text{ open}\}$ as a neighbourhood base for any other element g of $\langle \phi(U) \rangle$. This is easily seen to generate a topology. To show that the group operations remain continuous in this topology, the main task is to show that the conjugation operations $x \mapsto gxg^{-1}$ are continuous with respect to the neighbourhood base at the identity, in the sense that for every open neighbourhood W of the identity in U and every $g \in \langle \phi(U) \rangle$, there exists an open neighbourhood W' of the identity such that $\phi(W') \subset g\phi(W)g^{-1}$. But for $g \in \phi(U)$ this is clear from the injective local homomorphism properties of ϕ (after shrinking W small enough that $g\phi(W)g^{-1}$ will still fall in $\phi(U)$), and then an induction shows the same is true for g in any product set $\phi(U)^n$ of $\phi(U)$, and hence in all of $\langle \phi(U) \rangle$. (It is instructive to follow through this argument for the example given above.)

There is another characterisation of globalisability, due to Mal'cev [**Ma1941**], which is stronger than Lemma 15.2.2, but this strengthening is usually not needed in applications. Call a local group G *globally associative* if, whenever $g_1, \ldots, g_n \in G$ and there are two ways to associate the product $g_1 \ldots g_n$ which are individually well-defined, then the value obtained by these two associations are equal to each other. This implies, but is stronger than, local associativity (which only covers the cases $n \leq 3$).

Proposition 15.2.4 (Globalisation criterion). *Let G be a symmetric local group. Then G is isomorphic to (a restriction of) an open symmetric neighbourhood of the identity in a global topological group if and only if it is globally associative.*

By "restriction of an open symmetric neighbourhood of the identity U", I mean the local group formed from U by restricting the set $\Omega \subset U \times U$ of admissible products for the local group law to some open neighbourhood of $\{1\} \times U \cup U \times \{1\}$ in $U \times U$.

Proof. The "only if" direction is clear, so now suppose that G is a globally associative symmetric local group. Let us call a formal product $g_1 \ldots g_n$ with $g_1, \ldots, g_n \in G$ *weakly well-defined* if there is at least one way to associate the product so that it can be defined in G (this is opposed to actual well-definedness of $g_1 \ldots g_n$, which requires *all* associations to be well-defined). By global associativity, the product $g_1 \ldots g_n$ has a unique evaluation in G whenever it is weakly well-defined.

Let $F = (F, *)$ be the (discrete) free group generated by the elements of G (now viewed merely as a discrete set), thus each element of F can be expressed as a formal product $g_1^{*\epsilon_1} * \ldots * g_n^{*\epsilon_n}$ of elements $g_1, \ldots, g_n$ in G and their formal inverses $g_1^{*-1}, \ldots, g_n^{*-1}$, where $\epsilon_1, \ldots, \epsilon_n \in \{-1, +1\}$, and G can be viewed (only as a set, not as a local group) as a subset of F. Let N be the set of elements in F that have at least one representation (not necessarily reduced) of the form $g_1^{*\epsilon_1} * \ldots * g_n^{*\epsilon_n}$ such that $g_1, \ldots, g_n \in G$ and $\epsilon_1, \ldots, \epsilon_n \in \{-1, +1\}$ with $g_1^{\epsilon_1} \ldots g_n^{\epsilon_n}$ weakly well-defined and evaluating to the identity in G. It is easy to see that N is a normal subgroup of F, and so we may form the quotient group F/N and the quotient map $\pi : F \to F/N$. We claim π is injective on G, and so G is isomorphic (as a discrete local group) to $\pi(G)$. To see this, suppose for contradiction that there are distinct $g, h \in G$ such that $\pi(g) = h$, thus $g * g_1^{*\epsilon_1} * \ldots * g_n^{*\epsilon_n} = h$ for some $g_1^{*\epsilon_1} * \ldots * g_n^{*\epsilon_n} \in N$. As this identity takes place in the free group F, this implies that the formal word $g * g_1^{*\epsilon_1} * \ldots * g_n^{*\epsilon_n}$ can be reduced to the generator h of G by a finite number of operations in which an adjacent pair of the form $k * k^{*-1}$ or $k^{*-1} * k$ for some $k \in G$, or a singleton of the form 1, is deleted. In particular, this implies that $gg_1^{\epsilon_1} \ldots g_n^{\epsilon_n}$ is weakly well-defined and evaluates to h in G. On the other hand, as $g_1^{*\epsilon_1} * \ldots * g_n^{*\epsilon_n} \in N$, we also see (by associating in a different way) that $gg_1^{\epsilon_1} \ldots g_n^{\epsilon_n}$ is weakly well-defined and evaluates to g in G. This contradicts global associativity, and the claim follows.

To conclude the proof, we need to place a topology on F/N that makes G homeomorphic to G. This can be done by taking the sets $x\pi(W)$, with W an open neighbourhood of the identity in G, as a neighbourhood base for each $x \in F/N$. By arguing as in the proof of Lemma 15.2.2 one can verify that this generates a topology that makes F/N a topological group, and makes G homeomorphic to $\pi(G)$; we leave the details as an exercise. $\square$

Exercise 15.2.2. Complete the proof of the above proposition.

Exercise 15.2.3. Use Proposition 15.2.4 to give an alternate proof of Proposition 15.2.2.

15.3. A nonglobalisable group

We now prove Theorem 15.2.1. We begin with a preliminary construction, which gives a local group that has a fixed small neighbourhood that cannot arise from a global group.

Proposition 15.3.1 (Preliminary counterexample). *For any $m \geq 1$, there exists an equivalence relation $\sim_m$ on the open unit ball $B(0,1)$ of $\ell^\infty(\mathbf{N})$ which gives $B(0,1)/\sim_m$ the structure of a local group, but such that $B(0, 1/m)/\sim_m$ is not isomorphic (even as a discrete local group) to a subset of a global group.*

Proof. We will use a probabilistic construction, mimicking the three-dimensional example $(-1, 1)^3/\sim$ from the introduction. Fix m, and let N be a large integer (depending on m) to be chosen later. We identify the N-dimensional cube $(-1, 1)^N$ with the unit ball in $\ell^\infty(\{1, \ldots, N\})$, which embeds in the unit ball in $\ell^\infty(\mathbf{N})$ via extension by zero. Let $b_1, \ldots, b_{N+1} \in \{-1, 1\}^N$ be randomly chosen corners of this cube. A simple application of the union bound shows that with probability approaching 1 as $N \to \infty$, we have $b_i \neq \pm b_j$ for all $i < j$, but also that for any $1 \leq i_1 < \cdots < i_{100m} \leq N$

and any choice of signs $\epsilon_1, \ldots, \epsilon_{100m}$, the vectors $\epsilon_1 b_{i_1}, \ldots, \epsilon_{100m} b_{i_{100m}}$ agree on at least one coordinate. As a corollary of this, we see that

$$\| \sum_{i=1}^{N+1} n_i b_i \|_{\ell^\infty} = \sum_{i=1}^{N+1} |n_i|$$

for any integers $n_1, \ldots, n_{N+1}$ with $\sum_{i=1}^{N+1} |n_i| \leq 100m$.

Let $\varepsilon > 0$ be a small number (depending on m, N) to be chosen later. We let a_i be an element of $(-1, 1)^N$ which is within distance ε of b_i, then we have

(15.4)
$$\| \sum_{i=1}^{N+1} n_i a_i \|_{\ell^\infty} = \sum_{i=1}^{N+1} |n_i| + O(\varepsilon)$$

(allowing implied constants to depend on m, N) for $n_1, \ldots, n_{N+1}$ as above. By generically perturbing the $a_1, \ldots, a_{N+1}$, we may assume that they are *noncommensurable* in the sense that they span a dense subset of $\mathbf{R}^N$. In particular, the $a_1, \ldots, a_{N+1}$ are linearly independent over $\mathbf{Z}$.

We now define an equivalence relation $\sim_m$ on $B(0, 1)$ by defining $f \sim_m g$ whenever

(15.5)
$$f - g = \frac{1}{2m} \sum_{i=1}^{N+1} n_i a_i$$

for some integers $n_1, \ldots, n_N$ with $\sum_{i=1}^{N+1} |n_i| \leq 100m$. Since $\|f - g\|_{\ell^\infty} \leq 2$, we see from (15.4) (if ε is small enough) that the equation (15.5) can only be true if $\sum_{i=1}^{N+1} |n_i| \leq 4m$. As a consequence, we see that $\sim_m$ is a equivalence relation. We can then form the quotient space $B(0, 1)/ \sim_m$. Observe that if $f_1, f_2, g_1, g_2 \in B(0, 1)$ are such that $f_1 \sim_m g_1$, $f_2 \sim_m g_2$, and $f_1 + f_2, g_1 + g_2 \in B(0, 1)$, then we also have $f_1 + f_2 \sim_m g_1 + g_2$. Thus $B(0, 1)/ \sim_m$ has an addition operation $+$, defined on those equivalence classes $[f], [g] \in B(0, 1)/ \sim_m$ for which $f + g \in B(0, 1)$ for at least one representative f, g of $[f], [g]$, respectively. One easily verifies that this gives $B(0, 1)/ \sim_m$ the structure of a local group.

Now suppose for the sake of contradiction that $B(0, 1/m)/ \sim_m$ is isomorphic to a restriction of a global topological group G. Since $0 \sim \frac{1}{m} a_i$ for all $i = 1, \ldots, N+1$, we thus have a map $\phi \colon B(0, 1/m) \to G$ which annihilates all of the $\frac{1}{2m} a_i$, and is locally additive in the sense that $\phi(f + g) = \phi(f) \cdot \phi(g)$ whenever $f, g, f + g \in B(0, 1/m)$. In particular, we see that all the elements of $\phi(B(0, 1/2m))$ commute with each other. Furthermore, the kernel $\{f \in B(0, 1/m) : \phi(f) = 1\}$ is precisely equal to the set

(15.6)
$$\left\{ \frac{1}{2m} \sum_{i=1}^{N+1} n_i a_i : \sum_{i=1}^{N+1} |n_i| \leq 4m \right\} \cap B(0, 1/m).$$

As the $a_1, \ldots, a_{N+1}$ span a dense subset of $\mathbf{R}^n$, we can find integers $n_1, \ldots, n_{N+1}$ such that

$$n_1 a_1 + \cdots + n_{N+1} a_{N+1} \in B(0,1)$$

and

$$\sum_{i=1}^{N+1} |n_i| > 4m.$$

We claim that $\frac{1}{2m}(n_1 a_1 + \cdots + n_{N+1} a_{N+1})$ lies in the kernel of ϕ, contradicting the description (15.6) of that description (and the linear independence of the a_i). To see this, we observe for a sufficiently large natural number M that the local homomorphism property (and the commutativity) gives

$$\phi\left(\frac{1}{2mM}(n_1 a_1 + \cdots + n_{N+1} a_{N+1})\right) = \prod_{i=1}^{N+1} \phi\left(\frac{1}{2mM} a_{N+1}\right)^{n_{N+1}}$$

and hence (by further application of local homomorphism and commutativity)

$$\phi\left(\frac{1}{2m}(n_1 a_1 + \cdots + n_{N+1} a_{N+1})\right) = \prod_{i=1}^{N+1} \phi\left(\frac{1}{2mM} a_{N+1}\right)^{M n_{N+1}}.$$

Yet, by more application of the local homomorphism property, we have

$$\phi\left(\frac{1}{2mM} a_{N+1}\right)^M = \phi\left(\frac{1}{2m} a_{N+1}\right) = 1,$$

and the claim follows. $\square$

Now we glue together the examples in Proposition 15.3.1 to establish Theorem 15.2.1. We work in the space $\ell^\infty(\mathbf{N}^2)$, the elements of which we can think of as a sequence $(f_m)_{m \in \mathbf{N}}$ of uniformly bounded functions $f_m \in \ell^\infty(\mathbf{N})$. The unit ball in this space can then be identified (as a set) with the product $B(0,1)^{\mathbf{N}}$, where $B(0,1)$ is the unit ball in $\ell^\infty(\mathbf{N})$, though we caution that the topology on $B(0,1)^{\mathbf{N}}$ is *not* given by the product topology (or the box topology).

We can combine the equivalence relations $\sim_m$ on $B(0,1)$ to a relation $\sim$ on $B(0,1)^{\mathbf{N}}$, defined by setting $(f_m)_{m \in \mathbf{N}} \sim (g_m)_{m \in \mathbf{N}}$ iff $f_m \sim_m g_m$ for all $m \in \mathbf{N}$. This is clearly an equivalence relation on $B(0,1)^{\mathbf{N}}$, and so we can create the quotient space $B(0,1)^{\mathbf{N}}/\sim$ with the quotient topology. One easily verifies that this gives a local group. The sets $B(0,1/m)^{\mathbf{N}}/\sim$ form a neighbourhood base of the identity, but none of these sets is isomorphic (even as a discrete local group) to a subset of a global group, as it contains a copy of $B(0,1/m)/\sim_m$, and the claim follows.

Chapter 16

Central extensions of Lie groups, and cocycle averaging

The theory of Hilbert's fifth problem sprawls across many subfields of mathematics: Lie theory, representation theory, group theory, nonabelian Fourier analysis, point-set topology, and even a little bit of group cohomology. The latter aspect of this theory is what this section will be focused on. The general question that comes into play here is the *extension problem*: Given two (topological or Lie) groups H and K, what is the structure of the possible groups G that are formed by extending H by K? In other words, given a short exact sequence

$$0 \to K \to G \to H \to 0,$$

to what extent is the structure of G determined by that of H and K?

As an example of why understanding the extension problem would help in structural theory, let us consider the task of classifying the structure of a Lie group G. First, we factor out the connected component G° of the identity as

$$0 \to G^\circ \to G \to G/G^\circ \to 0;$$

as Lie groups are locally connected, G/G° is discrete. Thus, to understand general Lie groups, it suffices to understand the extensions of discrete groups by connected Lie groups.

Next, to study a connected Lie group G, we can consider the conjugation action $g\colon X \mapsto gXg^{-1}$ on the Lie algebra $\mathfrak{g}$, which gives the *adjoint representation* $\mathrm{Ad}\colon G \to GL(\mathfrak{g})$. The kernel of this representation consists of all the group elements g that commute with all elements of the Lie algebra, and thus (by connectedness) is the center $Z(G)$ of G. The adjoint representation is then faithful on the quotient $G/Z(G)$. The short exact sequence

$$0 \to Z(G) \to G \to G/Z(G) \to 0$$

then describes G as a central extension (by the abelian Lie group $Z(G)$) of $G/Z(G)$, which is a connected Lie group with a faithful finite-dimensional linear representation.

This suggests a route to Hilbert's fifth problem, at least in the case of connected groups G. Let G be a connected locally compact group that we hope to demonstrate is isomorphic to a Lie group. As discussed in Chapter 3, we first form the space $L(G)$ of one-parameter subgroups of G (which should, eventually, become the Lie algebra of G). Hopefully, $L(G)$ has the structure of a vector space. The group G acts on $L(G)$ by conjugation; this action should be both continuous and linear, giving an "adjoint representation" $\mathrm{Ad}\colon G \to GL(L(G))$. The kernel of this representation should then be the center $Z(G)$ of G. The quotient $G/Z(G)$ is locally compact and has a faithful linear representation, and is thus a Lie group by von Neumann's theorem (Theorem 3.0.16). The group $Z(G)$ is locally compact abelian, and so it should be a relatively easy task to establish that it is also a Lie group. To finish the job, one needs the following result:

Theorem 16.0.2 (Central extensions of Lie are Lie). *Let G be a locally compact group which is a central extension of a Lie group H by an abelian Lie group K. Then G is also isomorphic to a Lie group.*

This result can be obtained by combining a result of Kuranishi [**Ku1950**] with a result of Gleason [**Gl1950**]. It is superceded by the subsequent theory of Hilbert's fifth problem (cf. Exercise 17.2.1 below), but can be proven without using such machinery. The point here is that while G is initially only a topological group, the smooth structures of H and K can be combined (after a little bit of cohomology) to create the smooth structure on G required to upgrade G from a topological group to a Lie group. One of the main ideas here is to improve the behaviour of a cocycle by averaging it; this basic trick is helpful elsewhere in the theory, resolving a number of cohomological issues in topological group theory. The result can be generalised to show, in fact, that arbitrary (topological) extensions of Lie groups by Lie groups remain Lie; this was shown in [**Gl1950**]. However, the above special case of this result is already sufficient (in conjunction with the rest of the theory, of course) to resolve Hilbert's fifth problem.

Remark 16.0.3. We have shown in the above discussion that every connected Lie group is a central extension (by an abelian Lie group) of a Lie group with a faithful continuous linear representation. It is natural to ask whether this central extension is necessary. Unfortunately, not every connected Lie group admits a faithful continuous linear representation; see Exercise 1.1.1. (On the other hand, the group G in that example is certainly isomorphic to the extension of the linear group $\mathbf{R}^2$ by the abelian group $\mathbf{R}/\mathbf{Z}$.)

16.1. A little group cohomology

Let us first ignore the topological or Lie structure, consider the (central) extension problem for discrete groups only. Thus, let us suppose we have a (discrete) group $G = (G, \cdot)$ which is a central extension of a group H by a group K. We view K as a central subgroup of G (which we write additively to emphasise its abelian nature), and use $\pi \colon G \to H$ to denote the projection map. If $k \in K$ and $g \in G$, we write $g + k$ for $kg = gk$ to emphasise the central nature of K.

It may help to view G as a principal K-bundle over H, with K being thought of as the "vertical" component of G and H as the "horizontal" component. Thus G is the union of "vertical fibres" $\pi^{-1}(h)$, $h \in H$ indexed by the horizontal group H, each of which is a coset (or a *torsor*) of the vertical group H.

As central extensions are not unique, we will need to specify some additional data beyond H and K. One way to view this data is to specify a *section* of the extension, that is to say a map $\phi \colon H \to G$ that is a right-inverse for the projection map π; thus ϕ selects one element $\phi(h)$ from each fibre $\pi^{-1}(h)$ of the projection map (which is also a coset of K). Such a section can be always chosen using the axiom of choice, though of course there is no guarantee of any measurability, continuity, or smoothness properties of such a section if the groups involved have the relevant measure-theoretic, topological, or smooth structure.

Note that we do not necessarily require the section ϕ to map the group identity 1_H of H to the group identity 1_G of G, though in practice it is usually not difficult to impose this constraint if desired.

The group G can then be described, as a set, as the disjoint union of its fibres $\phi(h) + K$ for $h \in H$:

$$G = \biguplus_{h \in H} \phi(h) + K.$$

Thus each element $g \in G$ can be uniquely expressed as $\phi(h) + k$ for some $h \in H$ and $k \in K$, and can thus be viewed as a system of coordinates of G (identifying it as a set with $H \times K$). Now we turn to the group operations $()^{-1} \colon G \to G$ and $\cdot \colon G \times G \to G$. If $h_1, h_2 \in H$, then the product $\phi(h_1)\phi(h_2)$

must lie in the fibre of $h_1 h_2$, and so we have

$$\phi(h_1)\phi(h_2) = \phi(h_1 h_2) + \psi(h_1, h_2)$$

for some function $\psi \colon H \times H \to K$. By the centrality of K, this describes the product law for general elements of G in the $H \times K$ coordinates:

$$(16.1) \qquad (\phi(h_1) + k_1)(\phi(h_2) + k_2) = \phi(h_1 h_2) + (k_1 + k_2 + \psi(h_1, h_2)).$$

Using the associative law for G, we see that ψ must obey the *cocycle equation*

$$(16.2) \qquad \psi(h_1, h_2) + \psi(h_1 h_2, h_3) = \psi(h_1, h_2 h_3) + \psi(h_2, h_3)$$

for all $h_1, h_2, h_3 \in K$; we refer to functions ψ that obey this equation as *cocycles* (the reason for this terminology being explained in [**Ta2009**, §1.13]). The space of all such cocyles is denoted $Z^2(H, K)$ (or $Z^2_{\text{disc}}(H, K)$, if we wish to emphasise that we are working for now in the discrete category, as opposed to the measurable, topological, or smooth category); this is an abelian group with respect to pointwise addition. Using (16.1), we also have a description of the group identity 1_G of G in coordinates:

$$(16.3) \qquad\qquad 1_G = \phi(1_H) - \psi(1_H, 1_H).$$

Indeed, this can be seen by noting that 1_G is the unique solution of the group equation $g \cdot g = g$. Similarly, we can describe the inverse operation in $H \times K$ coordinates:

$$(16.4) \qquad (\phi(h) + k)^{-1} = \phi(h^{-1}) + (-k - \psi(h, h^{-1}) - \psi(1_H, 1_H)).$$

Thus we see that the cocycle ψ, together with the group structures on H and K, capture all the group-theoretic structure of G. Conversely, given any cocycle $\psi \in Z^2(H, K)$, one can place a group structure on the set $H \times K$ by declaring the multiplication law as

$$(h_1, k_1) \cdot (h_2, k_2) := (h_1 h_2, k_1 + k_2 + \psi(h_1, h_2)),$$

the identity element as $(1_H, -\psi(1_H, 1_H))$, and the inversion law as

$$(h, k)^{-1} := \left(h^{-1}, -k - \psi(h, h^{-1}) - \psi(1_H, 1_H) \right).$$

Exercise 16.1.1. Using the cocycle equation (16.2), show that the above operations do indeed yield group structure on $H \times K$ (i.e., the group axioms are obeyed). If ψ arises as the cocycle associated to a section of a group extension G, we then see from (16.1), (16.3), (16.4) that this group structure we have just placed on $H \times K$ is isomorphic to that on G.

Thus we see that once we select a section, we can describe a central extension of K by H (up to group isomorphism) as a cocycle, and conversely every cocycle arises in this manner. However, we have some freedom in deciding how to select this section. Given one section $\phi\colon H \to G$, any other section $\phi'\colon H \to G$ takes the form

$$\phi'(h) = \phi(h) + f(h)$$

for some function $f\colon H \to K$; conversely, every such function f can be used to shift a section ϕ to a new section ϕ'. We refer to such functions f as *gauge functions*. The cocycles ψ, ψ' associated to ϕ, ϕ' are related by the *gauge transformation*

$$\psi'(h_1, h_2) = \psi(h_1, h_2) + df(h_1, h_2)$$

where $df\colon H \times H \to K$ is the function

$$df(h_1, h_2) \coloneqq f(h_1) + f(h_2) - f(h_1 h_2).$$

We refer to functions of the form df as *coboundaries*, and denote the space of all such coboundaries as $B^2(H, K)$ (or $B^2_{\mathrm{disc}}(H, K)$, if we want to emphasise the discrete nature of these coboundaries). One easily verifies that all coboundaries are cocycles, and so $B^2(H, K)$ is a subgroup of $Z^2(H, K)$. We then define the *second group cohomology* $H^2(H, K)$ (or $H^2_{\mathrm{disc}}(H, K)$) to be the quotient group

$$H^2(H, K) \coloneqq Z^2(H, K)/B^2(H, K),$$

and refer to elements of $H^2(H, K)$ as *cohomology classes*. (There are higher order group cohomologies, which also have some relevance for the extension problem, but will not be needed here; see [**Ta2009**, §1.13] for further discussion. The first group cohomology $H^1(H, K) = \mathrm{Hom}(H, K)$ is just the space of homomorphisms from H to K; again, this has some relevance for the extension problem but will not be needed here.)

We call two cocycles *cohomologous* if they differ by a coboundary (i.e., they lie in the same cohomology class). Thus we see that different sections of a single central group extension provide cohomologous cocycles. Conversely, if two cocycles are cohomologous, then the group structures on $H \times K$ given by these cocycles are easily seen to be isomorphic; furthermore, if we restrict group isomorphism to fix each fibre of π, this is the only way in which group structures generated by such cocycles are isomorphic. Thus, we see that up to group isomorphism, central group extensions are described by cohomology classes.

Remark 16.1.1. All constant cocycles are coboundaries, and so if desired one can always normalise a cocycle $\psi\colon H \times H \to K$ (up to coboundaries) so that $\psi(1_H, 1_H) = 0$. This can lead to some minor simplifications to some of the cocycle formulae (such as (16.3), (16.4)).

The trivial cohomology class of course contains the trivial cocycle, which in turn generates the direct product $H \times K$. Nontrivial cohomology classes generate "skew" products that will not be isomorphic to the direct product $H \times K$ (at least if we insist on fixing each fibre). In particular, we see that if the second group cohomology $H^2(H, K)$ is trivial, then the only central extensions of H by K are (up to isomorphism) the direct product extension.

In general, we do not expect the cohomology group to be trivial: not every cocycle is a coboundary. A simple example is provided by viewing the integer group $\mathbf{Z}$ as a central extension of a cyclic group $\mathbf{Z}/N\mathbf{Z}$ by the subgroup $N\mathbf{Z}$ using the short exact sequence

$$0 \to N\mathbf{Z} \to \mathbf{Z} \to \mathbf{Z}/N\mathbf{Z} \to 0.$$

Clearly $\mathbf{Z}$ is not isomorphic to $N\mathbf{Z} \times \mathbf{Z}/N\mathbf{Z}$ (the former group is torsion-free, while the latter group is not), and so $H^2(\mathbf{Z}/N\mathbf{Z}, N\mathbf{Z})$ is nontrivial. If we use the section $\phi\colon \mathbf{Z}/N\mathbf{Z} \to \mathbf{Z}$ that maps $i \bmod N$ to i for $i = 0, \ldots, N-1$, then the associated cocycle $\psi(i \bmod N, j \bmod N)$ is the familiar "carry bit" one learns about in primary school, which equals N when $i, j = 0, 1, \ldots, N-1$ and $i + j \geq N$, but vanishes for other $i, j = 0, 1, \ldots, N-1$.

However, when the "vertical group" K is sufficiently "Euclidean" in nature, one can start deploying the method of *averaging* to improve the behaviour of cocycles. Here is a simple example of the averaging method in action:

Lemma 16.1.2. *Let H be a finite group, and let K be a real vector space. Then $H^2(H, K)$ is trivial.*

Proof. We need to show that every cocycle is a coboundary. Accordingly, let $\psi \in Z^2(H, K)$ be a cocycle, thus

$$\psi(h_1, h_2) + \psi(h_1 h_2, h_3) = \psi(h_1, h_2 h_3) + \psi(h_2, h_3)$$

for all $h_1, h_2, h_3 \in H$. We sum this equation over all $h_3 \in H$, and then divide by the cardinality $|H|$ of H; note that this averaging operation relies on the finite nature of H and the vector space nature of K. We obtain as a consequence

$$\psi(h_1, h_2) + f(h_1 h_2) = f(h_1) + f(h_2)$$

where

$$f(h) := \mathbf{E}_{h_3 \in H} \psi(h, h_3) = \frac{1}{|H|} \sum_{h_3 \in H} \psi(h, h_3)$$

is the averaging of ψ in the second variable. We thus have $\psi = df$, and so ψ is a coboundary as desired. $\square$

It is instructive to compare this argument against the nontriviality of $H^2(\mathbf{Z}/N\mathbf{Z}, N\mathbf{Z})$ mentioned earlier. While $\mathbf{Z}/N\mathbf{Z}$ is still finite, the problem here is that $N\mathbf{Z}$ is not a vector space and so one cannot divide by N to "straighten" the cocycle. However, the averaging argument can still achieve some simplification to such cocycles:

Exercise 16.1.2. Let H be a finite group, and let $\psi \in Z^2(H, \mathbf{Z})$ be a cocycle. Show that ψ is cohomologous to a cocycle taking values in $\{-1, 0, +1\}$. (*Hint:* Average the cocycle as before using the reals $\mathbf{R}$, then round to the nearest integer.)

Now we turn from discrete group theory to topological group theory. Now H, K, G are required to be topological groups rather than discrete groups, with the projection map $\pi \colon G \to H$ continuous. (In particular, if H is Hausdorff, then K must be closed.) In this case, it is no longer the case that every (discrete) cocycle $\psi \in Z^2_{\text{disc}}(H, K)$ gives rise to a topological group, because the group structure on $H \times K$ given by ψ need not be continuous with respect to the product topology of $H \times K$. However, if the cocycle ψ is *continuous*, then it is clear that the group operations will be continuous, and so we will in fact generate a topological group. Conversely, if we have a section $\phi \colon H \to G$ which is continuous, then it is not difficult to verify that the cocycle generated is also continuous. This leads to a slightly different group cohomology, using the space $Z^2_{top}(H, K)$ of continuous cocycles, and also the space $B^2_{top}(H, K)$ of coboundaries df arising from continuous gauge functions f.

However, such "global" cohomology contains some nontrivial global topological obstructions that limit its usefulness for the extension problem. One particularly fundamental such obstruction is that one does not expect global continuous sections $\phi \colon H \to G$ of a group extension to exist in general, unless H or K have a particularly simple topology (e.g., if they are contractible or simply connected). For instance, with the short exact sequence

$$0 \to \mathbf{Z} \to \mathbf{R} \to \mathbf{R}/\mathbf{Z} \to 0$$

there is no way to continuously lift the horizontal group $\mathbf{R}/\mathbf{Z}$ back up to the real line $\mathbf{R}$, due to the presence of nontrivial *monodromy*. While these global obstructions are quite interesting from an algebraic topology perspective, they are not of central importance in the theory surrounding Hilbert's fifth problem, which is instead more concerned with the *local* topological and Lie structure of groups.

Because of this, we shall work with *local* sections and cocycles instead of global ones. A *local section* in a topological group extension

$$0 \to K \to G \to H \to 0$$

is a continuous map $\phi \colon U \to G$ from an open neighbourhood U of the identity 1_H in H to G which is a right inverse of π on U, thus $\phi(h) \in \pi^{-1}(h)$ for all $h \in U$. (It is not yet obvious why local sections exist at all — there may still be *local* obstructions to trivialising the fibre bundle — but certainly this task should be easier than that of locating *global* continuous sections.)

Similarly, a *local cocycle* is a continuous map $\psi \colon U \times U \to K$ defined on some open neighbourhood U of the identity in H that obeys the cocycle equation (16.2) whenever $h_1, h_2, h_3 \in U$ are such that $h_1 h_2, h_2 h_3 \in U$. We consider two local sections $\phi \colon U \to G$, $\phi' \colon U' \to G$ to be *locally identical* if there exists a neighbourhood U'' of the identity contained in both U and U' such that ϕ and ϕ' agree on U''. Similarly, we consider two local cocycles $\psi \colon U \times U \to G$ and $\psi' \colon U' \times U' \to G$ to be locally identical if there is a neighbourhood U'' of the identity in $U \cap U'$ such that ψ and ψ' agree on $U'' \times U''$. We let $Z^2_{top,loc}(H, K)$ be the space of local cocycles, modulo local identity; this is easily seen to be an abelian group.

One easily verifies that every local section induces a local cocycle (perhaps after shrinking the open neighbourhood U slightly), which is well-defined up to local identity. Conversely, the computations used to show Exercise 16.1.1 show that every local cocycle creates a *local* group structure on $H \times K$.

If U is an open neighbourhood the identity of H, $f \colon U \to K$ is a continuous gauge function, and U' is a smaller neighbourhood such that $(U')^2 \subset U$, we can define the *local coboundary* $df \colon U' \times U' \to K$ by the usual formula

$$df(h_1, h_2) := f(h_1) + f(h_2) - f(h_1 h_2).$$

This is well-defined up to local identity. We let $B^2_{top,loc}(H, K)$ be the space of local coboundaries, modulo local identity; this is a subgroup of $Z^2_{top,loc}(H, K)$. Thus, one can form the local topological group cohomology $H^2_{top,loc}(H, K) := Z^2_{top,loc}(H, K)/B^2_{top,loc}(H, K)$. We say that two local cocycles are *locally cohomologous* if they differ by a local coboundary.

The relevance of local cohomology to the Lie group extension problem can be seen from the following lemma.

Lemma 16.1.3. *Let G be a central (topological) group extension of a Lie group H by a Lie group K. If there is a local section $\phi \colon U \to G$ whose associated local cocycle $\psi \colon U' \times U' \to K$ is locally cohomologous to a smooth local cocycle $\psi' \colon U'' \times U'' \to K$, then G is also isomorphic to a Lie group.*

Note that the notion of smoothness of a (local) cocycle $\psi \colon U' \times U' \to K$ makes sense, because there are smooth structures in place for both H (and hence U') and K. In contrast, one cannot initially talk about a smooth (local) section $\phi \colon U \to G$, because G initially only has a topological structure and not a smooth one.

Proof. By hypothesis, we locally have $\psi = \psi' + df$ for some continuous gauge function $f \colon U''' \to K$. Rotating ϕ by f, we thus obtain another local section $\phi' \colon U'''' \to G$ whose associated local cocycle is smooth. We use this section to identify $\pi^{-1}(U^{(5)})$ as a topological space with $U^{(5)} \times K$ for some sufficiently small neighbourhood $U^{(5)}$ of the origin. We can then give $\pi^{-1}(U^{(5)})$ a smooth structure induced from the product smooth structure on $U^{(5)} \times K$. By (16.1), (16.4) (and (16.3)), we see that the group operations are smooth on a neighbourhood of the identity in G.

To finish the job and give G the structure of a Lie group, we need to extend this smooth structure to the rest of G in a manner which preserves the smooth nature of the group operations. First, by using the local coordinates of $U^{(5)} \times K$ we see that G is locally connected. Thus, if G° is the connected component of the identity, then the quotient group G/G°.

Next, we extend the smooth structure on a neighbourhood of the identity on G to the rest of G by (say) left-invariance; it is easy to see that these smooth coordinate patches are compatible with each other. A continuity argument then shows that the group operations are smooth on G°. Each element g of G acts via conjugation by a continuous homomorphism $x \mapsto gxg^{-1}$ on the Lie group G°; by applying Cartan's theorem to the graph of this homomorphism as in the preceding post, we see that that such homomorphisms are smooth. From this we can then conclude that the group operations are smooth on all of G. $\square$

16.2. Proof of theorem

Lemma 16.1.3 suggests a strategy to prove Theorem 16.0.2: First, obtain a local section of G, extract the associated local cocycle, then "straighten" it to a smooth cocycle. This will indeed be how we proceed. In obtaining the local section and in smoothing the cocycle we will take advantage of the averaging argument, adapted to the Lie algebra setting.

We begin by constructing the section, following [**Gl1950**]. The idea is to use the zero set of a certain continuous function $F \colon G \to V$ from G into a finite-dimensional vector space V. We will use topological arguments to force this function to have a zero at every fibre, and will give F some "vertical nondegeneracy" to prevent it from having more than one zero on any given fibre, at least locally.

Let's see how this works in the special case when the abelian Lie group K is compact, so that (by the Peter-Weyl theorem, as discussed in this post) it can be modeled as a closed subgroup of a unitary group $U(n)$, thus we have an injective continuous homomorphism $\rho \colon K \to U(n)$. We view $U(n)$ as a subset of the vector space $\mathbf{C}^{n^2}$ of $n \times n$ complex matrices.

The group G is locally compact and also Hausdorff (being the extension of one Hausdorff group by another). Thus, we may then apply the Tietze extension theorem, and extend the map $\rho \colon K \to \mathbf{C}^{n^2}$ to a continuous (and compactly supported) function $F_0 \colon G \to \mathbf{C}^{n^2}$ defined on all of G.

Note that while the original map ρ was a homomorphism, thus

(16.5) $$\rho(k + k') = \rho(k)\rho(k')$$

for all $k, k' \in K$, the extension F_0 need not obey any analogous symmetry in general. However, this can be rectified by an averaging argument. Define the function $F_1 \colon G \to \mathbf{C}^{n^2}$ by the formula

$$F_1(g) := \int_K \rho(k')^{-1} F_0(g + k') \, d\mu_K(k'),$$

where μ_K is normalised Haar measure on K. Then F_1 is still continuous and still extends ρ; but, unlike F_0, it now obeys the *equivariance* property

$$F_1(g + k) = \rho(k) F_1(g)$$

for $g \in G$ and $k \in K$, as can easily be seen from (16.5).

The function F_1 is a better extension than F_0, but suffers from another defect; while ρ took values in the closed group $\rho(K)$, F_1 takes values in $\mathbf{C}^{n^2}$ (for instance, F_1 is going to be compactly supported). There may well be global topological reasons that prevent F_1 from taking values in $\rho(K)$ at all points, but as long as we are willing to work *locally*, we can repair this issue as follows. Using the inverse function theorem, one can find a manifold $W \subset \mathbf{C}^{n^2}$ going through the identity matrix I, which is transverse to $\rho(K)$ in the sense that every matrix M that is sufficiently close to the identity can be uniquely (and continuously) decomposed as $\pi_1(M)\pi_2(M)$, where $\pi_1(M) \in \rho(K)$ and $\pi_2(M) \in W$. (Indeed, to build W, one can just use a complementary subspace to the tangent space (or Lie algebra) of $\rho(K)$ at the identity.) For g sufficiently close to the identity 1_G, we may thus factor

$$F_1(g) = F_2(g) F_3(g)$$

where $F_2(g) \in \rho(K)$ and $F_3(g) \in W$. By construction, F_2 will be continuous and take values in $\rho(K)$ near 1_G, and equal ρ on K near 1_G; it will also obey the equivariance

$$F_2(g + k) = \rho(k) F_2(g)$$

for $g \in G$ and $k \in K$ that are sufficiently close to the identity. From this, we see that for each $h \in H$ that is sufficiently close to the identity, there exists a unique solution $\phi(h)$ in $\pi^{-1}(h)$ to the equation $F_2(g) = I$ that is also close to the identity; from the continuity of F_2 we see that ϕ is continuous near the identity and is thus a local section of G as required.

Now we argue in the general case, in which the abelian Lie group K need not be compact. If K is connected, then it is isomorphic to a Euclidean space quotiented by a discrete subgroup. In particular, it can be quotiented down to a torus in a manner which is a local homeomorphism near the origin. Embedding that torus in a unitary group $U(n)$, we obtain a continuous homomorphism $\rho\colon K \to U(n)$ which is still *locally* injective and has compact image. One can then repeat the previous arguments to obtain a local section; we omit the details.

Finally, if K is not connected, we can work with the connected component K° of the identity, which is locally the same as K, and repeat the previous argument again to obtain a local section.

Using this local section, we obtain a local cocycle $\psi\colon U \times U \to K$; shifting by a constant, we may assume that $\psi(1_H, 1_H) = 0$. Near the identity, the abelian Lie group K can be locally identified with a Euclidean space $\mathbf{R}^n$, so (shrinking U if necessary) ψ induces a $\mathbf{R}^n$-valued local cocycle $\tilde\psi\colon U \times U \to \mathbf{R}^n$.

We will smooth this cocycle by an averaging argument akin to the one used in Lemma 16.1.2. Instead of a discrete averaging, though, we will now use a nontrivial left-invariant Haar measure μ_H on H (the existence of which is guaranteed by Theorem 4.1.6). As we only have a local cocycle, though, we will need to only average using a smooth, compactly supported function $\eta\colon H \to \mathbf{R}^+$ supported on a small neighbourhood U' of U (small enough that $(U')^3 \subset U$), normalised so that $\int_H \eta(h_3) \, d\mu_H(h_3) = 1$. For any $h_1, h_2, h_3 \in U'$, we have the cocycle equation

$$\tilde\psi(h_1, h_2) + \tilde\psi(h_1 h_2, h_3) = \tilde\psi(h_1, h_2 h_3) + \tilde\psi(h_2, h_3);$$

we average this against $\eta(h_3) \, d\mu_H(h_3)$ and conclude that

$$\tilde\psi(h_1, h_2) + F(h_1 h_2) = F(h_1) + F(h_2) + \Psi(h_1, h_2)$$

for $h_1, h_2 \in U'$, where $F\colon (U')^2 \to \mathbf{R}^n$ is the function

$$F(h) := \int_H \tilde\psi(h, h_3) \eta(h_3) \, d\mu_H(h_3)$$

and $\Psi\colon U' \times U' \to \mathbf{R}^n$ is the function

$$\Psi(h_1, h_2) := \int_H (\tilde\psi(h_1, h_2 h_3) - \tilde\psi(h_1, h_3)) \eta(h_3) \, d\mu_H(h_3).$$

The function F is continuous, and so $\tilde\psi$ is locally cohomologous to Ψ. Using the left-invariance of μ_H, we can rewrite Ψ as

$$\Psi(h_1, h_2) := \int_H \tilde\psi(h_1, h_3)(\eta(h_2^{-1} h_3) - \eta(h_3)) \, d\mu_H(h_3),$$

from which it becomes clear that Ψ is smooth in the h_2 variable. A similar averaging argument (now against a bump function on a right-invariant measure) then shows that Ψ is in turn locally cohomologous to another local cocycle $\Psi' : U'' \times U'' \to \mathbf{R}^n$ which is now smooth in both the h_1 and h_2 variables. Pulling $\mathbf{R}^n$ back to K, we conclude that ψ is locally cohomologous to a smooth local cocycle, and Theorem 16.0.2 now follows from Lemma 16.1.3.

The Hilbert-Smith conjecture

The classical formulation of *Hilbert's fifth problem* asks whether topological groups that have the *topological* structure of a manifold, are necessarily Lie groups. This is indeed the case, thanks to following theorem of Gleason [**Gl1952**] and Montgomery and Zippin [**MoZi1952**]:

Theorem 17.0.1 (Hilbert's fifth problem). *Let G be a topological group which is locally Euclidean. Then G is isomorphic to a Lie group.*

This theorem was proven in Section 6.3. There is, however, a generalisation of Hilbert's fifth problem which remains open, namely the *Hilbert-Smith conjecture*, in which it is a space acted on by the group which has the manifold structure, rather than the group itself:

Conjecture 17.0.2 (Hilbert-Smith conjecture). *Let G be a locally compact topological group which acts continuously and faithfully (or effectively) on a connected finite-dimensional manifold X. Then G is isomorphic to a Lie group.*

Note that Conjecture 17.0.2 easily implies Theorem 17.0.1 as one can pass to the connected component G° of a locally Euclidean group (which is clearly locally compact), and then look at the action of G° on itself by left-multiplication.

The hypothesis that the action is faithful (i.e., each nonidentity group element $g \in G \backslash \{\text{id}\}$ acts nontrivially on X) cannot be completely eliminated, as any group G will have a trivial action on any space X. The requirement that G be locally compact is similarly necessary: Consider, for instance, the diffeomorphism group $\mathrm{Diff}(S^1)$ of, say, the unit circle S^1, which acts on S^1 but is infinite-dimensional and is not locally compact (with, say, the uniform topology). Finally, the connectedness of X is also important: The infinite torus $G = (\mathbf{R}/\mathbf{Z})^{\mathbf{N}}$ (with the product topology) acts faithfully on the disconnected manifold $X := \mathbf{R}/\mathbf{Z} \times \mathbf{N}$ by the action

$$(g_n)_{n \in \mathbf{N}}(\theta, m) := (\theta + g_m, m).$$

Note that $(\mathbf{R}/\mathbf{Z})^{\mathbf{N}}$ contains the p-adic group $\mathbf{Z}_p$ as an embedded subgroup (identifying a p-adic integer m with $(\frac{m}{p^n} \bmod 1)_{n \in \mathbf{N}}$), so this also gives a faithful action of $\mathbf{Z}_p$ on X.

The conjecture in full generality remains open. However, there are a number of partial results. For instance, it was observed by Montgomery and Zippin [**MoZi1974**] that the conjecture is true for *transitive* actions; see Section 6.4. Another partial result is the reduction of the Hilbert-Smith conjecture to the p-adic case. Indeed, it is known that Conjecture 17.0.2 is equivalent to

Conjecture 17.0.3 (Hilbert-Smith conjecture for p-adic actions). *It is not possible for a p-adic group $\mathbf{Z}_p$ to act continuously and effectively on a connected finite-dimensional manifold X.*

The reduction to the p-adic case follows from the structural theory of locally compact groups (specifically, the Gleason-Yamabe theorem, Theorem 1.1.13) and some results of Newman [**Ne1931**] that sharply restrict the ability of periodic actions on a manifold X to be close to the identity. This argument will also be given below (following the presentation in [**Le1997**]).

Very recently, the three-dimensional case of Conjecture 17.0.3 (and hence the three-dimensional case of Conjecture 17.0.2) was settled in [**Pa2013**], by topological methods; however, we will not discuss the proof of this conjecture here.

17.1. Periodic actions of prime order

We now study periodic actions $T: X \to X$ on a manifold X of some prime order p, thus $T^p = \mathrm{id}$.

The basic observation to exploit here is that of *rigidity*: A periodic action (or more precisely, the orbits of this action) cannot be too close to the identity, without actually *being* the identity. More precisely, we have the following theorem of Newman [**Ne1931**]:

Theorem 17.1.1 (Newman's first theorem). *Let U be an open subset of $\mathbf{R}^n$ containing the closed unit ball B, and let $T: U \to U$ be a homeomorphism of some prime period $p \geq 1$. Suppose that for every $x \in U$, the orbit $\{T^n x : n = 0, 1, \ldots, p-1\}$ has diameter strictly less than 1. Then $T(0) = 0$.*

Note that some result like this must be needed in order to establish the Hilbert-Smith conjecture. Suppose, for instance, that one could find a nontrivial transformation T of some period p on the unit ball B that acts trivially on the boundary of that ball. Then by placing infinitely many disjoint copies of that ball into $\mathbf{R}^n$, and considering maps that are equal to some power of T on each such ball, and on the identity outside all the balls, we can obtain a faithful action of $(\mathbf{Z}/p\mathbf{Z})^{\mathbf{N}}$ on $\mathbf{R}^n$, contradicting the Hilbert-Smith conjecture.

To prove the theorem we will need some basic *degree theory*. Given a continuously differentiable map $\Phi \colon B \to \mathbf{R}^n$, we know from *Sard's theorem* that almost every point x in $\mathbf{R}^n$ is a regular point, in that the preimage $\Phi^{-1}(\{x\})$ is finite and avoids the boundary ∂B of B, with $\nabla \Phi(x)$ being nondegenerate at each x. We can define the *degree* of Φ at the regular point x to be the number of preimages with $\nabla \Phi$ orientation-preserving, minus the number of preimages with $\nabla \Phi$ orientation-reversing. One can show that this degree extends to a constant integer-valued function on each connected component U of $\mathbf{R}^n \backslash \Phi(\partial B)$; indeed, one can define the degree $\deg(U)$ on such a component analytically by the formula

$$\deg(U) = \int_B \Phi^* \omega$$

for any volume form ω on U of total mass 1 (one can show that this definition is independent of the choice of ω). This definition is stable under uniform convergence of Φ, and thus can be used to also define the degree for maps Φ that are merely continuous rather than continuously differentiable.

Proof. Suppose for contradiction that $T(0) \neq 0$. We use an averaging argument, combined with *degree theory*. Let $\Phi \colon U \to \mathbf{R}^n$ be the map

$$\Phi(x) := \frac{1}{p} \sum_{n=0}^{p-1} T^n x.$$

Then from construction, Φ is continuous and T-invariant (thus $\Phi(x) = \Phi(Tx)$ for all $x \in U$) and we have $|\Phi(x) - x| < 1$ for all $x \in U$. In particular, Φ is nonzero on the boundary ∂B of B, and $\Phi(\partial B)$ is contractible to ∂B in $\mathbf{R}^n \backslash 0$ (by taking convex combinations of Φ and the identity). As such, the degree of Φ on B near the origin is equal to 1.

A similar argument shows that the homeomorphism T has degree 1 on B near the origin and, in particular, is orientation-preserving rather than orientation reversing.

On the other hand, it can be shown that the degree of Φ must be divisible by p, leading to a contradiction. This is easiest to see in the case when Φ is continuously differentiable, for then by Sard's theorem we may find a regular point x arbitrarily close to 0. On the other hand, since $T(0)$ is a positive distance away from 0, we see that for x sufficiently close to 0, x is not a fixed point of T, and thus (as T has prime order) all elements in the preimage $\Phi^{-1}(\{x\})$ are not fixed points of T either. Thus $\Phi^{-1}(\{x\})$ can be partitioned into a finite number of disjoint orbits of T of cardinality p. As T is orientation preserving and x is a regular point of Φ, each of these orbits contributes $+p$ or $-p$ to the degree, giving the claim.

The case when Φ is not continuously differentiable is trickier, as the degree is not as easily computed in this case. One way to proceed is to perturb T (or more precisely, the graph $\{(x, Tx, \ldots, T^{p-1}x) : x \in U\}$ in U^p) to be piecewise linear near the preimage of 0 (while preserving the periodicity properties of the graph), so that degree can be computed by hand; this is the approach taken in Newman's original paper [**Ne1931**]. Another is to use the machinery of *singular homology*, which is more general and flexible than degree theory; this is the approach taken in [**Sm1941**], [**Dr1969**]. □

Note that the above argument shows not only that T fixes the origin 0, but must also fix an open neighbourhood of the origin, by translating B slightly. One can then extend this open neighbourhood to the entire space by the following variant of Theorem 17.1.1.

Theorem 17.1.2 (Newman's second theorem). *Let X be a connected manifold, and let $T\colon X \to X$ be a homeomorphism of some prime order p that fixes a nonempty open set U. Then T is the identity.*

Proof. We need to show that T fixes all points in X, and not just U. Suppose for the sake of contradiction that T just fixes some of the points in X and not others. By a continuity argument, and applying a homeomorphic change of variables if necessary, we can find a coordinate chart containing the ball B, where T fixes all $x = (x_1, \ldots, x_n) \in B$ with $x_1 \leq 0$, but does not fix all $x \in B$ with $x_1 > 0$. By shrinking B we may assume that the entire orbit $B, TB, \ldots, T^{p-1}B$ stays inside the coordinate chart (and can thus be viewed as a subset of $\mathbf{R}^n$). Then the map Φ can be defined as before. This map is the identity on the left hemisphere $\{x \in B : x_1 \leq 0\}$. On the right hemisphere $\{x \in B : x_1 > 0\}$, one observes for x small enough that $x, Tx, \ldots, T^{p-1}x$ must all stay on the right-hemisphere (as they must lie in B and cannot enter the left-hemisphere, where T is the identity) and so Φ stays on the right. This implies that Φ has degree 1 near 0 on a small ball around the origin, but as before one can argue that the degree must in fact be divisible by p, leading again to a contradiction. □

17.2. Reduction to the p-adic case

We are now ready to prove Conjecture 17.0.2 assuming Conjecture 17.0.3. Let G be a locally compact group acting continuously and faithfully on a connected manifold X; we wish to show that G is Lie.

We first make some basic reductions. We will need the following fact:

Exercise 17.2.1. Show that the extension of a Lie group by another Lie group is again isomorphic to a Lie group. (*Hint:* Use Corollary 5.3.3.) This result was first established in [**Gl1951**]. Note that this result supersedes Theorem 16.0.2, though it uses more of the theory of Hilbert's fifth problem in its proof.

From the Gleason-Yamabe theorem (Theorem 1.1.13), every locally compact group G contains an open subgroup which is an extension of a Lie group by a compact subgroup of G. Since a group with a Lie group as an open subgroup is again Lie (because all outer automorphisms of Lie groups are smooth), it thus suffices by Exercise 17.2.1 to prove the claim when G is compact. In particular, all orbits of G on X are also compact.

Let B be a small ball in chart of X around some origin x_0. By continuity, there is some neighbourhood U of the identity in G such that $gB \ni x_0$ for all $g \in U$. By the Peter-Weyl theorem, there is a compact normal subgroup G' of G in U with G/G' linear (and hence Lie). The set $G'B$ is then a G'-invariant manifold, which is precompact and connected (because all the shifts gB are connected and share a common point). If we let G'' be the subgroup of G' that fixes $G'B$, then G'' is a compact normal subgroup of G', and G'/G'' acts faithfully on $G'B$. By Newman's first theorem (Theorem 17.1.1), we see that if U is small enough, then G'/G'' cannot contain any elements of prime order, and hence cannot contain any nontrivial periodic elements whatsoever; by Conjecture 17.0.3, it also cannot contain a continuously embedded copy of $\mathbf{Z}_p$ for any p. We claim that this forces G'/G'' to be trivial.

As $G'B$ is precompact, the space of $C(\overline{G'B} \to \overline{G'B})$ of continuous maps from $\overline{G'B}$ to itself (with the compact-open topology) is first countable, which makes G'/G'' first-countable as well (since G'/G'' is homeomorphic to a subspace of $C(\overline{G'B} \to \overline{G'B})$). The claim now follows from

Lemma 17.2.1. *Let G be a compact first-countable group which does not contain any nontrivial periodic elements or a continuously embedded copy of $\mathbf{Z}_p$ for any p. Then G is trivial.*

Proof. Every element g of G is contained in a compact *abelian* subgroup of G, namely the closed group $\overline{\langle g \rangle}$ generated by g. Thus we may assume without loss of generality that G is abelian.

As G is both compact and first countable, it can be written (using Exercise 4.2.9) as the inverse limit $\lim_{\leftarrow} G_n$ of a countable sequence of compact abelian Lie groups G_n, with surjective continuous projection homomorphisms $\pi_{n+1 \to n} \colon G_{n+1} \to G_n$ between these Lie groups. (Note that without first countability, one might only be an inverse limit of a *net* of Lie groups rather than a sequence, as one can see, for instance, with the example $(\mathbf{R}/\mathbf{Z})^{\mathbf{R}}$.)

Suppose for contradiction that at least one of the G_n, say G_1, is nontrivial. It is a standard fact that every compact abelian Lie group is isomorphic to the direct product of a torus and a finite group. (Indeed, in the connected case one can inspect the kernel of the exponential map, and then one can extend to the general case by viewing a compact Lie group as an extension of a finite group by a connected compact Lie group); in particular, the periodic points (i.e., points of finite order) are dense, and so there exists an element g_1 of G_1 of finite nontrivial order. By raising g_1 to a suitable power, we may assume that g_1 has some prime order p.

We now claim inductively that for each $n = 1, 2, \ldots$, g_1 can be lifted to an element $g_n \in G_n$ of some order p^{k_n}, where $1 = k_1 \leq k_2 \leq \ldots$ are a nondecreasing set of integers. Indeed, suppose inductively that we have already lifted g_1 up to $g_n \in G_n$ with order p^{k_n}. The preimage of g_n in G_{n+1} is then a dense subset of the preimage of $\langle g_n \rangle$, which is a compact abelian Lie group, and thus contains an element g'_{n+1} of some finite order. As g_n has order p^{k_n}, g'_{n+1} must have an order divisible by p^{k_n}, and thus of the form $p^{k_{n+1}} q$ for some q coprime to p and some $k_{n+1} \geq k_n$. By raising g'_{n+1} to a multiple of q that equals 1 mod p^{k_n}, we may eliminate q and obtain a preimage g_{n+1} of order $p^{k_{n+1}}$, and the claim follows.

There are now two cases, depending on whether k_n goes to infinity or not. If the k_n stay bounded, then they converge to a limit k, and the inverse limit of the g_n is then an element g of G of finite order p^k, a contradiction. But if the k_n are unbounded, then G contains a continuously embedded copy of the inverse limit of the $\mathbf{Z}/p^{k_n}\mathbf{Z}$, which is $\mathbf{Z}_p$, and again we have a contradiction. $\qquad\square$

Since G'/G'' and G/G' are both Lie groups, G/G'' is Lie too. Thus it suffices to show that G'' is Lie. Without loss of generality, we may therefore replace G by G'' and assume that B is fixed by G.

Now let Σ be the closure of the interior of the set of fixed points of G. Then Σ is nonempty, and is also clearly closed. We claim that Σ is open; by connectedness of X, this implies that $\Sigma = X$, which by faithfulness of G implies that G is trivial, giving the claim. Indeed, let $x \in \Sigma$, and let B be a small ball around x. As before, GB is then a connected G-invariant

σ-compact manifold, and so if G' is the subgroup of G that fixes GB, then G/G' is a compact Lie group that acts faithfully on GB fixing a nontrivial open subset of GB. By Newman's theorem (Theorem 17.1.2), G/G' cannot contain any periodic elements; by Conjecture 17.0.3, it also cannot contain any copy of $\mathbf{Z}_p$. By Lemma 17.2.1, G/G' is trivial, and so G fixes all of B. Thus gives the desired openness of Σ as required.

Remark 17.2.2. An alternate derivation of Conjecture 17.0.2 from Conjecture 17.0.3, suggested to the author by John Pardon, goes as follows. Again we may assume G to be compact. If G is NSS, then it is Lie (e.g., by Peter-Weyl) and we are done, so suppose G is not NSS. By Newman's theorem, we can find a compact neighbourhood of the identity in G whose action on X contains no periodic transformations other than the identity. As G is NSS, we may thus find a nontrivial compact subgroup G' inside this neighbourhood, which has no nontrivial periodic elements. Also, as the space of homeomorphisms on M is metrisable, it is first countable; as the action maps the compact Hausdorff space G' continuously and injectively into the Hausdorff space of homeomorphisms on M, G' is homeomorphic to its image and is thus also first countable. By Lemma 17.2.1, G' then contains a copy of $\mathbf{Z}_p$, and the claim follows.

Remark 17.2.3. Whereas actions of the finite group $\mathbf{Z}/p\mathbf{Z}$ on manifolds can be analysed by degree theory, it appears that actions of p-adic groups $\mathbf{Z}_p$ require more sophisticated homological tools; the p-adic analogues of the averaged map Φ now typically have infinite preimages and it is no longer obvious how to compute the degree of such maps. Nevertheless, some progress has been made along these lines under some additional regularity hypotheses on the action, such as Lipschitz continuity; see, for instance, [**ReSc97**]. Note that if $T\colon X \to X$ is the action of the generator of $\mathbf{Z}_p$, then the powers $T^{p^n}\colon X \to X$ will converge to the identity locally uniformly as $n \to \infty$. This is already enough to rule out generating maps T that are smooth but nontrivial by use of Taylor expansion; see [**MoZi1974**]. (In fact, this type of argument even works for C^1 actions.)

Chapter 18

The Peter-Weyl theorem and nonabelian Fourier analysis

Let G be a compact group. (Throughout this chapter, all topological groups are assumed to be Hausdorff.) Then G has a number of *unitary representations*, i.e., continuous homomorphisms $\rho\colon G \to U(H)$ to the group $U(H)$ of unitary operators on a Hilbert space H, equipped with the strong operator topology. In particular, one has the *left-regular representation* $\tau\colon G \to U(L^2(G))$, where we equip G with its normalised Haar measure μ (and the Borel σ-algebra) to form the Hilbert space $L^2(G)$, and τ is the translation operation

$$\tau(g)f(x) := f(g^{-1}x).$$

We call two unitary representations $\rho\colon G \to U(H)$ and $\rho'\colon G \to U(H')$ *isomorphic* if one has $\rho'(g) = U\rho(g)U^{-1}$ for some unitary transformation $U\colon H \to H'$, in which case we write $\rho \equiv \rho'$.

Given two unitary representations $\rho\colon G \to U(H)$ and $\rho'\colon G \to U(H')$, one can form their direct sum $\rho \oplus \rho'\colon G \to U(H \oplus H')$ in the obvious manner: $\rho \oplus \rho'(g)(v) := (\rho(g)v, \rho'(g)v)$. Conversely, if a unitary representation $\rho\colon G \to U(H)$ has a closed invariant subspace $V \subset H$ of H (thus $\rho(g)V \subset V$ for all $g \in G$), then the orthogonal complement $V^\perp$ is also invariant, leading to a decomposition $\rho \equiv \rho \restriction_V \oplus \rho \restriction_{V^\perp}$ of ρ into the *subrepresentations* $\rho \restriction_V\colon G \to U(V)$, $\rho \restriction_{V^\perp}\colon G \to U(V^\perp)$. Accordingly, we will call a unitary representation $\rho: G \to U(H)$ *irreducible* if H is nontrivial (i.e., $H \neq \{0\}$) and there are no nontrivial invariant subspaces (i.e., no invariant subspaces other than $\{0\}$ and H); the irreducible representations play a role in the subject analogous to those of prime numbers in multiplicative number theory. By the principle of infinite descent, every finite-dimensional unitary representation is then expressible (perhaps nonuniquely) as the direct sum of irreducible representations.

The *Peter-Weyl theorem* asserts, among other things, that the same claim is true for the regular representation:

Theorem 18.0.4 (Peter-Weyl theorem). *Let G be a compact group. Then the regular representation $\tau: G \to U(L^2(G))$ is isomorphic to the direct sum of irreducible representations. In fact, one has $\tau \equiv \bigoplus_{\xi \in \hat{G}} \rho_\xi^{\oplus \dim(V_\xi)}$, where $(\rho_\xi)_{\xi \in \hat{G}}$ is an enumeration of the irreducible finite-dimensional unitary representations $\rho_\xi: G \to U(V_\xi)$ of G (up to isomorphism). (It is not difficult to see that such an enumeration exists.)*

In the case when G is abelian, the Peter-Weyl theorem is a consequence of the *Plancherel theorem*; in that case, the irreducible representations are all one-dimensional, and are thus indexed by the space $\hat{G}$ of *characters* $\xi: G \to \mathbf{R}/\mathbf{Z}$ (i.e., continuous homomorphisms into the unit circle $\mathbf{R}/\mathbf{Z}$), known as the *Pontryagin dual* of G; see, e.g., [**Ta2010**, §1.12]. Conversely, the Peter-Weyl theorem can be used to deduce the Plancherel theorem for compact groups, as well as other basic results in Fourier analysis on these groups, such as the Fourier inversion formula. A baby version of this theorem (Theorem 4.2.2) is also an essential component in the solution to Hilbert's fifth problem, as discussed in Section 4. In this section, we will upgrade Theorem 4.2.2 to the full Peter-Weyl theorem.

18.1. Proof of the Peter-Weyl theorem

Henceforth, in this section, G is a fixed compact group.

Let $\rho: G \to U(H)$ and $\rho': G \to U(H')$ be unitary representations. An *(linear) equivariant map* $T: H \to H'$ is defined to be a continuous linear transformation such that $T\rho(g) = \rho'(g)T$ for all $g \in G$.

A fundamental fact in representation theory, known as *Schur's lemma*, asserts (roughly speaking) that equivariant maps cannot mix irreducible representations together unless they are isomorphic. More precisely:

Lemma 18.1.1 (Schur's lemma for unitary representations). *Suppose that $\rho: G \to U(H)$ and $\rho': G \to U(H')$ are irreducible unitary representations, and let $T: H \to H'$ be an equivariant map. Then T is either the zero* *transformation, or a constant multiple of an isomorphism. In particular, if $\rho \not\equiv \rho'$, then there are no nontrivial equivariant maps between H and H'.*

Proof. The adjoint map $T^*: H' \to H$ of the equivariant map T is also equivariant, and thus so is $T^*T: H \to H$. As T^*T is also a bounded self-adjoint operator, we can apply the *spectral theorem* to it. Observe that any closed invariant subspace of T^*T is G-invariant, and is thus either $\{0\}$ or H. By the spectral theorem, this forces T^*T to be a constant multiple of the identity. Similarly for TT^*. This forces T to either be zero or a constant multiple of a unitary map, and the claim follows. (Thanks to Frederick Goodman for this proof.) $\square$

Schur's lemma has many foundational applications in the subject. For instance, we have the following generalisation of the well-known fact that eigenvectors of a unitary operator with distinct eigenvalues are necessarily orthogonal:

Corollary 18.1.2. *Let $\rho \downharpoonright_V: G \to U(V)$ and $\rho \downharpoonright_W: G \to U(W)$ be two irreducible subrepresentations of a unitary representation $\rho: G \to U(H)$. Then one either has $\rho \downharpoonright_V \equiv \rho \downharpoonright_W$ or $V \perp W$.*

Proof. Apply Schur's lemma to the orthogonal projection from W to V. $\square$

Another application shows that finite-dimensional *linear* representations can be canonically identified (up to constants) with finite-dimensional unitary representations:

Corollary 18.1.3. *Let $\rho: G \to GL(V)$ be a linear representation on a finite-dimensional space V. Then there exists a Hermitian inner product $\langle , \rangle$ on V that makes this representation unitary. Furthermore, if V is irreducible, then this inner product is unique up to constants.*

Proof. To show existence of the Hermitian inner product that unitarises ρ, take an arbitrary Hermitian inner product $\langle , \rangle_0$ and then form the average

$$\langle v, w \rangle := \int_G \langle \rho(g)v, \rho(g)w \rangle_0 \, d\mu(g)$$

(this is the "Weyl averaging trick", which crucially exploits compactness of G). Then one easily checks (using the fact that V is finite-dimensional and thus locally compact) that $\langle , \rangle$ is also Hermitian, and that ρ is unitary with respect to this inner product, as desired. (This part of the argument does not use finite dimensionality.)

To show uniqueness up to constants, assume that one has two such inner products $\langle , \rangle$, $\langle , \rangle'$ on V, and apply Schur's lemma to the identity map between the two Hilbert spaces $(V, \langle, \rangle)$ and $(V, \langle, \rangle')$. (Here, finite dimensionality is used to establish   □

A third application of Schur's lemma allows us to express the trace of a linear operator as an average:

Corollary 18.1.4. *Let* $\rho \colon G \to GL(H)$ *be an irreducible unitary representation on a nontrivial finite-dimensional space* H, *and let* $T \colon H \to H$ *be a linear transformation. Then*

$$\frac{1}{\dim(H)} \operatorname{tr}_H(T) I_H = \int_G \rho(g) T \rho(g)^* \, d\mu(g),$$

where $I_H \colon H \to H$ *is the identity operator.*

Proof. The right-hand side is equivariant, and hence by Schur's lemma is a multiple of the identity. Taking traces, we see that the right-hand side also has the same trace as T. The claim follows.   □

Let us now consider the irreducible subrepresentations $\rho \mid_V \colon G \to U(V)$ of the left-regular representation $\rho \colon G \to U(L^2(G))$. From Corollary 18.1.2, we know that those subrepresentations coming from different isomorphism classes in $\hat{G}$ are orthogonal, so we now focus attention on those subrepresentations coming from a single class $\xi \in \hat{G}$. Define the ξ-*isotypic component* $L^2(G)_\xi$ of the regular representation to be the finite-dimensional subspace of $L^2(G)$ spanned by the functions of the form

$$f_{\xi,v,w} \colon g \mapsto \langle v, \rho_\xi(g)w \rangle_{V_\xi}$$

where v, w are arbitrary vectors in V_ξ. This is clearly a left-invariant subspace of $L^2(G)$ (in fact, it is bi-invariant, a point which we will return to later), and thus induces a subrepresentation of the left-regular representation. In fact, it captures precisely all the subrepresentations of the left-regular representation that are isomorphic to ρ_ξ:

Proposition 18.1.5. *Let* $\xi \in \hat{G}$. *Then every irreducible subrepresentation* $\tau \mid_V \colon G \to U(V)$ *of the left-regular representation* $\tau \colon G \to U(L^2(G))$ *that is isomorphic to* ρ_ξ *is a subrepresentation of* $L^2(G)_\xi$. *Conversely,* $L^2(G)_\xi$ *is isomorphic to the direct sum* $\rho_\xi^{\dim(V_\xi)}$ *of* $\dim(V_\xi)$ *copies of* $\rho_\xi \colon G \to U(V_\xi)$. *(In particular,* $L^2(G)_\xi$ *has dimension* $\dim(V_\xi)^2$.)

Proof. Let $\tau \mid_V \colon G \to U(V)$ be a subrepresentation of the left-regular representation that is isomorphic to ρ_ξ. Thus, we have an equivariant isometry $\iota \colon V_\xi \to L^2(G)$ whose image is V; it has an adjoint $\iota^* \colon L^2(G) \to V_\xi$.

Let $v \in V_\xi$ and $K \in L^2(G)$. The convolution

$$\iota(v) * K(g) := \int_G \iota(v)(gh) K(h^{-1}) \, d\mu(h)$$

can be re-arranged as

$$\int_G \tau(g^{-1})(\iota(v))(h)\overline{\tilde{K}(h)} \, d\mu(h)$$
$$= \langle \tau(g^{-1})(\iota(v)), \tilde{K} \rangle_{L^2(G)}$$
$$= \langle \iota(\rho_\xi(g^{-1})v), \tilde{K} \rangle_{L^2(G)}$$
$$= \langle \rho_\xi(g^{-1})v, \iota^*\tilde{K} \rangle_{V_\xi}$$
$$= \langle v, \rho_\xi(g)\iota^*\tilde{K} \rangle_{V_\xi}$$

where

$$\tilde{K}(g) := \overline{K(g^{-1})}.$$

In particular, we see that $\iota(v) * K \in L^2(G)_\xi$ for every K. Letting K be a sequence (or net) of approximations to the identity, we conclude that $\iota(v) \in L^2(G)_\xi$ as well, and so $V \subset L^2(G)_\xi$, which is the first claim.

To prove the converse claim, write $n := \dim(V_\xi)$, and let $e_1, \ldots, e_n$ be an orthonormal basis for V_ξ. Observe that we may then decompose $L^2(G)_\xi$ as the direct sum of the spaces

$$L^2(G)_{\xi,e_i} := \{f_{\xi,v,e_i} : v \in V_\xi\}$$

for $i = 1, \ldots, n$. The claim follows. $\square$

From Corollary 18.1.2, the ξ-isotypic components $L^2(G)_\xi$ for $\xi \in \hat{G}$ are pairwise orthogonal, and so we can form the direct sum $\bigoplus_{\xi \in \hat{G}} L^2(G)_\xi \equiv \bigoplus_{\xi \in \hat{G}} \rho_\xi^{\oplus \dim(G)}$, which is an invariant subspace of $L^2(G)$ that contains all the finite-dimensional irreducible subrepresentations (and hence also all the finite-dimensional representations, period). The essence of the Peter-Weyl theorem is then the assertion that this direct sum in fact occupies all of $L^2(G)$:

Proposition 18.1.6. *We have $L^2(G) = \bigoplus_{\xi \in \hat{G}} L^2(G)_\xi$.*

Proof. Suppose this is not the case. Taking orthogonal complements, we conclude that there exists a nontrivial $f \in L^2(G)$ which is orthogonal to all $L^2(G)_\xi$ and is, in particular, orthogonal to all finite-dimensional subrepresentations of $L^2(G)$.

Now let $K \in L^2(G)$ be an arbitrary self-adjoint kernel, thus $\overline{K(g^{-1})} = K(g)$ for all $g \in G$. The convolution operator $T: f \mapsto f * K$ is then a self-adjoint Hilbert-Schmidt operator and is thus compact. (Here, we have crucially used the compactness of G.) By the spectral theorem, the cokernel $\ker(T)^\perp$ of this operator then splits as the direct sum of finite-dimensional eigenspaces. As T is equivariant, all these eigenspaces are invariant, and thus orthogonal to f; thus f must lie in the kernel of T, and thus $f * K$ vanishes for all self-adjoint $K \in L^2(G)$. Using a sequence (or net) of approximations to the identity, we conclude that f vanishes also, a contradiction. $\square$

Theorem 18.0.4 follows by combining this proposition with 18.1.5.

18.2. Nonabelian Fourier analysis

Given $\xi \in \hat{G}$, the space $HS(V_\xi)$ of linear transformations from V_ξ to V_ξ is a finite-dimensional Hilbert space, with the Hilbert-Schmidt inner product $\langle S, T \rangle_{HS(V_\xi)} := \mathrm{tr}_{V_\xi} ST^*$; it has a unitary action of G as defined by $\rho_{HS(V_\xi)}(g): T \mapsto \rho_\xi(g)T$. For any $T \in HS(V_\xi)$, the function $g \mapsto \langle T, \rho(g) \rangle_{HS(V_\xi)}$ can be easily seen to lie in $L^2(G)_\xi$, giving rise to a map $\iota_\xi: HS(V_\xi) \to L^2(G)_\xi$. It is easy to see that this map is equivariant.

Proposition 18.2.1. *For each $\xi \in \hat{G}$, the map $\dim(V_\xi)^{1/2} \iota_\xi$ is unitary.*

Proof. As $HS(V_\xi)$ and $L^2(G)_\xi$ are finite-dimensional spaces with the same dimension $\dim(V_\xi)^2$, it suffices to show that this map is an isometry, thus we need to show that

$$\langle \mathcal{F}_\xi^*(S), \mathcal{F}_\xi^*(T) \rangle_{L^2(G)} = \frac{1}{\dim(V_\xi)} \langle S, T \rangle_{HS(V_\xi)}$$

for all $S, T \in HS(V_\xi)$. By bilinearity, we may reduce to the case when S, T are rank one operators

$$S := ab^*; \quad T := cd^*$$

for some $a, b, c, d \in V_\xi$, where $b^*: V_\xi \to \mathbf{C}$ is the dual vector $b^*: v \mapsto \langle v, b \rangle$ to b, and similarly for d. Then we have

$$\langle S, T \rangle_{HS(V_\xi)} = \mathrm{tr}_{V_\xi} ab^* dc^* = (c^* a)(b^* d)$$

and

$$\langle \mathcal{F}_\xi^*(S), \mathcal{F}_\xi^*(T) \rangle_{L^2(G)} = \int_G (b^* \rho_\xi(g) a)\overline{(d^* \rho_\xi(g) c)} \, d\mu(G).$$

The latter expression can be rewritten as

$$\int_G \langle \rho_\xi(g)^* db^* \rho_\xi(g), ca^* \rangle_{HS(V_\xi)} \, d\mu(g).$$

Applying Fubini's theorem, followed by Corollary 18.1.4, this simplifies to

$$\left\langle \frac{1}{\dim(V_\xi)} \mathrm{tr}(db^*) I_{V_\xi}, ca^* \right\rangle_{HS(V_\xi)},$$

which simplifies to $\frac{1}{\dim(V_\xi)}(c^* a)(b^* d)$, and the claim follows. $\square$

As a corollary of the above proposition, the orthogonal projection of a function $f \in L^2(G)$ to $L^2(G)_\xi$ can be expressed as

$$\dim(V_\xi)\iota_\xi\iota_\xi^* f.$$

We call

$$\hat{f}(\xi) := \iota_\xi^* f = \int_G f(g)\rho(g) \, d\mu(g) \in HS(V_\xi)$$

the *Fourier coefficient* of f at ξ, thus the projection of f to $L^2(G)_\xi$ is the function

$$g \mapsto \dim(V_\xi)\langle \hat{f}(\xi), \rho(g)\rangle$$

which has an $L^2(G)$ norm of $\dim(V_\xi)^{1/2}\|\hat{f}(\xi)\|_{HS(V_\xi)}$. From the Peter-Weyl theorem we thus obtain the *Fourier inversion formula*

$$f(g) = \sum_{\xi \in \hat{G}} \dim(V_\xi)\langle \hat{f}(\xi), \rho(g)\rangle$$

and the *Plancherel identity*

$$\|f\|_{L^2(G)}^2 = \sum_{\xi \in \hat{G}} \dim(V_\xi)\|\hat{f}(\xi)\|_{HS(V_\xi)}^2.$$

We can write these identities more compactly as an isomorphism

(18.1) $$L^2(G) \equiv \bigoplus_{\xi \in \hat{G}} \dim(V_\xi) \cdot HS(V_\xi)$$

where the dilation $c \cdot H$ of a Hilbert space H is formed by using the inner product $\langle v, w\rangle_{c \cdot H} := c\langle v, w\rangle_H$. This is an isomorphism not only of Hilbert spaces, but of the left-action of G. Indeed, it is an isomorphism of the bi-action of $G \times G$ on both the left and right of both $L^2(G)$ and $HS(V_\xi)$, defined by

$$\rho_{L^2(G), G \times G}(g, h)(f)(x) := f(g^{-1}xh)$$

and

$$\rho_{\xi, G \times G}(g, h)(T) := \rho(g)T\rho(h)^*.$$

It is easy to see that each of the $HS(V_\xi)$ are irreducible with respect to the $G \times G$ action. Indeed, first observe from Proposition 18.2.1 that ι_ξ^* is surjective, and thus $\rho_\xi(g) \in HS(V_\xi)$ must span all of $HS(V_\xi)$. Thus, any bi-invariant subspace of $HS(V_\xi)$ must also be invariant with respect to left and right multiplication by arbitrary elements of $HS(V_\xi)$ and, in particular, by rank one operators; from this one easily sees that there are no nontrivial bi-invariant subspaces. Thus we can view the Peter-Weyl theorem as also describing the irreducible decomposition of $L^2(G)$ into $G \times G$-irreducible components.

Remark 18.2.2. In view of (18.1), it is natural to view $\hat{G}$ as being the "spectrum" of G, with each "frequency" $\xi \in \hat{G}$ occurring with "multiplicity" $\dim(V_\xi)$.

In the abelian case, any eigenspace of one unitary operator $\rho(g)$ is automatically an invariant subspace of all other $\rho(h)$, which quickly implies (from the spectral theorem) that all irreducible finite-dimensional unitary representations must be one-dimensional, in which case we see that the above formulae collapse to the usual Fourier inversion and Plancherel theorems for compact abelian groups.

In the case of a finite group G, we can take dimensions in (18.1) to obtain the identity

$$|G| = \sum_{\xi \in \hat{G}} \dim(V_\xi)^2.$$

In the finite abelian case we see, in particular, that G and $\hat{G}$ have the same cardinality.

Direct computation also shows other basic Fourier identities, such as the convolution identity

$$\widehat{f_1 * f_2}(\xi) = \hat{f}_1(\xi)\hat{f}_2(\xi)$$

for $f_1, f_2 \in L^2(G)$, thus partially diagonalising convolution into multiplication of linear operators on finite-dimensional vector spaces V_ξ. (Of course, one cannot expect complete diagonalisation in the nonabelian case, since convolution would then also be nonabelian, whereas diagonalised operators must always commute with each other.)

Call a function $f \in L^2(G)$ a *class function* if it is conjugation-invariant, thus $f(gxg^{-1}) = f(x)$ for all $x, g \in G$. It is easy to see that this is equivalent to each of the Fourier coefficients $\hat{f}(\xi)$ also being conjugation-invariant: $\rho_\xi(g)\hat{f}(\xi)\rho_\xi(g)^* = \hat{f}(\xi)$. By Lemma 18.1.4, this is in turn equivalent to $\hat{f}(\xi)$ being equal to a multiple of the identity:

$$\hat{f}(\xi) = \frac{1}{\dim(V_\xi)} \operatorname{tr}(\hat{f}(\xi))I_{V_\xi} = \frac{1}{\dim(V_\xi)} \langle f, \chi_\xi \rangle_{L^2(G)} I_{V_\xi}$$

where the *character* $\chi_\xi \in L^2(G)$ of the representation ρ_ξ is given by the formula

$$\chi_\xi(g) := \operatorname{tr}_{V_\xi} \rho_\xi(g).$$

The Plancherel identity then simplifies to

$$f = \sum_{\xi \in \hat{G}} \langle f, \chi_\xi \rangle_{L^2(G)} \chi_\xi,$$

thus the χ_ξ form an orthonormal basis for the space $L^2(G)^G$ of class functions. Analogously to (18.1), we have

$$L^2(G)^G \equiv \bigoplus_{\xi \in \hat{G}} \mathbf{C}.$$

(In particular, in the case of finite groups G, $\hat{G}$ has the same cardinality as the space of conjugacy classes of G.)

Characters are a fundamentally important tool in analysing finite-dimensional representations V of G that are not necessarily irreducible; indeed, if V decomposes into irreducibles as $\bigoplus_{\xi \in \hat{G}} V_\xi^{\oplus m_\xi}$, then the character $\chi_V(g) := \mathrm{tr}_V(\rho_g)$ then similarly splits as

$$\chi_V = \sum_{\xi \in \hat{G}} m_\xi \chi_\xi$$

and so the multiplicities m_ξ of each component V_ξ in V can be given by the formula

$$m_\xi = \langle \chi_V, \chi_\xi \rangle_{L^2(G)}.$$

In particular, these multiplicities are unique: all decompositions of V into irreducibles have the same multiplicities.

Remark 18.2.3. Representation theory becomes much more complicated once one leaves the compact case; convolution operators $f \mapsto f * K$ are no longer compact, and can now admit continuous spectrum in addition to pure point spectrum. Furthermore, even when one has pure point spectrum, the eigenspaces can now be infinite-dimensional. Thus, one must now grapple with infinite-dimensional irreducible representations, as well as continuous combinations of representations that cannot be readily resolved into irreducible components. Nevertheless, in the important case of *locally compact groups*, it is still the case that there are "enough" irreducible unitary representations to recover a significant portion of the above theory. The fundamental theorem here is the *Gelfand-Raikov theorem*, which asserts that given any nontrivial group element g in a locally compact group, there exists an irreducible unitary representation (possibly infinite-dimensional) on which g acts nontrivially. Very roughly speaking, this theorem is first proven by observing that g acts nontrivially on the regular representation, which (by the *Gelfand-Naimark-Segal (GNS) construction*) gives a state on the *-algebra of measures on G that distinguishes the Dirac mass δ_g at g from the Dirac mass δ_0 from the origin. Applying the *Krein-Milman theorem*, one then finds an *extreme* state with this property; applying the GNS construction, one then obtains the desired irreducible representation.

Chapter 19

Polynomial bounds via nonstandard analysis

As discussed in Chapter 7, nonstandard analysis is useful in allowing one to import tools from infinitary (or qualitative) mathematics in order to establish results in finitary (or quantitative) mathematics. One drawback, though, to using nonstandard analysis methods is that the bounds one obtains by such methods are usually *ineffective*: in particular, the conclusions of a nonstandard analysis argument may involve an unspecified constant C that is known to be finite but for which no explicit bound is obviously available[1].

Because of this fact, it would seem that quantitative bounds, such as polynomial type bounds $X \leq CY^C$ that show that one quantity X is controlled in a polynomial fashion by another quantity Y, are not easily obtainable through the ineffective methods of nonstandard analysis. Actually, this is not the case; as I will demonstrate by an example below, nonstandard analysis can certainly yield polynomial type bounds. The catch is that the exponent C in such bounds will be ineffective; but nevertheless such bounds are still good enough for many applications.

Let us now illustrate this by reproving a lemma from a paper of Chang [**Ch2003**, Lemma 2.14], which was recently pointed out to me by Van Vu. Chang's paper is focused primarily on the sum-product problem, but she

---

[1]In many cases, a bound can eventually be worked out by performing *proof mining* on the argument and, in particular, by carefully unpacking the *proofs* of all the various results from infinitary mathematics that were used in the argument, as opposed to simply using them as "black boxes", but this is a time-consuming task and the bounds that one eventually obtains tend to be quite poor (e.g., tower exponential or Ackermann type bounds are not uncommon).

uses a quantitative lemma from algebraic geometry which is of independent interest. To motivate the lemma, let us first establish a qualitative version:

Lemma 19.0.4 (Qualitative solvability)**.** *Let* $P_1, \dots, P_r \colon \mathbf{C}^d \to \mathbf{C}$ *be a finite number of polynomials in several variables with rational coefficients. If there is a complex solution* $z = (z_1, \dots, z_d) \in \mathbf{C}^d$ *to the simultaneous system of equations*

$$P_1(z) = \cdots = P_r(z) = 0,$$

then there also exists a solution $z \in \overline{\mathbf{Q}}^d$ *whose coefficients are algebraic numbers (i.e., they lie in the algebraic closure* $\mathbf{Q}$ *of the rationals).*

Proof. Suppose there was no solution to $P_1(z) = \cdots = P_r(z) = 0$ over $\overline{\mathbf{Q}}$. Applying *Hilbert's nullstellensatz* (which is available as $\overline{\mathbf{Q}}$ is algebraically closed), we conclude the existence of some polynomials $Q_1, \dots, Q_r$ (with coefficients in $\overline{\mathbf{Q}}$) such that

$$P_1 Q_1 + \cdots + P_r Q_r = 1$$

as polynomials. In particular, we have

$$P_1(z)Q_1(z) + \cdots + P_r(z)Q_r(z) = 1$$

for all $z \in \mathbf{C}^d$. This shows that there is no solution to $P_1(z) = \cdots = P_r(z) = 0$ over $\mathbf{C}$, as required. $\square$

Remark 19.0.5. Observe that in the above argument, one could replace $\mathbf{Q}$ and $\mathbf{C}$ by any other pair of fields, with the latter containing the algebraic closure of the former, and still obtain the same result.

The above lemma asserts that if a system of rational equations is solvable at all, then it is solvable with some algebraic solution. But it gives no bound on the complexity of that solution in terms of the complexity of the original equation. Chang's lemma provides such a bound. If $H \geq 1$ is an integer, let us say that an algebraic number has *height* at most H if its minimal polynomial (after clearing denominators) consists of integers of magnitude at most H.

Lemma 19.0.6 (Quantitative solvability)**.** *Let* $P_1, \dots, P_r \colon \mathbf{C}^d \to \mathbf{C}$ *be a finite number of polynomials of degree at most* D *with rational coefficients, each of height at most* H*. If there is a complex solution* $z = (z_1, \dots, z_d) \in \mathbf{C}^d$ *to the simultaneous system of equations*

$$P_1(z) = \cdots = P_r(z) = 0,$$

then there also exists a solution $z \in \overline{\mathbf{Q}}^d$ *whose coefficients are algebraic numbers of degree at most* C *and height at most* CH^C*, where* $C = C_{D,d,r}$ *depends only on* D*,* d *and* r*.*

Chang proves this lemma by essentially establishing a quantitative version of the nullstellensatz, via elementary elimination theory (somewhat similar, actually, to the approach I took to the nullstellensatz in [**Ta2008**, §1.15]). She also notes that one could also establish the result through the machinery of *Gröbner bases*. In each of these arguments, it was not possible to use Lemma 19.0.4 (or the closely related nullstellensatz) as a black box; one actually had to unpack one of the proofs of that lemma or nullstellensatz to get the polynomial bound. However, using nonstandard analysis, it is possible to get such polynomial bounds (albeit with an ineffective value of the constant C) directly from Lemma 19.0.4 (or more precisely, the generalisation in Remark 19.0.5) *without* having to inspect the proof, and instead simply using it as a black box, thus providing a "soft" proof of Lemma 19.0.6 that is an alternative to the "hard" proofs mentioned above.

Here's how the proof works. Informally, the idea is that Lemma 19.0.6 should follow from Lemma 19.0.4 after replacing the field of rationals $\mathbf{Q}$ with "the field of rationals of polynomially bounded height". Unfortunately, the latter object does not really make sense as a field in standard analysis; nevertheless, it is a perfectly sensible object in nonstandard analysis, and this allows the above informal argument to be made rigorous.

We turn to the details. As is common whenever one uses nonstandard analysis to prove finitary results, we use a "compactness and contradiction" argument (or more precisely, an "ultralimit and contradiction" argument). Suppose for contradiction that Lemma 19.0.6 failed. Carefully negating the quantifiers (and using the axiom of choice), we conclude that there exists D, d, r such that for each natural number n, there is a positive integer $H^{(n)}$ and a family $P_1^{(n)}, \ldots, P_r^{(n)} \colon \mathbf{C}^d \to \mathbf{C}$ of polynomials of degree at most D and rational coefficients of height at most $H^{(n)}$, such that there exist at least one complex solution $z^{(n)} \in \mathbf{C}^d$ to

$$(19.1) \qquad P_1^{(n)}(z^{(n)}) = \cdots = P_r(z^{(n)}) = 0,$$

but such that there does not exist any such solution whose coefficients are algebraic numbers of degree at most n and height at most $n(H^{(n)})^n$.

Now we take ultralimits, as in Chapter 7. Let $\alpha \in \beta\mathbf{N}\backslash\mathbf{N}$ be a nonprincipal ultrafilter. For each $i = 1, \ldots, r$, the ultralimit

$$P_i := \lim_{n \to \alpha} P_i^{(n)}$$

of the (standard) polynomials $P_i^{(n)}$ is a nonstandard polynomial $P_i \colon {}^*\mathbf{C}^d \to {}^*\mathbf{C}$ of degree at most D, whose coefficients now lie in the nonstandard rationals ${}^*\mathbf{Q}$. Actually, due to the height restriction, we can say more. Let $H := \lim_{n \to \alpha} H^{(n)} \in {}^*\mathbf{N}$ be the ultralimit of the $H^{(n)}$, this is a nonstandard natural number (which will almost certainly be unbounded, but we will not need to use this). Let us say that a nonstandard integer a is *of polynomial size* if we have $|a| \leq CH^C$ for some standard natural number C, and say that a nonstandard rational number a/b is *of polynomial height* if a, b are of polynomial size. Let $\mathbf{Q}_{\mathrm{poly}(H)}$ be the collection of all nonstandard rationals of polynomial height. (In the language of nonstandard analysis, $\mathbf{Q}_{\mathrm{poly}(H)}$ is an *external* set rather than an internal one, because it is not itself an ultra-product of standard sets; but this will not be relevant for the argument that follows.) It is easy to see that $\mathbf{Q}_{\mathrm{poly}(H)}$ is a field, basically because the sum or product of two integers of polynomial size, remains of polynomial size. By construction, it is clear that the coefficients of P_i are nonstandard rationals of polynomial height, and thus $P_1, \ldots, P_r$ are defined over $\mathbf{Q}_{\mathrm{poly}(H)}$.

Meanwhile, if we let $z := \lim_{n \to \alpha} z^{(n)} \in {}^*\mathbf{C}^d$ be the ultralimit of the solutions $z^{(n)}$ in (19.1), we have

$$P_1(z) = \cdots = P_r(z) = 0,$$

thus $P_1, \ldots, P_r$ are solvable in ${}^*\mathbf{C}$. Applying Lemma 19.0.4 (or more precisely, the generalisation in Remark 19.0.5), we see that $P_1, \ldots, P_r$ are also solvable in $\overline{\mathbf{Q}_{\mathrm{poly}(H)}}$. (Note that as $\mathbf{C}$ is algebraically closed, ${}^*\mathbf{C}$ is also (by Łoś's theorem), and so ${}^*\mathbf{C}$ contains $\overline{\mathbf{Q}_{\mathrm{poly}(H)}}$.) Thus, there exists $w \in \overline{\mathbf{Q}_{\mathrm{poly}(H)}}^d$ with

$$P_1(w) = \cdots = P_r(w) = 0.$$

As $\overline{\mathbf{Q}_{\mathrm{poly}(H)}}^d$ lies in ${}^*\mathbf{C}^d$, we can write w as an ultralimit $w = \lim_{n \to \alpha} w^{(n)}$ of standard complex vectors $w^{(n)} \in \mathbf{C}^d$. By construction, the coefficients of w each obey a nontrivial polynomial equation of degree at most C and whose coefficients are nonstandard integers of magnitude at most CH^C, for some standard natural number C. Undoing the ultralimit, we conclude that for n sufficiently close to p, the coefficients of $w^{(n)}$ obey a nontrivial polynomial equation of degree at most C whose coefficients are *standard* integers of magnitude at most $C(H^{(n)})^C$. In particular, these coefficients have height at most $C(H^{(n)})^C$. Also, we have

$$P_1^{(n)}(w^{(n)}) = \cdots = P_r^{(n)}(w^{(n)}) = 0.$$

But for n larger than C, this contradicts the construction of the $P_i^{(n)}$, and the claim follows. (Note that as p is nonprincipal, any neighbourhood of p in $\mathbf{N}$ will contain arbitrarily large natural numbers.)

Remark 19.0.7. The same argument actually gives a slightly stronger version of Lemma 19.0.6, namely that the integer coefficients used to define the algebraic solution z can be taken to be polynomials in the coefficients of $P_1, \ldots, P_r$, with degree and coefficients bounded by $C_{D,d,r}$.

Remark 19.0.8. A related application of nonstandard analysis to quantitative algebraic geometry was given in [**Sc1989**]. (Thanks to Matthias Aschenbrenner for this reference.)

Chapter 20

Loeb measure and the triangle removal lemma

Formally, a *measure space* is a triple $(X, \mathcal{B}, \mu)$, where X is a set, $\mathcal{B}$ is a σ-algebra of subsets of X, and $\mu \colon \mathcal{B} \to [0, +\infty]$ is a countably additive unsigned measure on $\mathcal{B}$. If the measure $\mu(X)$ of the total space is one, then the measure space becomes a *probability space*. If a nonnegative function $f \colon X \to [0, +\infty]$ is $\mathcal{B}$-*measurable* (or *measurable* for short), one can then form the integral $\int_X f \, d\mu \in [0, +\infty]$ by the usual abstract measure-theoretic construction (as discussed for instance in [**Ta2011**, §1.4]).

A measure space is *complete* if every subset of a null set (i.e., a measurable set of measure zero) is also a null set. Not all measure spaces are complete, but one can always form the *completion* $(X, \overline{\mathcal{B}}, \mu)$ of a measure space $(X, \mathcal{B}, \mu)$ by enlarging the σ-algebra $\mathcal{B}$ to the space of all sets which are equal to a measurable set outside of a null set, and extending the measure μ appropriately.

Given two (σ-finite) measure spaces $(X, \mathcal{B}_X, \mu_X)$ and $(Y, \mathcal{B}_Y, \mu_Y)$, one can form the *product space* $(X \times Y, \mathcal{B}_X \times \mathcal{B}_Y, \mu_X \times \mu_Y)$. This is a measure space whose domain is the Cartesian product $X \times Y$, the σ-algebra $\mathcal{B}_X \times \mathcal{B}_Y$ is generated by the "rectangles" $A \times B$ with $A \in \mathcal{B}_X$, $B \in \mathcal{B}_Y$, and the measure $\mu_X \times \mu_Y$ is the unique measure on $\mathcal{B}_X \times \mathcal{B}_Y$ obeying the identity

$$\mu_X \times \mu_Y(A \times B) = \mu_X(A)\mu_Y(B).$$

See, for instance, [**Ta2011**, §1.7] for a formal construction of product measure[1]. One of the fundamental theorems concerning product measure is *Tonelli's theorem* (which is basically the unsigned version of the more well-known *Fubini theorem*), which asserts that if $f \colon X \times Y \to [0, +\infty]$ is $\mathcal{B}_X \times \mathcal{B}_Y$

---

[1]There are technical difficulties with the theory when X or Y is not σ-finite, but in this section we will only be dealing with probability spaces, which are clearly σ-finite, so this difficulty will not concern us.

measurable, then the integral expressions

$$\int_X (\int_Y f(x,y)\ d\mu_Y(y))\ d\mu_X(x),$$

$$\int_Y (\int_X f(x,y)\ d\mu_X(x))\ d\mu_Y(y),$$

and

$$\int_{X\times Y} f(x,y)\ d\mu_{X\times Y}(x,y),$$

all exist (thus all integrands are almost-everywhere well-defined and measurable with respect to the appropriate σ-algebras), and are all equal to each other; see, e.g., [**Ta2011**, Theorem 1.7.10].

Any finite nonempty set V can be turned into a probability space $(V, 2^V, \mu_V)$ by endowing it with the discrete σ-algebra $2^V := \{A : A \subset V\}$ of all subsets of V, and the normalised counting measure

$$\mu(A) := \frac{|A|}{|V|},$$

where $|A|$ denotes the cardinality of A. In this discrete setting, the probability space is automatically complete, and every function $f \colon V \to [0, +\infty]$ is measurable, with the integral simply being the average:

$$\int_V f\ d\mu_V = \frac{1}{|V|} \sum_{v \in V} f(v).$$

Of course, Tonelli's theorem is obvious for these discrete spaces; the deeper content of that theorem is only apparent at the level of continuous measure spaces.

Among other things, this probability space structure on finite sets can be used to describe various statistics of dense graphs. Recall that a graph $G = (V, E)$ is a finite vertex set V, together with a set of edges E, which we will think of as a symmetric subset of the Cartesian product $V \times V$. (If one wishes, one can prohibit loops in E, so that E is disjoint from the diagonal $V^\Delta := \{(v,v) \colon v \in V\}$ of $V \times V$, but this will not make much difference for the discussion below.) Then, if V is nonempty, and ignoring some minor errors coming from the diagonal V^Δ, the *edge density* of the graph is essentially

$$e(G) := \mu_{V\times V}(E) = \int_{V\times V} 1_E(v,w)\ d\mu_{V\times V}(v,w),$$

the *triangle density* of the graph is basically

$$t(G) := \int_{V\times V\times V} 1_E(u,v) 1_E(v,w) 1_E(w,u)\ d\mu_{V\times V\times V}(u,v,w),$$

and so forth.

In [**Ru1978**], Ruzsa and Szemerédi established the *triangle removal lemma* concerning triangle densities, which informally asserts that a graph with few triangles can be made completely triangle-free by removing a small number of edges:

Lemma 20.0.9 (Triangle removal lemma). *Let $G = (V, E)$ be a graph on a nonempty finite set V, such that $t(G) \leq \delta$ for some $\delta > 0$. Then there exists a subgraph $G' = (V, E')$ of G with $t(G') = 0$, such that $e(G \backslash G') = o_{\delta \to 0}(1)$, where $o_{\delta \to 0}(1)$ denotes a quantity bounded by $c(\delta)$ for some function $c(\delta)$ of δ that goes to zero as $\delta \to 0$.*

The original proof of the triangle removal lemma was a "finitary" one, and proceeded via the *Szemerédi regularity lemma* [**Sz1978**]. It has a number of consequences; for instance, as already noted in that paper, the triangle removal lemma implies as a corollary the famous theorem of Roth [**Ro1953**] that subsets of **Z** of positive upper density contain infinitely many arithmetic progressions of length three.

It is, however, also possible to establish this lemma by infinitary means. There are at least three basic approaches for this. One is via a correspondence principle between questions about dense finite graphs, and questions about exchangeable random infinite graphs, as was pursued in [**Ta2007**]. A second (closely related to the first) is to use the machinery of graph limits, as developed in [**LoSz2006**], [**BoChLoSoVe2008**]. The third is via nonstandard analysis (or equivalently, by using ultraproducts), as was pursued in [**ElSz2006**]. These three approaches differ in the technical details of their execution, but the net effect of all of these approaches is broadly the same, in that they both convert statements about large dense graphs (such as the triangle removal lemma) to measure theoretic statements on infinitary measure spaces. (This is analogous to how the Furstenberg correspondence principle converts combinatorial statements about dense sets of integers into ergodic-theoretic statements on measure-preserving systems, as discussed, for instance, in [**Ta2009**].)

In this section we will illustrate the nonstandard analysis approach from [**ElSz2006**] by providing a nonstandard proof of the triangle removal lemma. The main technical tool used here (besides the basic machinery of nonstandard analysis) is that *Loeb measure*, [**Lo1975**] which gives a probability space structure $(V, \mathcal{B}_V, \mu_V)$ to *nonstandard* finite nonempty sets $V = \prod_{n \to \alpha} V_n$ that is an infinitary analogue of the discrete probability space structures $V = (V, 2^V, \mu_V)$ one has on standard finite nonempty sets. The nonstandard analogue of quantities such as triangle densities then become the integrals of various nonstandard functions with respect to Loeb measure. With this approach, the epsilons and deltas that are so prevalent in the finitary approach to these subjects disappear almost completely; but to compensate for this, one now must pay much more attention to questions of measurability, which were automatic in the finitary setting but now require some care in the infinitary one.

The nonstandard analysis approaches are also related to the regularity lemma approach; see [**Ta2011c**, §4.4] for a proof of the regularity lemma using Loeb measure.

As usual, the nonstandard approach offers a complexity tradeoff: there is more effort expended in building the foundational mathematical structures of the argument (in this case, ultraproducts and Loeb measure), but once these foundations are completed, the actual arguments are shorter than their finitary counterparts. In the case of the triangle removal lemma, this tradeoff does not lead to a particularly significant reduction in complexity (and arguably leads, in fact, to an increase in the length of the arguments, when written out in full), but the gain becomes more apparent when proving more complicated results, such as the hypergraph removal lemma, in which the initial investment in foundations leads to a greater savings in net complexity, as can be seen in [**ElSz2006**].

20.1. Loeb measure

We use the usual setup of nonstandard analysis (as reviewed in Chapter 7). Thus, we will fix a nonprincipal ultrafilter $\alpha \in \beta\mathbf{N}\backslash\mathbf{N}$ on the natural numbers $\mathbf{N}$.

Consider a nonstandard finite nonempty set V, i.e., an ultraproduct $V = \prod_{n\to\alpha} V_n$ of standard finite nonempty sets V_n. Define an *internal subset* of V to be a subset of V of the form $A = \prod_{n\to\alpha} A_n$, where each A_n is a subset of V_n. It is easy to see that the collection $\mathcal{A}_V$ of all internal subsets of V is a Boolean algebra. In general, though, $\mathcal{A}_V$ will not be a σ-algebra. For instance, suppose that the V_n are the standard discrete intervals $V_n := [1, n] := \{i \in \mathbf{N} : i \le n\}$, then V is the nonstandard discrete interval $V = [1, N] := \{i \in {}^*\mathbf{N} : i \le N\}$, where N is the unbounded nonstandard natural number $N := \lim_{n\to\alpha} n$. For any standard integer m, the subinterval $[1, N/m]$ is an internal subset of V; but the intersection

$$[1, o(N)] := \bigcap_{m\in\mathbf{N}} [1, N/m] = \{i \in {}^*\mathbf{N} : i = o(N)\}$$

is not an internal subset of V. (This can be seen, for instance, by noting that all nonempty internal subsets of $[1, N]$ have a maximal element, whereas $[1, o(N)]$ does not.)

Given any internal subset $A = \prod_{n\to\alpha} A_n$ of V, we can define the cardinality $|A|$ of A, which is the nonstandard natural number $|A| := \lim_{n\to\alpha} |A_n|$. We then have the nonstandard density $\frac{|A|}{|V|}$, which is a nonstandard real number between 0 and 1. By Exercise 7.4.1, this bounded nonstandard real number $\frac{|A|}{|V|}$ has a unique *standard part* $\text{st}(\frac{|A|}{|V|})$, which is a standard real number in $[0, 1]$ such that

$$\frac{|A|}{|V|} = \text{st}(\frac{|A|}{|V|}) + o(1),$$

where $o(1)$ denotes a nonstandard infinitesimal (i.e., a nonstandard number which is smaller in magnitude than any standard $\varepsilon > 0$).

In [**Lo1975**], Loeb observed that this standard density can be extended to a complete probability measure:

Theorem 20.1.1 (Construction of Loeb measure). *Let V be a nonstandard finite nonempty set. Then there exists a complete probability space $(V, \mathcal{L}_V, \mu_V)$, with the following properties:*

- *(Internal sets are Loeb measurable) If A is an internal subset of V, then $A \in \mathcal{L}_V$ and*

$$\mu_V(A) = \text{st}(\frac{|A|}{|V|}).$$

- *(Loeb measurable sets are almost internal) If E is a subset of V, then E is Loeb measurable if and only if, for every standard $\varepsilon > 0$, there exists internal subsets $A, B_1, B_2, \dots$ of V such that*

$$E \Delta A \subset \bigcup_{n=1}^{\infty} B_n$$

and

$$\sum_{n=1}^{\infty} \mu_V(B_n) \leq \varepsilon.$$

Proof. The map $\mu_V \colon A \mapsto \text{st}(\frac{|A|}{|V|})$ is a finitely additive probability measure on $\mathcal{A}_V$. We claim that this map μ_V is in fact a *pre-measure* on $\mathcal{A}_V$, thus one has

(20.1) $$\mu_V(A) = \sum_{n=1}^{\infty} \mu_V(A_n)$$

whenever A is an internal set that is partitioned into a disjoint sequence of internal sets A_n. But the countable sequence of sets $A \backslash (A_1 \cup \dots A_n)$ are internal, and have empty intersection, so by the countable saturation property of ultraproducts (Lemma 7.2.6), one of the $A \backslash (A_1 \cup \dots \cup A_n)$ must be empty. The pre-measure property (20.1) then follows from the finite additivity of μ_V.

Invoking the *Hahn-Kolmogorov extension theorem* (see, e.g., [**Ta2011**, Theorem 1.7.8]), we conclude that μ_V extends to a countably additive probability measure on the σ-algebra $\langle \mathcal{A}_V \rangle$ generated by the internal sets. This measure need not be complete, but we can then pass to the completion $\mathcal{L}_V := \overline{\langle \mathcal{A}_V \rangle}$ of that σ-algebra. This probability space certainly obeys the first property. The "only if" portion of second property asserts that all Loeb measurable sets differ from an internal set by sets of arbitrarily small outer measure, but this is easily seen since the space of all sets that have this property is easily verified to be a complete σ-algebra that contain the algebra of internal sets. The "if" portion follows easily from the fact that $\mathcal{L}_V$ is a complete σ-algebra containing the internal sets. (These facts are very similar to the more familiar facts that a bounded subset of a Euclidean space is Lebesgue measurable if and only if it differs from an elementary set by a set of arbitrarily small outer measure.) $\square$

Now we turn to the analogue of Tonelli's theorem for Loeb measure, which will be a fundamental tool when it comes to prove the triangle removal lemma. Let V, W be two nonstandard finite nonempty sets, then $V \times W$ is also a nonstandard finite nonempty set. We then have three Loeb probability spaces

$$(V, \mathcal{L}_V, \mu_V),$$

$$(W, \mathcal{L}_W, \mu_W),$$

and

(20.2) $(V \times W, \mathcal{L}_{V \times W}, \mu_{V \times W}),$

and we also have the product space

(20.3) $(V \times W, \mathcal{L}_V \times \mathcal{L}_W, \mu_V \times \mu_W).$

It is then natural to ask how the two probability spaces (20.2) and (20.3) are related. There is one easy relationship, which shows that (20.2) extends (20.3):

Exercise 20.1.1. Show that (20.2) is a refinement of (20.3), thus $\mathcal{L}_V \times \mathcal{L}_W$, and $\mu_{V \times W}$ extends $\mu_V \times \mu_W$. (*Hint:* First recall why the product of Lebesgue measurable sets is Lebesgue measurable, and mimic that proof to show that the product of a $\mathcal{L}_V$-measurable set and a $\mathcal{L}_W$-measurable set is $\mathcal{L}_{V \times W}$-measurable, and that the two measures $\mu_{V \times W}$ and $\mu_V \times \mu_W$ agree in this case.)

In the converse direction, (20.2) enjoys the type of Tonelli theorem that (20.3) does:

Theorem 20.1.2 (Tonelli theorem for Loeb measure). *Let V, W be two nonstandard finite nonempty sets, and let $f: V \times W \to [0, +\infty]$ be an unsigned $\mathcal{L}_{V \times W}$-measurable function. Then the expressions*

$$\text{(20.4)} \qquad \int_V (\int_W f(v, w) \, d\mu_W(w)) \, d\mu_V(v),$$

$$\text{(20.5)} \qquad \int_W (\int_V f(v, w) \, d\mu_W(w)) \, d\mu_V(v),$$

and

$$\text{(20.6)} \qquad \int_{V \times W} f(v, w) \, d\mu_{V \times W}(v, w),$$

are well-defined (thus all integrands are almost everywhere well-defined and appropriately measurable) and equal to each other.

This result is sometimes referred to as the *Keisler-Fubini theorem*.

Proof. By the monotone convergence theorem it suffices to verify this when f is a simple function; by linearity we may then take f to be an indicator function $f = 1_E$. Using Theorem 20.1.1 and an approximation argument (and many further applications of monotone convergence) we may assume without loss of generality that E is an internal set. We then have

$$\int_{V \times W} f(v, w) \, d\mu_{V \times W}(v, w) = \text{st}(\frac{|E|}{|V||W|})$$

and for every $v \in V$, we have

$$\int_W f(v, w) \, d\mu_W(w) = \text{st}\left(\frac{|E_v|}{|W|}\right),$$

where E_v is the internal set

$$E_v := \{w \in W : (v, w) \in E\}.$$

Let n be a standard natural number, then we can partition V into the internal sets $V = V_1 \cup \cdots \cup V_n$, where

$$V_i := \left\{v \in V : \frac{i-1}{n} < \frac{|E_v|}{|W|} \le \frac{i}{n}\right\}.$$

On each V_i, we have

$$\text{(20.7)} \qquad \int_W f(v, w) \, d\mu_W(w) = \frac{i}{n} + O\left(\frac{1}{n}\right)$$

and

$$\text{(20.8)} \qquad \frac{|E_v|}{|W|} = \frac{i}{n} + O\left(\frac{1}{n}\right).$$

From (20.7), we see that the upper and lower integrals of $\int_W f(v,w)\, d\mu_W(w)$ are both of the form

$$\sum_{i=1}^{n} \frac{i}{n}\frac{|V_i|}{|V|} + O\left(\frac{1}{n}\right).$$

Meanwhile, using the nonstandard double counting identity

$$\frac{1}{|V|}\sum_{v\in V}\frac{|E_v|}{|W|} = \frac{|E|}{|V||W|}$$

(where all arithmetic operations are interpreted in the nonstandard sense, of course) and (20.8), we see that

$$\frac{|E|}{|V||W|} = \sum_{i=1}^{n}\frac{i}{n}\frac{|V_i|}{|V|} + O\left(\frac{1}{n}\right).$$

Thus we see that the upper and lower integrals of $\int_W f(v,w)\, d\mu_W(w)$ are equal to $\frac{|E|}{|V||W|} + O\left(\frac{1}{n}\right)$ for every standard n. Sending n to infinity, we conclude that $\int_W f(v,w)\, d\mu_W(w)$ is measurable, and that

$$\int_V \left(\int_W f(v,w)\, d\mu_W(w)\right) d\mu_V(v) = \mathrm{st}\left(\frac{|E|}{|V||W|}\right)$$

showing that (20.4) and (20.6) are well-defined and equal. A similar argument holds for (20.5) and (20.6), and the claim follows. $\qquad\square$

Remark 20.1.3. It is well known that the product of two Lebesgue measure spaces $\mathbf{R}^n, \mathbf{R}^m$, upon completion, becomes the Lebesgue measure space on $\mathbf{R}^{n+m}$. Drawing the analogy between Loeb measure and Lebesgue measure, it is then natural to ask whether (20.2) is simply the completion of (20.3). But while (20.2) certainly contains the completion of (20.3), it is a significantly larger space in general (a fact first observed by Doug Hoover). Indeed, suppose $V = \prod_{n\to\alpha} V_n$, $W = \prod_{n\to\alpha} W_n$, where the cardinality of V_n, W_n goes to infinity at some reasonable rate, e.g., $|V_n|, |W_n| \geq n$ for all n. For each n, let E_n be a random subset of $V_n \times W_n$, with each element of $V_n \times W_n$ having an independent probability of $1/2$ of lying in E_n. Then, as is well known, the sequence of sets E_n is almost surely *asymptotically regular* in the sense that almost surely, we have the bound

$$\sup_{A_n\subset V_n, B_n\subset W_n} \frac{||E_n \cap (A_n \times B_n)| - \frac{1}{2}|A_n||B_n||}{|V_n||W_n|} \to 0$$

as $n \to \infty$. Let us condition on the event that this asymptotic regularity holds. Taking ultralimits, we conclude that the internal set $E := \prod_{n\to p} E_n$ obeys the property

$$\mu_{V\times W}(E \cap (A \times B)) = \frac{1}{2}\mu_{V\times W}(A \times B)$$

for all internal $A \subset V, B \subset W$; in particular, E has Loeb measure $1/2$. Using Theorem 20.1.1 we conclude that

$$\mu_{V \times W}(E \cap F) = \frac{1}{2}\mu_{V \times W}(F)$$

for all $\mathcal{L}_V \times \mathcal{L}_W$-measurable F, which implies, in particular, that E cannot be $\mathcal{L}_V \times \mathcal{L}_W$-measurable. (Indeed, $1_E - \frac{1}{2}$ is "anti-measurable" in the sense that it is orthogonal to all functions in $L^2(\mathcal{L}_V \times \mathcal{L}_W)$; or equivalently, we have the conditional expectation formula $\mathbf{E}(1_E | \mathcal{L}_V \times \mathcal{L}_W) = \frac{1}{2}$ almost everywhere.)

Intuitively, a $\mathcal{L}_V \times \mathcal{L}_W$-measurable set corresponds to a subset of $V \times W$ that is of "almost bounded complexity", in that it can be approximated by a bounded Boolean combination of Cartesian products. In contrast, $\mathcal{L}_{V \times W}$-measurable sets (such as the set E given above) have no bound on their complexity.

20.2. The triangle removal lemma

Now we can prove the triangle removal lemma, Lemma 20.0.9. We will deduce it from the following nonstandard (and tripartite) counterpart (a special case of a result established in [**Ta2007**]):

Lemma 20.2.1 (Nonstandard triangle removal lemma). *Let V be a nonstandard finite nonempty set, and let $E_{12}, E_{23}, E_{31} \subset V \times V$ be Loeb-measurable subsets of $V \times V$ which are almost triangle-free in the sense that*

$$(20.9) \quad \int_{V \times V \times V} 1_{E_{12}}(u,v) 1_{E_{23}}(v,w) 1_{E_{31}}(w,u) \, d\mu_{V \times V \times V}(u,v,w) = 0.$$

Then for any standard $\varepsilon > 0$, there exists a internal subsets $F_{ij} \subset V \times V$ for $ij = 12, 23, 31$ with $\mu_{V \times V}(E_{ij} \backslash F_{ij}) < \varepsilon$, which are completely triangle-free in the sense that

$$(20.10) \quad 1_{F_{12}}(u,v) 1_{F_{23}}(v,w) 1_{F_{31}}(w,u) = 0$$

for all $u, v, w \in V$.

Let us first see why Lemma 20.2.1 implies Lemma 20.0.9. We use the usual "compactness and contradiction" argument. Suppose for contradiction that Lemma 1.2 failed. Carefully negating the quantifiers, we can find a (standard) $\varepsilon > 0$, and a sequence $G_n = (V_n, E_n)$ of graphs with $t(G_n) \leq 1/n$, such that for each n, there does *not* exist a subgraph $G'_n = (V_n, E'_n)$ of n with $|E_n \backslash E'_n| \leq \varepsilon |V_n|^2$ with $t(G'_n) = 0$. Clearly we may assume the V_n are nonempty.

We form the ultraproduct $G = (V, E)$ of the G_n, thus $V = \prod_{n \to \alpha} V_n$ and $E = \prod_{n \to \alpha} E_n$. By construction, E is a symmetric internal subset of

$V \times V$ and we have

$$\int_{V \times V \times V} 1_E(u,v) 1_E(v,w) 1_E(w,u) \, d\mu_{V \times V \times V}(u,v,w) = \operatorname{st} \lim_{n \to \alpha} t(G_n) = 0.$$

Thus, by Lemma 20.2.1, we may find internal subsets F_{12}, F_{23}, F_{31} of $V \times V$ with $\mu_{V \times V}(E \backslash F_{ij}) < \varepsilon/6$ (say) for $ij = 12, 23, 31$ such that (20.10) holds for all $u, v, w \in V$. By letting E' be the intersection of all E with all the F_{ij} and their reflections, we see that E' is a symmetric internal subset of E with $\mu_{V \times V}(E \backslash E') < \varepsilon$, and we still have

$$1_{E'}(u,v) 1_{E'}(v,w) 1_{E'}(w,u) = 0$$

for all $u, v, w \in V$. If we write $E' = \lim_{n \to \alpha} E_n'$ for some sets E_n', then for n sufficiently close to p, one has E_n' a symmetric subset of E_n with

$$\mu_{V_n \times V_n}(E_n \backslash E_n') < \varepsilon$$

and

$$1_{E_n'}(u,v) 1_{E_n'}(v,w) 1_{E_n'}(w,u) = 0.$$

If we then set $G_n' := (V_n, E_n)$, we thus have $|E_n \backslash E_n'| \le \varepsilon |V_n|^2$ and $t(G_n') = 0$, which contradicts the construction of G_n by taking n sufficiently large.

Now we prove Lemma 20.2.1. The idea (similar to that used to prove the Furstenberg recurrence theorem, as discussed, for instance, in [**Ta2009**, §2.10]) is to first prove the lemma for very simple examples of sets E_{ij}, and then work one's way towards the general case. Readers who are familiar with the traditional proof of the triangle removal lemma using the regularity lemma will see strong similarities between that argument and the one given here (and, on some level, they are essentially the same argument).

To begin with, we suppose first that the E_{ij} are all *elementary sets*, in the sense that they are finite Boolean combinations of products of internal sets. (At the finitary level, this corresponds to graphs that are bounded combinations of bipartite graphs.) This implies that there is an internal partition $V = V_1 \cup \cdots \cup V_n$ of the vertex set V, such that each E_{ij} is the union of some of the $V_a \times V_b$.

Let F_{ij} be the union of all the $V_a \times V_b$ in E_{ij} for which V_a and V_b have positive Loeb measure; then $\mu_{V \times V}(E_{ij} \backslash F_{ij}) = 0$. We claim that (20.10) holds for all $u, v, w \in V$, which gives Theorem 20.2.1 in this case. Indeed, if $u \in V_a, v \in V_b, w \in V_c$ were such that (20.10) failed, then E_{12} would contain $V_a \times V_b$, E_{23} would contain $V_b \times V_c$, and E_{31} would contain $V_c \times V_a$. The integrand in (20.9) is then equal to 1 on $V_a \times V_b \times V_c$, which has Loeb measure $\mu_V(V_a)\mu_V(V_b)\mu_V(V_c)$ which is nonzero, contradicting (20.9). This gives Theorem 20.2.1 in the elementary set case.

Next, we increase the level of generality by assuming that the E_{ij} are all $\overline{\mathcal{L}_V \times \mathcal{L}_V}$-measurable. (The finitary equivalent of this is a little difficult to pin down; roughly speaking, it is dealing with graphs that are not quite bounded combinations of bounded graphs, but can be well approximated by such bounded combinations; a good example is the *half-graph*, which is a bipartite graph between two copies of $\{1, \ldots, N\}$, which joins an edge between the first copy of i and the second copy of j iff $i < j$.) Then each E_{ij} can be approximated to within an error of $\varepsilon/3$ in $\mu_{V \times V}$ by elementary sets. In particular, we can find a finite partition $V = V_1 \cup \cdots \cup V_n$ of V, and sets E'_{ij} that are unions of some of the $V_a \times V_b$, such that $\mu_{V \times V}(E_{ij} \Delta E'_{ij}) < \varepsilon/3$.

Let F_{ij} be the union of all the $V_a \times V_b$ contained in E'_{ij} such that V_a, V_b have positive Loeb measure, and such that

$$\mu_{V \times V}(E_{ij} \cap (V_a \times V_b)) > \frac{2}{3}\mu_{V \times V}(V_a \times V_b).$$

Then the F_{ij} are internal subsets of $V \times V$, and $\mu_{V \times V}(E_{ij} \backslash F_{ij}) < \varepsilon$.

We now claim that the F_{ij} obey (20.10) for all u, v, w, which gives Theorem 20.2.1 in this case. Indeed, if $u \in V_a, v \in V_b, w \in V_c$ were such that (20.10) failed, then E_{12} occupies more than $\frac{2}{3}$ of $V_a \times V_b$, and thus

$$\int_{V_a \times V_b \times V_c} 1_{E_{12}}(u, v) \, d\mu_{V \times V \times V}(u, v, w) > \frac{2}{3}\mu_{V \times V \times V}(V_a \times V_b \times V_c).$$

Similarly for $1_{E_{23}}(v, w)$ and $1_{E_{31}}(w, u)$. From the inclusion-exclusion formula, we conclude that

$$\int_{V_a \times V_b \times V_c} 1_{E_{12}}(u, v) 1_{E_{23}}(v, w) 1_{E_{31}}(w, u) \, d\mu_{V \times V \times V}(u, v, w) > 0,$$

contradicting (20.9), and the claim follows.

Finally, we turn to the general case, when the E_{ij} are merely $\mathcal{L}_{V \times V}$-measurable. Here, we split

$$1_{E_{ij}} = f_{ij} + g_{ij}$$

where $f_{ij} := \mathbf{E}(1_{E_{ij}} | \overline{\mathcal{L}_V \times \mathcal{L}_V})$ is the conditional expectation of $1_{E_{ij}}$ onto $\overline{\mathcal{L}_V \times \mathcal{L}_V}$, and $g_{ij} := 1_{E_{ij}} - f_{ij}$ is the remainder. We observe that each $g_{ij}(u, v)$ is orthogonal to any tensor product $f(u)g(v)$ with f, g bounded and $\mathcal{L}_V$-measurable. From this and Tonelli's theorem for Loeb measure (Theorem 20.1.2) we conclude that each of the g_{ij} make a zero contribution to (20.9), and thus

$$\int_{V \times V \times V} f_{12}(u, v) f_{23}(v, w) f_{31}(w, u) \, d\mu_{V \times V \times V}(u, v, w) = 0.$$

Now let $E'_{ij} := \{(u, v) \in V \times V : f_{ij}(u, v) \geq \varepsilon/2\}$, then the E'_{ij} are $\overline{\mathcal{L}_V \times \mathcal{L}_V}$-measurable, and we have

$$\int_{V \times V \times V} 1_{E'_{12}}(u, v) 1_{E'_{23}}(v, w) 1_{E'_{31}}(w, u) \, d\mu_{V \times V \times V}(u, v, w) = 0.$$

Also, we have

$$
\mu_{V \times V}(E_{ij} \backslash E'_{ij}) = \int_{V \times V} 1_{E_{ij}}(1 - 1_{E'_{ij}})
$$
$$
= \int_{V \times V} f_{ij}(1 - 1_{E'_{ij}})
$$
$$
\leq \varepsilon/2.
$$

Applying the already established cases of Theorem 20.2.1, we can find internal sets F_{ij} obeying (20.10) with $\mu_{V \times V}(E'_{ij} \backslash F_{ij}) < \varepsilon/2$, and hence $\mu_{V \times V}(E_{ij} \backslash F_{ij}) < \varepsilon$, and Theorem 20.2.1 follows.

Remark 20.2.2. The full hypergraph removal lemma can be proven using similar techniques, but with a longer tower of generalisations than the three cases given here; see [**Ta2007**] or [**ElSz2006**].

Chapter 21

Two notes on Lie groups

In this chapter we record two small miscellaneous facts about Lie groups.

The first fact concerns the *exponential map* exp: $\mathfrak{g} \to G$ from a Lie algebra $\mathfrak{g}$ of a Lie group G to that group. (For this discussion we will only consider finite-dimensional Lie groups and Lie algebras over the reals **R**.) A basic fact in the subject is that the exponential map is *locally* a homeomorphism: There is a neighbourhood of the origin in $\mathfrak{g}$ that is mapped homeomorphically by the exponential map to a neighbourhood of the identity in G. This local homeomorphism property is the foundation of an important dictionary between Lie groups and Lie algebras.

It is natural to ask whether the exponential map is globally a homeomorphism, and not just locally; in particular, whether the exponential map remains both injective and surjective. For instance, this is the case for connected, simply connected, nilpotent Lie groups (as can be seen from the Baker-Campbell-Hausdorff formula from Section 2.5).

The circle group S^1, which has **R** as its Lie algebra, already shows that global injectivity fails for any group that contains a circle subgroup, which is a huge class of examples (including, for instance, the positive dimensional compact Lie groups, or nonsimply-connected Lie groups). Surjectivity also obviously fails for disconnected groups, since the Lie algebra is necessarily connected, and so the image under the exponential map must be connected also. However, even for connected Lie groups, surjectivity can fail. To see this, first observe that if the exponential map was surjective, then every group element $g \in G$ would have a square root (i.e., an element $h \in G$ with

$h^2 = g$), since $\exp(x)$ has $\exp(x/2)$ as a square root for any $x \in \mathfrak{g}$. However, there exist elements in connected Lie groups without square roots. A simple example is provided by the matrix

$$g = \begin{pmatrix} -4 & 0 \\ 0 & -1/4 \end{pmatrix}$$

in the connected Lie group $SL_2(\mathbf{R})$. This matrix has eigenvalues $-4, -1/4$. Thus, if $h \in SL_2(\mathbf{R})$ is a square root of g, we see (from the Jordan normal form) that it must have at least one eigenvalue in $\{-2i, +2i\}$, and at least one eigenvalue in $\{-i/2, i/2\}$. On the other hand, as h has real coefficients, the complex eigenvalues must come in conjugate pairs $\{a + bi, a - bi\}$. Since h can only have at most 2 eigenvalues, we obtain a contradiction.

However, there is an important case where surjectivity is recovered:

Proposition 21.0.3. *If G is a compact connected Lie group, then the exponential map is surjective.*

Proof. The idea here is to relate the exponential map in Lie theory to the exponential map in Riemannian geometry. We first observe that every compact Lie group G can be given the structure of a Riemannian manifold with a bi-invariant metric. This can be seen in one of two ways. First, one can put an arbitrary positive definite inner product on $\mathfrak{g}$ and average it against the adjoint action of G using Haar probability measure (which is available since G is compact); this gives an ad-invariant positive-definite inner product on $\mathfrak{g}$ that one can then translate by either left or right translation to give a bi-invariant Riemannian structure on G. Alternatively, one can use Theorem 4.2.4 to embed G in a unitary group $U(n)$, at which point one can induce a bi-invariant metric on G from the one on the space $M_n(\mathbf{C}) \equiv \mathbf{C}^{n^2}$ of $n \times n$ complex matrices.

As G is connected and compact and thus complete, we can apply the *Hopf-Rinow theorem* and conclude that any two points are connected by at least one geodesic, so that the *Riemannian* exponential map from $\mathfrak{g}$ to G formed by following geodesics from the origin is surjective. But one can check that the Lie exponential map and Riemannian exponential map agree; for instance, this can be seen by noting that the group structure naturally defines a connection on the tangent bundle which is both torsion-free and preserves the bi-invariant metric, and must therefore agree with the Levi-Civita metric. (Alternatively, one can embed into a unitary group $U(n)$ and observe that G is totally geodesic inside $U(n)$, because the geodesics in $U(n)$ can be described explicitly in terms of one-parameter subgroups.) The claim follows. $\qquad\square$

The other basic fact we will present here concerns the algebraic nature of Lie groups and Lie algebras. An important family of examples of Lie groups are the *algebraic groups* — algebraic varieties with a group law given by algebraic maps. Given that one can always automatically upgrade the smooth structure on a Lie group to analytic structure (by using the Baker-Campbell-Hausdorff formula), it is natural to ask whether one can upgrade the structure further to an algebraic structure. Unfortunately, this is not always the case. A prototypical example of this is given by the one-parameter subgroup

$$(21.1) \qquad G := \left\{ \begin{pmatrix} t & 0 \\ 0 & t^\alpha \end{pmatrix} : t \in \mathbf{R}^+ \right\}$$

of $GL_2(\mathbf{R})$. This is a Lie group for any exponent $\alpha \in \mathbf{R}$, but if α is irrational, then the curve that G traces out is not an algebraic subset of $GL_2(\mathbf{R})$ (as one can see by playing around with *Puiseux series*).

This is not a true counterexample to the claim that every Lie group can be given the structure of an algebraic group, because one can give G a different algebraic structure than one inherited from the ambient group $GL_2(\mathbf{R})$. Indeed, G is clearly isomorphic to the additive group $\mathbf{R}$, which is of course an algebraic group. However, a modification of the above construction works:

Proposition 21.0.4. *There exists a Lie group G that cannot be given the structure of an algebraic group.*

Proof. We use an example from [**TaYu2005**]. Consider the subgroup

$$G := \left\{ \begin{pmatrix} 1 & 0 & 0 \\ x & t & 0 \\ y & 0 & t^\alpha \end{pmatrix} : x, y \in \mathbf{R}; t \in \mathbf{R}^+ \right\}$$

of $GL_3(\mathbf{R})$, with α an irrational number. This is a three-dimensional (*metabelian*) Lie group, whose Lie algebra $\mathfrak{g} \subset \mathfrak{gl}_3(\mathbf{R})$ is spanned by the elements

$$X := \begin{pmatrix} 0 & 0 & 0 \\ 0 & 1 & 0 \\ 0 & 0 & \alpha \end{pmatrix},$$

$$Y := \begin{pmatrix} 0 & 0 & 0 \\ -1 & 0 & 0 \\ 0 & 0 & 0 \end{pmatrix},$$

$$Z := \begin{pmatrix} 0 & 0 & 0 \\ 0 & 0 & 0 \\ -\alpha & 0 & 0 \end{pmatrix},$$

with the Lie bracket given by

$$[Y, X] = -Y; [Z, X] = -\alpha Z; [Y, Z] = 0.$$

As such, we see that if we use the basis X, Y, Z to identify $\mathfrak{g}$ to $\mathbf{R}^3$, then the *adjoint representation* of G is the identity map.

If G is an algebraic group, it is easy to see that the adjoint representation $\mathrm{Ad}\colon G \to GL(\mathfrak{g})$ is also algebraic, and so $\mathrm{Ad}(G) = G$ is algebraic in $GL(\mathfrak{g})$. Specialising to our specific example, in which adjoint representation is the identity, we conclude that if G has *any* algebraic structure, then it must also be an algebraic subgroup of $GL_3(\mathbf{R})$; but G projects to the group (21.1) which is not algebraic, a contradiction.   □

A slight modification of the same argument also shows that not every Lie algebra is *algebraic*, in the sense that it is isomorphic to a Lie algebra of an algebraic group. (However, there are important classes of Lie algebras that are automatically algebraic, such as nilpotent or semisimple Lie algebras.)

Bibliography

[Ar1950] K. Arrow, *A Difficulty in the Concept of Social Welfare*, Journal of Political Economy **58** (1950), 328–346.

[BoChLoSoVe2008] C. Borgs, J. Chayes, L. Lovász, V. Sós, K. Vesztergombi, *Convergent sequences of dense graphs. I. Subgraph frequencies, metric properties and testing*, Adv. Math. **219** (2008), no. 6, 1801–1851.

[Bo1968] N. Bourbaki, Éléments de mathématique. Fasc. XXXIV. Groupes et algèbres de Lie. Chapitre IV: Groupes de Coxeter et systèmes de Tits. Chapitre V: Groupes engendrés par des réflexions. Chapitre VI: systèmes de racines. Actualités Scientifiques et Industrielles, No. 1337 Hermann, Paris 1968.

[BrGrTa2011] E. Breuillard, B. Green, T. Tao, *The structure of approximate groups*, Publ. Math. Inst. Hautes Études Sci. **116** (2012), 115–221.

[Ch2003] M. Chang, *Factorization in generalized arithmetic progressions and applications to the Erdos-Szemerédi sum-product problems*, Geom. Funct. Anal. **13** (2003), no. 4, 720–736.

[ChCo1996] J. Cheeger, T. Colding, *Lower bounds on Ricci curvature and the almost rigidity of warped products*, Ann. of Math. **144** (1996), no. 1, 189–237.

[ChGr1972] J. Cheeger, D. Gromoll, *On the structure of complete manifolds of nonnegative curvature*, Ann. of Math. **96** (1972), 413–443.

[CoMi1997] T. Colding, W. Minicozzi, *Harmonic functions on manifolds*, Ann. of Math. **146** (1997), no. 3, 725–747.

[CrSi2010] E. Croot, O. Sisask, *A probabilistic technique for finding almost-periods of convolutions*, Geom. Funct. Anal. **20** (2010), 1367–1396.

[Dr1969] A. Dress, *Newman's theorems on transformation groups*, Topology **8** (1969), 203–207.

[ElSz2006] G. Elek, B. Szegedy, *A measure-theoretic approach to the theory of dense hypergraphs*, Adv. Math. **231** (2012), 1731–1772.

[FuHa1991] W. Fulton, J. Harris, Representation theory. A first course. Graduate Texts in Mathematics, 129. Readings in Mathematics. Springer-Verlag, New York, 1991.

[Fr1973] G. A. Freiman, Foundations of a structural theory of set addition, American Mathematical Society, Providence, R. I., 1973. Translated from the Russian, Translations of Mathematical Monographs, Vol 37.

[Fr1936] H. Freudenthal, *Einige Sätze über topologische Gruppen*, Annals of Math. **37** (1936), 46–56.

[Gl1950] A. M. Gleason, *Spaces with a compact Lie group of transformations*, Proc. Amer. Math. Soc. **1** (1950), 35–43.

[Gl1952] A. M. Gleason, *Groups without small subgroups*, Ann. of Math. **56** (1952), 193–212.

[Gl1951] A. M. Gleason, *The structure of locally compact groups*, Duke Math. J. **18** (1951), 85–104.

[Go1938] K. Gödel, *Consistency of the axiom of choice and of the generalized continuum-hypothesis with the axioms of set theory*, Proc. Nat. Acad. Sci, **24** (1938), 556–557.

[Go2009] I. Goldbring, Nonstandard methods in lie theory. Ph.D. Thesis, University of Illinois at Urbana-Champaign, 2009.

[Go2010] I. Goldbring, *Hilbert's fifth problem for local groups*, Ann. of Math. **172** (2010), no. 2, 1269–1314.

[GoSa1964] E. D. Golod, I. R. Shafarevich, *On the class field tower*, Izv. Akad. Nauk SSSR Ser. Mat. **28** (1964) 261–272.

[GrRu2007] B. Green, I. Z. Ruzsa, *Freiman's theorem in an arbitrary abelian group*, J. Lond. Math. Soc. **75** (2007), 163–175.

[Gr1984] R.I. Grigorchuk, *Degrees of growth of finitely generated groups and the theory of invariant means*, Math. USSR Izv. **25** (1985), 259–300; Russian original: Izv. Akad. Nauk SSSR Sr. Mat. **48** (1984), 939–985.

[Gr1981] M. Gromov, *Groups of polynomial growth and expanding maps*, Inst. Hautes Études Sci. Publ. Math. No. **53** (1981), 53–73.

[HeRo1979] E. Hewitt, K. Ross, Abstract harmonic analysis. Vol. I. Structure of topological groups, integration theory, group representations. Second edition. Grundlehren der Mathematischen Wissenschaften [Fundamental Principles of Mathematical Sciences], 115. Springer-Verlag, Berlin-New York, 1979.

[Hi1990] J. Hirschfeld, *The nonstandard treatment of Hilbert's fifth problem*, Trans. Amer. Math. Soc. **321** (1990), no. 1, 379–400.

[Ho1991] B. Host, *Mixing of all orders and pairwise independent joinings of systems with singular spectrum*, Israel J. Math. **76** (1991), no. 3, 289–298.

[Hr2012] E. Hrushovski, *Stable group theory and approximate subgroups*, J. Amer. Math. Soc. **25** (2012), no. 1, 189–243.

[Jo1878] C. Jordan, *Mémoire sur les équations différentielles linéaires à intégrale algébrique*, J. Reine Angew. Math. **84** (1878), 89–215.

[KaPeTu2010] V. Kapovitch, A. Petrunin, W. Tuschmann, *Nilpotency, almost nonnegative curvature, and the gradient flow on Alexandrov spaces*, Ann. of Math. **171** (2010), no. 1, 343–373.

[KaWi2011] V. Kapovitch, B. Wilking, *Structure of fundamental groups of manifolds with Ricci curvature bounded below*, preprint.

[KiSo1972] A. Kirman, D. Sondermann, *Arrow's theorem, many agents, and invisible dictators*, Journal of Economic Theory **5** (1972), 267–277.

[Kl2010] B. Kleiner, *A new proof of Gromov's theorem on groups of polynomial growth*, J. Amer. Math. Soc., **23** (2010), 815–829.

[Ku1998] W. Kulpa, *Poincaré and domain invariance theorem*, Acta Univ. Carolin. Math. Phys. 39 (1998), no. 1-2, 127–136.

[Ku1950] M. Kuranishi, *On Euclidean local groups satisfying certain conditions*, Proc. Amer. Math. Soc. **1** (1950), 372–380.

[Le1997] J. S. Lee, *Totally disconnected groups, p-adic groups and the Hilbert-Smith conjecture*, Commun. Korean Math. Soc. **12** (1997), no. 3, 691–699.

[Lo1975] P. Loeb, *Conversion from nonstandard to standard measure spaces and applications in probability theory*, Trans. Amer. Math. Soc. **211** (1975), 113–122.

[Lo1955] J. Los, *Quelques remarques, théorèmes et problèmes sur les classes définissables d'algébres*, in: Mathematical Interpretation of Formal Systems, North-Holland, Amsterdam, 1955, 98–113.

[LoSz2006] L. Lovász, B. Szegedy, *Limits of dense graph sequences*, J. Combin. Theory Ser. B **96** (2006), no. 6, 933–957.

[Ma1941] A. Malcev, *Sur les groupes topologiques locaux et complets*, C. R. (Doklady) Acad. Sci. URSS (N.S.) **32**, (1941). 606–608.

[Mi1956] J. Milnor, *On manifolds homeomorphic to the 7-sphere*, Annals of Mathematics **64** (1956), 399–405.

[MoZi1952] D. Montgomery, L. Zippin, *Small subgroups of finite-dimensional groups*, Ann. of Math. **56** (1952), 213–241.

[MoZi1974] D. Montgomery, L. Zippin, Topological transformation groups. Reprint of the 1955 original. Robert E. Krieger Publishing Co., Huntington, N.Y., 1974. xi+289 pp.

[My1941] S. B. Myers, *Riemannian manifolds with positive mean curvature*, Duke Mathematical Journal **8** (1941), 401–404.

[Ne1931] M. H. A. Newman, *A theorem on periodic transformations of spaces*, Quart. J. Math **2** (1931), 1–9.

[Ol1996] P. Olver, *Non-associative local Lie groups*, J. Lie Theory **6** (1996), no. 1, 23–51.

[Pa2013] J. Pardon, *The Hilbert–Smith conjecture for three-manifolds*, J. Amer. Math. Soc. **26** (2013), no. 3, 879–899.

[Pe2006] P. Petersen, Riemannian geometry. Second edition. Graduate Texts in Mathematics, 171. Springer, New York, 2006.

[ReSc97] D. Repovš, E. Ščepin, *A proof of the Hilbert-Smith conjecture for actions by Lipschitz maps*, Math. Ann. **308** (1997), no. 2, 361–364.

[Ro1966] A. Robinson, Non-standard analysis. North-Holland Publishing Co., Amsterdam 1966.

[Ro1953] K.F. Roth, *On certain sets of integers*, J. London Math. Soc. **28** (1953), 245–252.

[Ru1962] W. Rudin, Fourier analysis on groups. Interscience Tracts in Pure and Applied Mathematics, No. 12 Interscience Publishers (a division of John Wiley and Sons), New York-London 1962 ix+285 pp.

[Ru1978] I. Ruzsa, E. Szemerédi, *Triple systems with no six points carrying three triangles*, Combinatorics (Proc. Fifth Hungarian Colloq., Keszthely, 1976), Vol. II, pp. 939-945, Colloq. Math. Soc. János Bolyai, 18, North-Holland, Amsterdam-New York, 1978.

[Sa2009] T. Sanders, *On a non-abelian Balog-Szemeredi-type lemma*, J. Aust. Math. Soc. **89** (2010), no. 1, 127–132.

[Sa2010] T. Sanders, *On the Bogolyubov-Ruzsa lemma*, Anal. PDE **5** (2012), 627–655.

[Sc1995] J. Schmid, *On the affine Bezout inequality*, Manuscripta Math. **88** (1995), no. 2, 225–232.

[Sc1989] K. Schmidt-Göttsch, *Polynomial bounds in polynomial rings over fields*, J. Algebra **125** (1989), no. 1, 164–180.

[Se1931] H. Seifert, *Konstruction drei dimensionaler geschlossener Raume*, Berichte Sachs. Akad. Leipzig, Math.-Phys. Kl. **83** (1931) 26–66.

[Se1964] J.-P. Serre, Lie algebras and Lie groups. 1964 lectures given at Harvard University. Corrected fifth printing of the second (1992) edition. Lecture Notes in Mathematics, 1500. Springer-Verlag, Berlin, 2006.

[ShTa2010] Y. Shalom, T. Tao, *A finitary version of Gromov's polynomial growth theorem*, Geom. Funct. Anal. **20** (2010), no. 6, 1502–1547.

[Sm1941] P. A. Smith, *Transformations of finite period. III. Newman's theorem*, Ann. of Math. **42** (1941), 446–458.

[Sz1978] E. Szemerédi, *Regular partitions of graphs*, in "Problemés Combinatoires et Théorie des Graphes, Proc. Colloque Inter. CNRS," (Bermond, Fournier, Las Vergnas, Sotteau, eds.), CNRS Paris, 1978, 399–401.

[Ta2007] T. Tao, *A correspondence principle between (hyper)graph theory and probability theory, and the (hyper)graph removal lemma*, J. Anal. Math. **103** (2007), 1–45.

[Ta2008] T. Tao, Structure and Randomness: pages from year one of a mathematical blog, American Mathematical Society, Providence RI, 2008.

[Ta2008b] T. Tao, *Product set estimates for non-commutative groups*, Combinatorica **28** (2008), 547–594.

[Ta2009] T. Tao, Poincaré's Legacies: pages from year two of a mathematical blog, Vol. I, American Mathematical Society, Providence RI, 2009.

[Ta2009b] T. Tao, Poincaré's Legacies: pages from year two of a mathematical blog, Vol. II, American Mathematical Society, Providence RI, 2009.

[Ta2010] T. Tao, An epsilon of room, Vol. I, American Mathematical Society, Providence RI, 2010.

[Ta2010b] T. Tao, An epsilon of room, Vol. II, American Mathematical Society, Providence RI, 2010.

[Ta2011] T. Tao, An introduction to measure theory, American Mathematical Society, Providence RI, 2011.

[Ta2011b] T. Tao, Higher order Fourier analysis, American Mathematical Society, Providence RI, 2011.

[Ta2011c] T. Tao, Compactness and contradiction, American Mathematical Society, Providence, RI, 2013.

[TaVu2006] T. Tao, V. Vu, Additive combinatorics. Cambridge Studies in Advanced Mathematics, 105. Cambridge University Press, Cambridge, 2006.

[TaYu2005] P. Tauvel, R. Yu, Lie algebras and algebraic groups, Springer Monographs in Mathematics, Springer-Verlag, Berlin, 2005.

[vDa1936] D. van Dantzig, *Zur topologischen Algebra. III, Brouwersche und Cantorsche Gruppen*, Compositio Math. **3** (1936), 408–426.

[vDrGo2010] L. van den Dries, I. Goldbring, *Globalizing locally compact local groups*, J. Lie Theory **20** (2010), no. 3, 519–524.

[vKa1833] E. R. van Kampen, *On the connection between the fundamental groups of some related spaces*, Amer. J. Math., **55** (1933), 261–267.

[vDrWi1984] L. van den Dries, A. J. Wilkie, *Gromov's theorem on groups of polynomial growth and elementary logic*, J. Algebra, **89** (1984), 349–374.

[vEsKo1964] W. T. van Est, Th. J.Korthagen, *Non-enlargible Lie algebras*, Nederl. Akad. Wetensch. Proc. Ser. A 67, Indag. Math. **26** (1964), 15–31.

[We1973] B. A. F. Wehrfritz, Infinite linear groups. An account of the group-theoretic properties of infinite groups of matrices. Ergebnisse der Matematik und ihrer Grenzgebiete, Band 76. Springer-Verlag, New York-Heidelberg, 1973.

[Ya1953] H. Yamabe, *On the conjecture of Iwasawa and Gleason*, Ann. of Math. **58**, (1953), 48–54.

[Ya1953b] H. Yamabe, *A generalization of a theorem of Gleason*, Ann. of Math. **58** (1953), 351–365.

[Ze1990] E. I. Zel'manov, *Solution of the restricted Burnside problem for groups of odd exponent*, Izv. Akad. Nauk SSSR Ser. Mat. **54** (1990), no. 1, 42–59.

Index